KT-385-418

UNIVERSITY OF SALFORD
DEPT. OF
PHYSICS

The Communications Satellite

The Communications Satellite

Mark Williamson

BSc CPhys MInstP CEng MIEE FBIS

Space Technology Consultant

Adam Hilger
Bristol and New York

British Library Cataloguing in Publication Data

Williamson, Mark
The communications satellite
1. Communications satellites
I. Title
621.38'0422

ISBN 0-85274-192-8

Library of Congress Cataloging-in-Publication Data

Williamson, Mark.
The Communications satellite/by Mark Williamson.
p.24 cm.
Includes bibliographical references.
ISBN 0-85274-192-8
1. Artificial satellites in telecommunication. I. Title.
TK5104.W55 1990
621.382'5—dc20 89-37838

Cover photograph: the G-Star III communications satellite being lifted from a thermal-vacuum chamber after testing. [GE Astro]

Published under the Adam Hilger imprint by IOP Publishing Ltd
Techno House, Redcliffe Way, Bristol BS1 6NX, England
335 East 45th Street, New York, NY 10017-3483, USA

Typeset by CGS Print Group Ltd, Poole Hall Industrial Estate, Ellesmere Port, South Wirral
Printed in Great Britain by Dotesios Printers Ltd, Trowbridge, Wiltshire

To my wife
Rita
a captive student of
space technology

Contents

Foreword

I am pleased to be associated with this book because, unlike many others appearing nowadays on the space stage, it really does fill a need.

A great deal has been written about the techniques of communications by satellite, but relatively little about the satellites themselves. Students—and the term includes all of us who have had to apply our minds to this topic—have not found it easy to locate clear explanations as to how communications satellites are designed, and most teachers have themselves had to resort to preparing their own notes, compiled from a variety of sources. This volume is a very intelligent attempt to fill the gap and to provide a logical account both of the hardware subsystems and the communications system.

There is, of course, nothing special about the satellite used for communications; indeed, virtually all satellites perform some communication function, and so the book will be equally helpful to those who wish to understand satellites in general.

At a time when the United Kingdom Government is unwilling to increase its spending on space, it may appear somewhat whimsical for an Englishman to insist on the importance of space technology. Whatever role the British finally decide to play, however, communications satellites, and satellites for other purposes, will inevitably play an increasingly important part in our lives, and an understanding of their capabilities and their limitations forms part of what the well educated person needs to know in the modern world. The knowledge is useful not only to the young aspiring engineer, but also to the established businessman with no pretensions to technical prowess. Space techniques are being used in an ever-widening range of activities, and it is essential to de-mystify the subject by providing explanations which allow non-specialists to make their own assessment of the extent to which these new techniques can help them.

Satellites by no means provide a universal remedy: they can sometimes offer unique services, but in most cases they are in competition with other techniques. The services which satellites can offer must be judged in terms of comparative cost and reliability—no one can today afford to be sentimental about space technology. It follows, therefore, that many potential users need to have a better understanding of the new possibilities—rather than be entirely in the hands of the enthusiastic specialists. I am sure that they will find an answer in this book.

It will, of course, also be of help to students, whether or not they intend to follow a career in space. Indeed, in the coming decades it will be hard to find a career which is not in some way involved in, influenced by or dependent upon space.

Let me also express the hope that Mark Williamson's work will help to re-activate official interest in space here in the United Kingdom, as well as preparing young Britons for the day of the awakening.

Roy Gibson†
25 December 1988
Capdella, Mallorca

†Roy Gibson was the first director general of the European Space Agency (1975–80) and the first director general of the British National Space Centre (1985–7). In 1987 he was appointed special adviser to the director general of Inmarsat.

Preface

My principal aim in writing this book is to fill a gap I perceive in the literature available to the student of spacecraft technology. Many books on satellite communications have been published in the last decade. In the main they cover the techniques involved in communicating by satellite, but say little of the hardware itself. This book, in contrast, is a 'what it is and how it works' guide to the communications satellite.

The desire to produce such a book stems originally from my frustration as a young engineer in not being able to find a book that explained clearly and concisely, and without resort to tedious pages of mathematical 'flute music', the basic workings of the communications satellite. The non-mathematical nature of this book should make it accessible to most readers with a technical interest in the field. Although the book is aimed primarily at the undergraduate and young engineer, it is by no means beyond the interested layman—or indeed the more experienced members of the engineering profession, few of whom can claim an in-depth knowledge of all the aspects of technology embodied within the communications satellite.

The book itself is divided into two main parts. After a discussion of satellite design philosophy, the first section covers the major spacecraft subsystems chapter by chapter. Sadly, most engineers are educated to adopt a rather blinkered and parochial view of their chosen specialisation. Although the chapters on subsystems present the problems inherent in the design of a particular subsystem and the relevant engineering solutions, running through them is the interactive nature of satellite design—almost nothing can be designed in isolation. This theme is continued through chapters covering both ends of the communications link—the satellite payload and the earth station. Later chapters deal with aspects of the industry which, although peripheral to the satellite hardware, are essential for a full appreciation of the design and operation of the communications satellite, namely launch vehicles, spacecraft reliability and satellite insurance. Part A concludes with a discussion of future trends in communications satellite technology, without which a text on the communications satellite would be incomplete.

Part B is an adjunct to the above in that it concentrates on the *raison d'être* of the communications satellite, that is the communications link it is designed to provide. Although the section is a case study of a particular type of communications satellite—that designed for direct broadcasting of television—the techniques can be applied to the communications system of any space vehicle, unmanned or manned. The section discusses the regulatory framework behind direct broadcasting by satellite (DBS), the fundamental resources of a satellite communications system and the techniques of interference reduction. It concludes with the derivation of a link budget for the uplink and downlink transmission paths, the calculation that reflects the ultimate viability of the system. This final chapter is believed to be unique in its coverage and depth of explanation of the most fundamental calculation in satellite communications.

Although the book as a whole is an introduction to the communications satellite, the subsystem chapters taken in isolation can be applied to almost any spacecraft. All spacecraft have thermal, power and attitude control requirements; it is only the application of the techniques which differ. In addition, all active spacecraft need a communications system, and all such systems are designed using link budget techniques similar to those outlined in part B.

No book can hope to be entirely definitive where such a wide-ranging subject is concerned. This book contains little of the design of the electronic control systems scattered liberally throughout the typical communications satellite; nor does it cover in detail the transmission and coding methods used in satellite communications. Both these topics are subjects in their own right and have had many excellent texts devoted to them. I hope the limitations I have placed on the contents do not detract from the usefulness of this volume and that it will enhance the reader's understanding of this important technology.

It is not too many years since the by-line 'Live via Satellite' marked a special event in the evening news programme. The fact that the label is now rarely seen shows how this technology has been absorbed into our everyday lives, making little more impact than an interview recorded in the local high street. This is *not* an indication that the technology is less important or impressive than it was; quite the contrary. It is simply a testimony to its maturity.

The communications satellite has reached a maturity afforded only to those aspects of technology which excite the interest of the mass media when they go wrong. This book is an introduction to the engineering design solutions intended to ensure the anonymity of the communications satellite.

Mark Williamson
Arlesey, Bedfordshire
January 1989

Acknowledgments

The author of a book which covers a multitude of different technologies is very rarely master of them all. Realising this from the start, I sought the assistance of a number of friends and colleagues who specialise in one or other of these fields and asked them to read sections of the original manuscript covering their speciality. I am indebted to them for the time they gave so freely and would like to acknowledge them here. Although they have helped to improve this book, they are not of course accountable for the text as it appears—any errors are entirely my own responsibility.

I would particularly like to thank Tom Keates, Alan Hutchinson and Mike Armstrong-Smith who, between them, read a large proportion of the manuscript. I am also grateful to Ron Cooper, Jehangir Pocha, Roger Franzen, Anthony Giles, Ian White and Karen Burt whose contributions were invaluable. Many thanks are also due to Geoff Statham, Paul Brooks, Brad Smith, Gerard Firmain, Brian Sparke, Alan Dean, Claude Bonnet, Philippe Rasse, Adrian Stear, Bob Parkinson, John Watson and my brother Hugh for their helpful comments.

I would also like to thank Roy Gibson for sparing the time (on Christmas Day note!) to write a foreword to this volume. And, *in memoriam*, I would like to pay tribute to the late Larry Blonstein, a kindred spirit in spreading the word on satellites: thanks for the inspiration and encouragement during those halcyon days at British Aerospace.

Finally, I would like to thank my wife, Rita, who has done everything from typing my first drafts and correcting the later versions to proof-reading the final manuscript. Her belief in the book has helped me through the times when the words refused to flow and the brain refused to function.

MW

Part A

The Communications Satellite

A1 Satellite Design

1.1 INTRODUCTION

Although spacecraft engineering has its roots in the aeronautical industry, spacecraft designers have an extra, severe and overriding constraint placed upon them: their product has to function in space. A typical satellite is required to operate for a period of up to 10 years, totally unattended, in a vacuum and with temperature extremes well outside the freezing and boiling points of water. All electronic and mechanical components are expected to operate perfectly without on-site servicing, adjustment, lubrication, parts replacement or replenishment of consumables. Little wonder that test programmes for spacecraft equipment are the most stringent ever devised.

The main purpose of this book is to describe and explain the principles which govern the design of the communications satellite and, by extension, most other satellites. Like other branches of high technology, the design process is largely evolutionary. The communications satellites of today are more complex in appearance and vastly superior in capability compared with those early examples of orbiting relay stations, but the key to their operation is rooted in the same basic physical principles. All orbiting spacecraft are governed by the laws of motion and gravitation, are required to operate in the hostile environment of space and, if they are to be of any practical use, must communicate with their operators on Earth.

Later chapters discuss the various subsystems of the satellite and its communications payload. This chapter introduces the types of satellite available.

1.2 SATELLITE DESIGN TYPES

Despite the perceived variety of satellites, displaying a panoply of extendable booms, panels and dishes, and used for meteorology, telephony, surveillance and TV relay, there are but two basic types of satellite design which are fundamentally different in nature and readily distinguishable, even to the casual observer. Figures A1.1 and A1.2 depict the two basic forms which result from an initial design decision concerning the method of stabilisation to be used. The essence of this decision is whether to spin the whole body of the satellite, to obtain gyroscopic stability like a spinning top, or to spin a number of wheels within the stationary body of the satellite. These methods result in the *spin-stabilised* and *body-stabilised* designs, respectively. The colloquial term for the former is the *spinner* and, due to its shape, it is sometimes known as *drum stabilised*. A better description of the body-stabilised satellite refers to the way the wheels are arranged within the body: they are usually aligned with the three orthogonal (mutually perpendicular) axes. This type of spacecraft is sometimes described as axially stabilised, but again the more common term is *three-axis* or alternatively *tri-axis stabilised*.

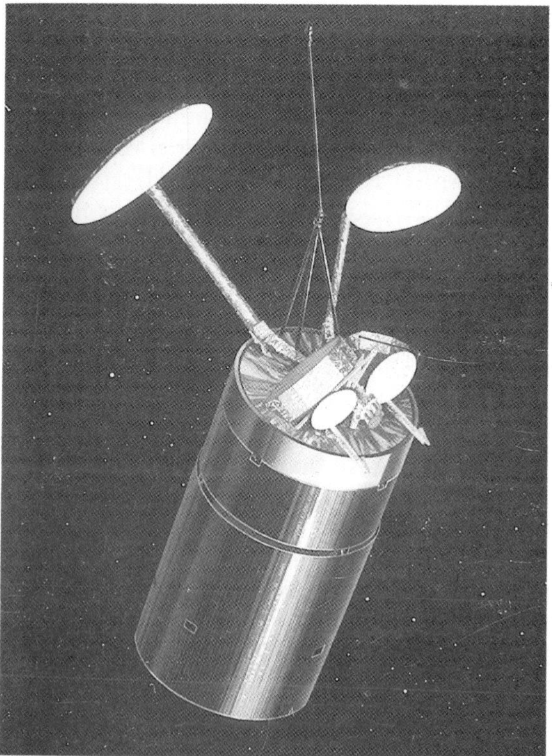

Figure A1.1 Intelsat VI—an example of a spin-stabilised satellite. [Hughes Aircraft Company]

Figure A1.2 Intelsat V—an example of a three-axis stabilised satellite. [Intelsat]

Historically, the choice for a particular satellite (between spin and three-axis stabilisation) has depended largely on the organisation submitting the winning bid for the contract to manufacture that satellite, since most prime contractors specialise in one or the other. In future it is more likely to be the size of the satellite that governs the choice, as we shall see.

Due to the complexities of the contract process, which can involve a number of aerospace companies or consortia bidding competitively, it is difficult to make a definitive statement on satellite genealogy. However, it is fair to say that most contemporary commercial spin-stabilised communications satellites are made by Hughes Aircraft Co. (USA), who have chosen to develop this technology. Perhaps recognising the difficulties in competing with the giant Hughes corporation, most other prime contractors have chosen to develop the three-axis-stabilised spacecraft. Examples in this category are Ford (USA), Aerospatiale (France) and British Aerospace (UK). The pros and cons of these competing design-types are discussed in box A1.1.

The method of stabilisation is of fundamental importance since it affects practically every other function of the satellite. All satellites have the same basic requirements, so the basic subsystems remain a necessity, but the dependence on the method of stabilisation manifests itself in the manner of the design solution. The most obvious manifestation is in the structure of the spacecraft itself. The cylindrical spinner and the cubic three-axis satellite present two very different volumes for the

BOX A1.1: SPIN STABILISATION VERSUS THREE-AXIS STABILISATION

The effect of the choice between the two basic design types on the other subsystems is discussed below with a view to identifying the best design.

Stability [Chapter A4]:

A spinning body has an inherent stability, especially if it is a wide flat disk. Unfortunately, launch vehicle shrouds restrict a satellite's width, and the need to provide power from large areas of solar cells tends to increase its height. As the power requirements increase, the drum becomes taller and its inherent stability decreases. The addition of a de-spun platform contributes to the satellite's instability by actively encouraging nutation, to the extent that these so-called two-body spinners require active nutation damping in both geostationary transfer orbit (GTO) and geostationary orbit (GEO) [see figure A4.9 and box A4.1]. In addition, as the fuel tanks are drained, the remaining propellant becomes increasingly prone to 'sloshing', which represents an energy loss to the spinning section and contributes to the destabilisation. So what begins as an inherently stable design results in practice in a demonstrably unstable configuration.

The penalties for the more complex three-axis attitude and orbital control system are that it uses more power and accounts for a greater proportion of the mass budget in small satellites.

Power [Chapter A5]:

However, the advantage lies with the three-axis satellite as far as power generation is concerned, since the size of the array—folded or rolled—is not fundamentally limited by payload fairings, and all the cells can be used together. Confining the cells to the drum surface of a spinner limits the area and therefore the power which may be derived from a particular array since only about a third of the cells can operate at a given time.

In both designs the power has to be coupled via slip rings from the rotating part of the spacecraft to the relatively stationary part. The difference lies in the speed of rotation: about once every second for the spinner and once every 24 h for the three-axis type. The two arrays of the three-axis designer to fill with the necessary plethora of subsystems, and require two differing structural designs to ensure that the spacecraft does not collapse under the forces of the launch.

1.3 THE MODULAR CONCEPT

The philosophy which unifies the disparate designs is that of the modular concept. It is invariably more efficient, in terms of both time

satellite offer at least some redundancy should one of the bearings fail. Although such failure is unlikely, the spin-stabilised design features a possible *single point failure* in the provision of both stability and power.

Thermal [Chapter A6]:

The spin-stabilised design does, however, offer an advantage to the thermal subsystem, since the solar cells absorb heat for only about one-third of the time but radiate all the time, making the average temperature of a spinner lower than a three-axis satellite. In fact the thermal subsystem for a spinner is much simpler overall and often largely passive in nature.

Propulsion [Chapter A3]:

The propulsion system is also simpler. A spinner requires fewer thrusters than a three-axis satellite since the rotation of the spacecraft moves the thrusters into a position where they can be pulsed in the desired direction. It is all a matter of timing: radial thrusters can fire at the correct moment to provide any desired torque. The three-axis satellite requires thrusters mounted in all three axes [see §A4.5].

Propellant feed on a spinner is by centrifugal force, so there is no need for complex pressurisation systems or propellant management devices [see box A3.2]. The spinner is generally simpler, since there are less components to malfunction, but redundancy is always built in and the three-axis craft is more flexible should a problem arise.

Payload [Chapter A7]:

The design of the spinner restricts the number and size of antennas since they must all be de-spun; the three-axis satellite offers several faces on which to mount antennas and their size is limited only by antenna technology, not the design of the spacecraft.

Since the communications capacity of a satellite is related directly to the available power, the available area for the solar arrays becomes a limiting factor. Greater power means greater communications capability. This suggests that spin stabilisation is only ideal when the satellite is small, low powered and uncomplicated.

and cost, to construct a number of functionally separate modules in isolation and bring them together at a relatively late stage to form the final article.

To draw an analogy with automobile manufacturing, the concept may be illustrated by the fact that car engines, tyres and radio-cassettes require very different production environments and varying degrees of skill and time for their manufacture. It would, for example, be inefficient to bring anything less than the completed power unit to the vehicle assembly line and more than unwise for the technician to solder radio components from the half-finished passenger seat!

As might be expected, the financial scale of satellite manufacturing is somewhat different, but the philosophy of modularity is essentially the same.

The communications satellite typically comprises two major, functionally separate, modules: the *service module* and the *payload module*. The relationship between the two modules is analogous to that between the chassis and power-train of a motor coach, and its upholstered passenger cabin. In both cases the former provides the motive power, stability and environmental control, while the latter carries the revenue-earning part of the load.

The payload of a communications satellite is the equipment which provides the communications service. The subsystems necessary to ensure the continual operation of the payload are housed within the service module, otherwise known as a *platform* since it provides the structure and stability required by the payload. The service module is known colloquially as the *bus* (as an omnibus carries people, it 'carries' the payload).

One major reason for building the two modules separately is that they contain technologies which have historically been concentrated upon by different industrial contractors. A typical spacecraft project, therefore, may have the service module built by one company and the payload by another. They will be tested separately and then brought together for the final *integration* and further testing in the completed configuration. Naturally, individual components of the two modules are provided by numerous subcontractors who integrate submodules assembled from increasingly insignificant components as the contracting tree branches further.

Given the understanding that no one company can manufacture every component of the satellite, the enduring desire of most major contractors is to supply the 'complete system'. The nearest any company has come to this is the position held by the Hughes Aircraft Co., who provide platforms, payloads and even the earth stations forming the other end of the communications link. Given an equivalent time and sufficient capital, the competition is likely to reach this position. In the meantime, realising that growth without direction is a risk-ridden policy, most of the world's satellite contractors have chosen to form consortia, which can show a united front of complementary capabilities to a potential customer. Thus, British Aerospace and Matra can be found grouped under the Satcom International banner, and Eurosatellite comprises AEG-Telefunken, Aerospatiale, Alcatel-Thomson Espace, ETCA and MBB/ERNO.

1.4 THE PRODUCT RANGE

In addition to encouraging everyone to 'do what they do best', the modular concept allows major manufacturers to develop a continuing sales market, whereupon further customers come forward to purchase the technology so expensively developed. Apart from enhancing a company's credibility, the introduction of a range of products using a major component is economically sound, especially if it has already been developed and can be identified as successful. There may also be a market for differently sized versions of the same product, which is certainly the case for vans and trucks. The philosophy is currently being proved for the service module of the communications satellite, despite the fact that it is far from being mass produced, in the usual sense of the word, and it is not an item which may be purchased 'off-the-shelf'.

Although satellites may contain similar components and almost identical subsystems, each one is individually designed to meet the customer's requirements, certainly as far as the payload is concerned. This is because different customers require differing services and different coverage areas on the surface of the Earth. For example, relaying telephone calls across the Atlantic Ocean between earth stations with 30 m antennas, and broadcasting television to 90 cm domestic 'dishes' in Switzerland or Luxemburg, represent two very different requirements. The former requires a satellite payload with low-power transmitting equipment operating into a simple spacecraft antenna providing global or hemispheric coverage, whereas the latter involves high-power transmitters and specially shaped antenna beams to confine the signal to the area concerned.

The service module, however, is less dependent on the particular service provided and can be used for any number of different applications. There are a number of mitigating factors, such as the provision of differing amounts of power and a variable requirement on the accuracy of *antenna pointing*, but the basic platform may be marketed as a product in its own right. Thus, for example, in Europe there is the Spacebus family (by Aerospatiale and MBB/ERNO) used in its '300' version for the TV-SAT, TDF and Tele-X direct broadcast satellites, and, in the United States, the HS376 (Hughes standard spinner) used for a long list of communications satellites including SBS, Westar, Aussat and Morelos. TDF-1 and an SBS HS376 are shown in figures A1.3 and A1.4 respectively.

Figure A1.3 TDF-1 three-axis-stabilised satellite in cleanroom prior to testing. Note folded solar array panels on side. [Aerospatiale]

Although more than one satellite will usually be built under each contract, the average production run will not exceed three or four. Although production runs for Intelsat spacecraft are invariably larger, a more typical customer may require just two, practically identical, operational spacecraft and a spare. This surely amounts to a very limited edition.

Figure A1.4 SBS spin-stabilised satellite—one of the successful HS376 series. Note the lower telescoping array panel shown in its deployed position and the reflective radiator panel on the upper section. [Hughes Aircraft Company]

1.5 IDEALISED DESIGNS

The form of the modular concept for both three-axis and spin-stabilised satellites is illustrated in figure A1.5. Figure A1.5(*a*) shows how the three-axis design is based around a central thrust structure, in this case a corrugated cylinder, which provides stiffness in the vertical axis to protect the spacecraft from the forces imparted by the launch vehicle. The thrust cylinder also acts as a container for the spacecraft's primary propulsion system—its apogee kick motor [discussed in chapter A3]. The remainder of the service module consists of a number of floors and panels arranged, together with various strengthening devices, to provide a rigid structure and offer mounting points for the service module equipment and its connecting wiring harness. Mounted above and around the service module are the payload and antenna modules; the latter may be structurally part of the payload module or a separate antenna floor or platform. The two vertical panels shown in the figure have the communications payload equipment, the *transponders*, mounted upon them.

Figure A1.5(*b*) shows the equivalent idealised design for the spin-stabilised satellite. In a similar manner to the three-axis satellite, the basis of the service module is a central thrust tube containing the primary propulsion system. In this case, however, the tube comprises a lower conical shell, an intermediate cylindrical shell and an upper cone.

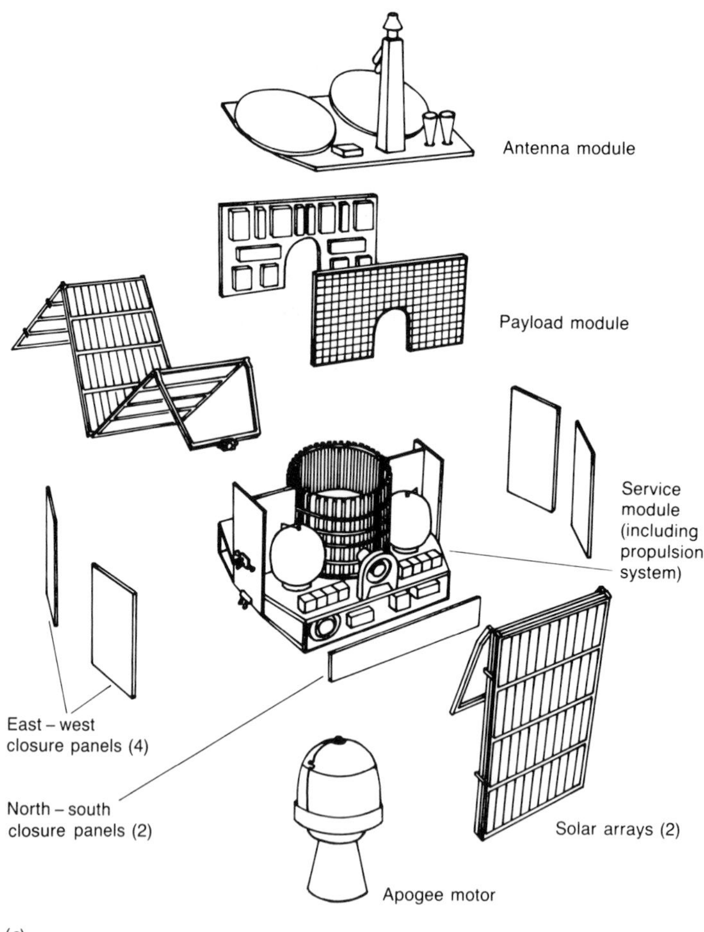

Figure A1.5 (*a*) Modular design of a three-axis-stabilised satellite; (*b*) modular design of a spin-stabilised satellite.

Various horizontal equipment platforms or shelves are supported between the thrust structure and the cylindrical outer hull of the satellite. The upper shelf holds the communications payload and, along with the antenna subsystem, may be termed the payload module. With this design the service and propulsion modules are spun to provide stability, but the antenna platform must remain stationary with respect to the Earth, otherwise the radio beams formed by the antennas would sweep across the Earth like searchlight beams. The antenna platform must therefore be *de-spun;* whether the shelf containing the communications transponder is de-spun too depends on the specific design.

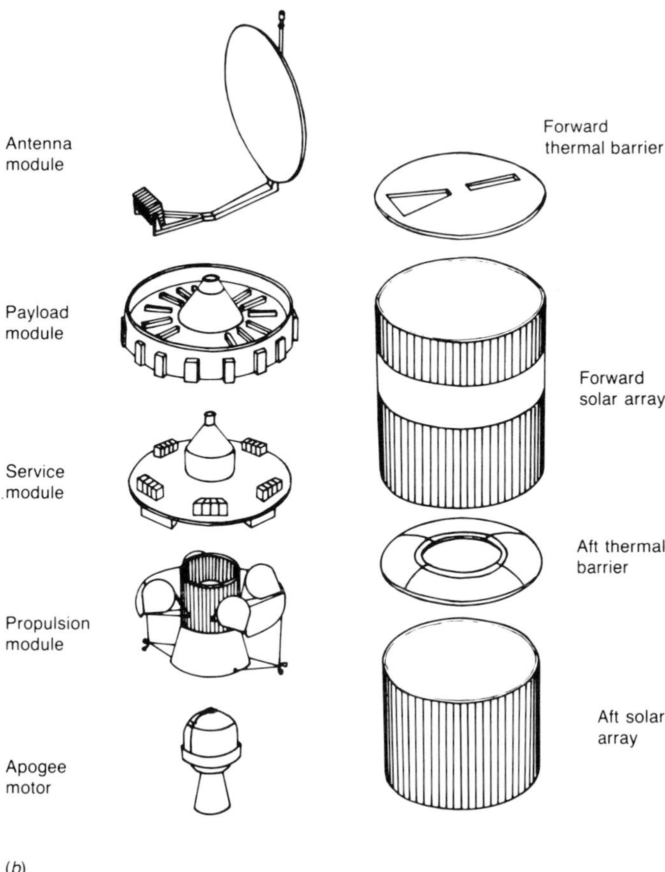

Figure A1.5 (continued)

1.6 DESIGN CONSTRAINTS

Having discussed the design philosophies behind the manufacture of the communications satellite, in part by comparison with the automotive industries, it is time to return to the restrictions which set it apart.

Perhaps more than any other product of advanced technology, the design of the spacecraft is limited by a large number of constraints, mostly due to the requirement that the spacecraft must be lifted into orbit. With typical communications satellites weighing 1 or 2 tonnes, an idea of the energy required can be gained from a knowledge of the velocity of a satellite in geostationary orbit, that is 3075 m s^{-1} or 11 070 km h^{-1} (although this is only part of the story [see box A10.1]). This is the main reason why launching satellites is a very costly business: every extra kilogram increases this cost by thousands of dollars. In fact, at the end of 1988, commercial launch prices were said to be between about $24 000 and $26 000 per kilogram [Dornheim 1988]. The cost of the satellite itself (in 1987 dollars) was, according to Hudson and Gartrell [1987], about $70 000 per kilogram. This means that the average 1 tonne satellite would cost the best part of $70 million.

The satellite is therefore severely *mass limited* and every effort must be made to save weight wherever possible. This has led, for example, to the widespread use of aluminium honeycomb panels and composite materials such as carbon-fibre-reinforced plastic (CFRP) [see chapter A2].

In addition to limiting the satellite's mass, the launcher constrains the spacecraft's size and shape; fitting the payload within the launcher shroud is as much an art as a science. Take the three-axis-stabilised satellite for instance: like the spinner, if its antenna subsystem is large, it must be folded to fit inside the shroud, but its solar arrays exhibit an even greater need for compression [see figure A10.7]. From the very beginning of the design process the *envelope* of the launcher must be considered. This process has been taken to its logical conclusion in the case of the Hughes spin-stabilised Syncom IV satellite, which has been designed to have the maximum diameter allowed by the Space Shuttle payload bay [see figure A10.9]. This, in effect, decreases the height of the drum and the overall cost of launching, since the shuttle launch costs are related to mass *and* length of payload bay utilised.

Another constraint is that of the power available to the satellite once in orbit. One major component of the communications satellite is the device which provides all electrical power for the lifetime of the spacecraft—the solar array. For this important component the design type of the satellite has a fundamental effect, requiring the arrays to be mounted in significantly different ways. On the spinner they are wrapped around the periphery of the drum, whereas the three-axis craft features array wings mounted on opposite faces of the spacecraft. These arrays are folded against the sides of the spacecraft for launch and are deployed in orbit.

Owing to the limited conversion efficiency of the individual photovoltaic cells, fairly large areas of cells are required to produce even modest amounts of power, which of course limits the overall power available for communications services. Power generated by the solar arrays may be stored in batteries for later use, but this is only a temporary storage medium. The number of batteries a satellite may carry is severely limited by their weight, so the only way to obtain more power is to increase the size of the solar array [see chapter A5].

It is in this respect that the spin-stabilised satellite is at a disadvantage because of the limited area of its drum surface [see box A1.1]. In fact, only about a third of the spinner's cells produce power at the same instant, since only that number face the sun at a time. The three-axis satellite, on the other hand, may deploy arrays much larger than the surface area of its body and, moreover, all of them may be used together.

Increasing the size of the arrays is, however, not the ideal solution it might at first appear. The main component of the satellite communications system [see chapter A7] is the travelling wave tube amplifier (TWTA), a particularly power-hungry device. Typical TWTAs are at best about 50% efficient in their conversion of DC power (from the arrays) to the radio-frequency (RF) power of the amplified signals. This means that one-half to two-thirds of the input power is effectively wasted as heat, the common by-product of any power amplification process. The heat produced *must* be radiated from the spacecraft to maintain its thermal balance [see chapter A6]. However, the amount of thermal energy which can be radiated in a given time is dependent upon the surface area of the radiator, which, in the case of a satellite, is severely restricted by the available surface area of the spacecraft. We are back to the size constraint.

This, in part, has led to the design of larger second-generation satellites which have larger surface areas available for both the mounting of solar cells, in the case of the spinner, and for the radiation of excess heat.

One final design constraint concerns the lifetime of the satellite itself. Apart from the natural tendency of components to wear out, life is limited by the amount of station-keeping fuel the spacecraft can carry, which again 'costs' mass. In orbit the satellite is affected by various gravitational and other forces which tend to move it from its desired position. It is returned to its nominal position by firing small rocket thrusters which consume the above-mentioned station-keeping fuel. Allowing for a certain amount of fuel to boost the satellite out of the geostationary arc at the end of its life (whimsically referred to as the graveyard burn), the depletion of fuel means the death of the satellite because it can no longer be effectively controlled. It *is* possible to calculate the amount of fuel required for a certain lifetime, but any unforeseen use of propellant, for instance to correct an inaccurate orbital placement by the launch vehicle, will decrease that figure.

1.7 CONFLICT AND INTERDEPENDENCE

With the design of complex systems, design constraints rarely combine in a complementary fashion. Applying restrictions to one factor will usually affect other factors in a detrimental way. The design of the communications satellite is no exception to this immutable law.

The constraints on the launch vehicle limit the size, shape and mass of the spacecraft, which limit *its* capacity for power generation, heat rejection and life extension. The structure must be strong enough to hold the satellite together but light enough to reduce launch costs and allow more useful payload to be carried. The solar arrays must be large to provide power for the communications payload and service subsystems, and the spacecraft must be large to provide sufficient radiator area, but the package must be compact to fit within the launcher shroud. In practical terms, of course, these constraints are more than mere concepts, and solutions must be found. The solutions emerge in an iterative fashion over a number of months or years, mainly through the efforts of a large team of design engineers each specialising in their own particular subsystem. It should be clear, however, that no subsystem can be designed in isolation; the interdependence of spacecraft subsystems is very great indeed.

The conflicting interests of the subsystem design engineers may be likened to a forum of political discussion—a blend of logical argument and 'party policy' laced with scientific fact, hearsay, threats and lies. In the end it invariably turns out that the most elegant design solution exploits the art of compromise.

REFERENCES

Dornheim M A 1988 *Aviat. Week Space Technol.* 19 Dec. 73–4

Hudson W R and Gartrell C F 1987 *Int. Astronautical Fed. Congr.* IAF-87-07

A2 Structure and Materials

2.1 INTRODUCTION

It will be apparent from the previous chapter that no satellite subsystem can be designed in isolation. The design of the structural subsystem, while no exception, is also inextricably linked to that of the vehicle employed to deliver it to orbit. In fact the primary constraints on a satellite design are those imposed by the launch vehicle, since it places a limit on the size, shape and mass of the satellite. In consequence, two of the main engineering concepts which guide the spacecraft designer in the initial stages and throughout a project are the *payload envelope* and the *mass budget*.

The payload envelope defines precisely the maximum volume available and constrains the size and shape of the satellite and its various appendages, such as antennas and solar arrays. [Example payload envelopes for the Ariane expendable launch vehicle and the Space Shuttle are shown in figures A10.6 and A10.8 and described further in chapter A10.]

The mass budget is a method of accounting: each subsystem engineering discipline designs within it and reports against it, so that throughout the project the overall spacecraft mass can be monitored. A simple example budget featuring the major subsystems is given in table A2.1. The figures, which are percentages, have been derived by averaging the subsystem masses of a number of actual spacecraft, both spinners and three-axis-stabilised satellites. A real mass budget would of course show the actual masses in kilograms and would incorporate a *balance mass*, typically about 1%, which must be allowed for when balancing the total spacecraft in preparation for launch. It should be noted, however, that this budget represents the *dry mass* of the spacecraft; the addition of propellant for orbital injection and station keeping can more than double the mass.

Table A2.1 Mass budget.

Subsystem	Three-axis (%)	Spinner (%)
Structure	18	21
Propulsion (AKM + RCS)	12	11
AOCS	7	5
Power	23	24
Thermal	4	5
TT&C	4	5
Payload (including antennas)	28	25
Electrical harness, etc	4	4
	100	100

The structure of the spacecraft accounts for about 20% of the total mass, a figure which tends to decrease for larger three-axis craft and increase for larger spinners as their diameters increase. Figure A2.1 shows that the cross-over between the two is at a dry mass of about 500 kg.

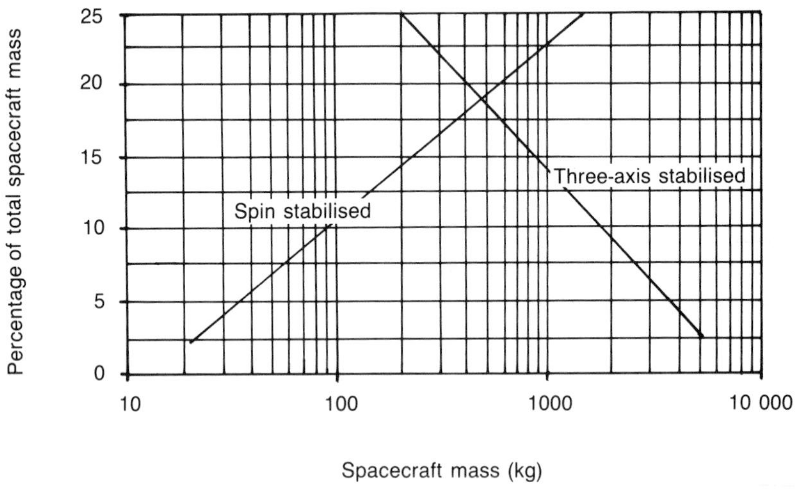

Figure A2.1 Variation of structure subsystem mass with overall spacecraft mass for spin-stabilised and three-axis-stabilised satellites.

The dimensional constraint and the mass constraint are of course intimately linked, since big satellites tend to weigh more. This leads to a fairly obvious conclusion: if the available launch envelope is to be filled with a useful package of communications equipment, the structure must contribute as little as possible to the overall mass—lead sheets on a cast-iron frame are definitely out! This makes the constituent materials of a satellite structure almost as important as the design of the structure itself.

The incentive for the continual development and use of low-density structural materials stems, in part, from the progressive drive to launch bigger and heavier satellites. The continual pressure from the payload community spurs on the launcher designers to produce more powerful vehicles with larger payload envelopes, but the spacecraft designers then have to ensure that the spacecraft which fills the envelope does not exceed the lifting capability of the launch vehicle. Part of the solution is to specify materials with ever-improving strength-to-weight ratios.

2.2 STRUCTURAL REQUIREMENTS

The structure of a spacecraft is required to provide accommodation for the subsystem equipment, support it in a manner which will maintain the alignment of its constituent parts and protect it against the rigours of the launch and the conditions of the orbital environment.

In terms of accommodation, the structure must provide sufficient volume for bulky items such as the apogee kick motor or propellant tanks and sufficient mounting area for the numerous boxes of electronics which constitute the payload. The placement of certain items is complicated by their need to radiate thermal energy, confining them to specific parts of the structure. Batteries and travelling wave tubes, for example, generate large amounts of heat which must be radiated to the external environment in order to maintain the temperature of the satellite [see chapter A6]. While providing space for the equipment, the structure must be designed to facilitate initial construction and continual access at all stages of the integration process.

When it comes to 'support', several components of the various subsystems require more than a panel to which to be bolted. Antennas and feedhorns, for example, must be accurately aligned to provide the desired coverage pattern on the Earth. Any relative movements due to mechanical loading or thermal distortion must be minimised by the design of a rigid support structure, or *antenna platform*, and the use of materials with low thermal coefficients of expansion. The spacecraft attitude control system, which controls the overall alignment of the antenna platform, can only operate effectively within a rigid supporting structure: the optical axes of its sensors must bear a known and constant relationship to the mechanical axes of its momentum wheels and reaction control thrusters [see chapter A4]. In order for the attitude and orbital control system (AOCS) to produce predictable results under operational conditions, the mechanical configuration of the spacecraft must ensure that the centre of mass, static and dynamic balance and inertia properties are maintained. Failure to address this requirement would result in wobbling satellites!

One of the primary requirements for a spacecraft structure is that it should protect and support the satellite against the mechanical loads experienced at launch, during the transfer orbit and throughout its orbital lifetime. The most severe force experienced by the satellite is that exerted by the launch vehicle as it imparts the thrust required to lift it into orbit. There are, however, other loads such as vibration, stage-separation shocks and further 'kicks' as second- and third-stage motors fire. Although the launch environment of the Shuttle is less severe, because it is a manned vehicle, satellites have typically been designed for compatability with both the Shuttle and expendable vehicles and the

BOX A2.1: ELECTROSTATIC DISCHARGE (ESD)

The phenomenon of ESD was first reported in the early 1970s when it was discovered that spacecraft surfaces were becoming charged. In 1979 the US Air Force launched the Scatha (spacecraft charging at high altitude) satellite to gather data. In fact it was electrostatic *discharge* that constituted the potential hazard, generally to electronic devices but particularly to astronomical detectors and the like. A vague indication of the severity of the hazard is represented by the Rosen scale, where 0 is 'no hazard', 1 is a 'nuisance' or outage of a second or less, 5 an outage of a few hours and 10 a catastrophe.

Although the scientific community recognised ESD as a real problem, satellite users, industrial contractors and project managers were sceptical about the seriousness of the threat [Kalweit 1981] because of the difficulties in proving a connection between ESD and the observed anomaly. Following more recent ESD occurrences, however, the phenomenon is now widely recognised. One satellite which experienced ESD problems was Telecom 1A in 1984: charging of the satellite's thermal protection and subsequent discharge caused the military communications package to switch off on several occasions.

Solutions

General: the outside surface of the satellite should be conductive and grounded to the structure to avoid differential charging. Treating the entire surface area may be too expensive, but particular attention should be paid to shaded areas, which are not discharged by photoemission.

Thermal control surfaces: conductive black paint for radiator surfaces and conductive coatings for optical solar reflectors (OSRs) should be used. The structure must be capable of withstanding the harshest launch regime.

In transfer orbit most satellites are temporarily spin stabilised, whether spinners or three-axis craft, and thus experience rotational stresses as well as further motor thrusts [see chapter A3]. In orbit the mechanical stresses are more benign, being limited to rotational stresses for a spinner and the occasional thruster firing for both types of stabilisation. All these loads must be specified and analysed during the design phase, using data gathered on previous missions, and the final structure must be subjected to simulated loads during the test phase to check the design.

Other important structural requirements include the provision of thermal and electrical paths. Some parts of the structure will be required to conduct heat, whereas others must be thermal barriers. The spacecraft should have an electrically conductive ground plane to help prevent electrostatic discharge [see box A2.1] and, in some cases, to provide a ground return (or *earth*) for the electrical distribution system, in a similar manner to a car body. This 'zero reference' is essential for the inter-

expense of conductive coatings led ESA to develop conductive flexible second-surface mirrors (SSMs): for example, Kapton-based SSM with conductive indium–tin oxide layer. However, current SSMs are of a more efficient non-conductive type, the emphasis being placed on making electronic components less susceptible to ESD at the outset [see chapter A6 for a description of OSRs and SSMs].

Solar arrays: conductive coatings are available but, mainly due to the labour-intensive interconnection required and the loss of about 3% of the power, they are only recommended for scientific satellites with stringent charge requirements. For other satellites the sunlit face of the arrays is discharged sufficiently by photoemission; the rear faces of 'wing' arrays do not benefit from this advantage and are thus more prone to ESD (carbon-fibre composites are recommended in these cases). The European manufacturers of solar arrays use cover glasses doped with cerium, which therefore have some inherent conductivity. In this respect they are superior to the quartz cover glasses used by US manufacturers.

Circuit protection: general EMC (electromagnetic compatibility) techniques, grounding and shielding should be used. Protective devices, usually filters and diode-like limiters, should be inserted in the input and output lines.

Conclusion

If the guidelines of ESD proofing are followed, the ESD hazard should be no more serious and no less predictable than other design problems (namely vibration, thermal loads, etc). The risk can be reduced to less than 1 on the Rosen scale.

operation of equipment with widely differing voltage requirements [see chapter A5]. A final requirement of the spacecraft structure is protection from the hostile environment of Earth orbit, that is mainly ultraviolet (UV) radiation, thermal radiation and micrometeoroids.

Before looking in detail at the space environment and the materials developed to fulfil the above requirements there, we shall investigate the design solutions applicable to the basic spacecraft structure.

2.3 STRUCTURAL CONFIGURATION

In general, spacecraft structures are derived from the basic structural elements of aeronautical practice, like the longeron and truss in figure A2.2, which formed the backbones of the earliest satellites. In more recent years satellite structures have followed their own line of evolution, from the simplest structural types to that based on the thrust tube or thrust cylinder.

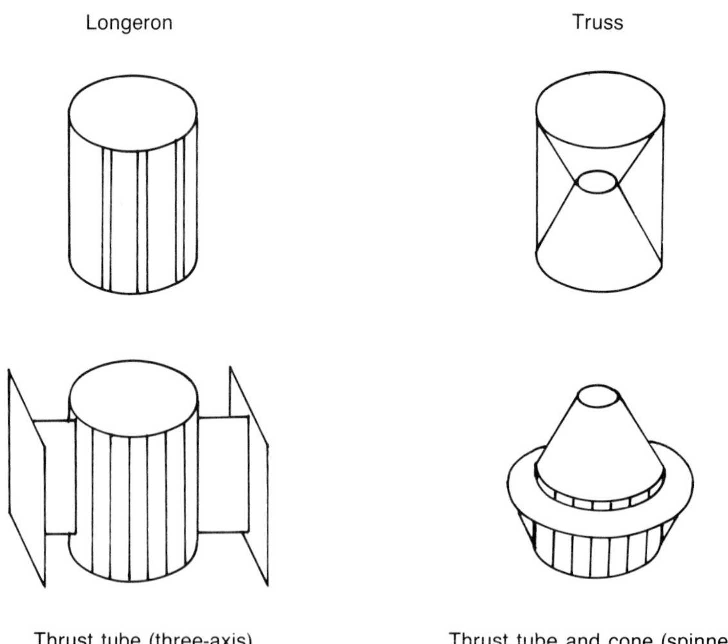

Figure A2.2 Derivation of spacecraft structures from basic aerospace structures.

The basic structure is designed to transfer all loads on subsystem equipment and minor structural elements ultimately to the launch vehicle interface, which for an expendable launch vehicle (ELV) is a simple horizontal ring forming an adaptor between the spacecraft and the launcher. On the other hand, it is undesirable for the vibration modes of the launch vehicle to be coupled back through the adaptor to the satellite. Dynamic coupling between the two is kept to a minimum by an analysis of resonant frequencies, which leads to an overall stiffness requirement for the satellite. Since mass must be kept to a minimum, however, the satellite's strength-to-weight ratio must be carefully optimised. We are back to the conflicting requirements of high strength and low weight introduced in chapter A1.

Figure A2.3 Orbital Test Satellite (OTS) structure model. [ESA]

In a process which ensures that the satellite structure is capable of surviving the mechanical loads inflicted by the launch vehicle, the structure can be designed conceptually upwards from the launch vehicle adaptor ring. In the most prevalent design, for both types of stabilisation, the *load path* from launcher to satellite is via the adaptor ring to a rigid, often corrugated, cone or cylinder [see figure A1.5]. Based on the philosophy of modularity introduced in chapter A1, the thrust cylinder becomes the backbone of the service module. The horizontal floors or platforms and the vertical panels, which constitute the rest of the service module's primary structure, are cantilevered off the central cylinder and braced by struts and *shear panels* in such a way that all loads on the primary structure are passed to the cylinder. The remaining structural elements, such as support brackets for propellant tanks, mountings for antenna reflectors, feed assemblies, etc, fit into the category of *secondary structure*, thus completing the load path from subsystem equipment, through secondary and primary structures, to the thrust cylinder. Figure A2.3 shows the Orbital Test Satellite (OTS) structure model as an example.

Figure A2.4 illustrates the integration process for an idealised three-axis-stabilised satellite. After the complex wiring looms have been added to the service module floors, they are joined to the thrust cylinder (*a*). The floors are then rigidised into a box structure by vertical shear panels or *shear webs* and external walls (*b*) and joined together using angle brackets and attachment *cleats*. Secondary struts attached to *hard points* on the panels tie the primary structure together and *closure panels* (added later) complete the stiff cellular construction of the box.

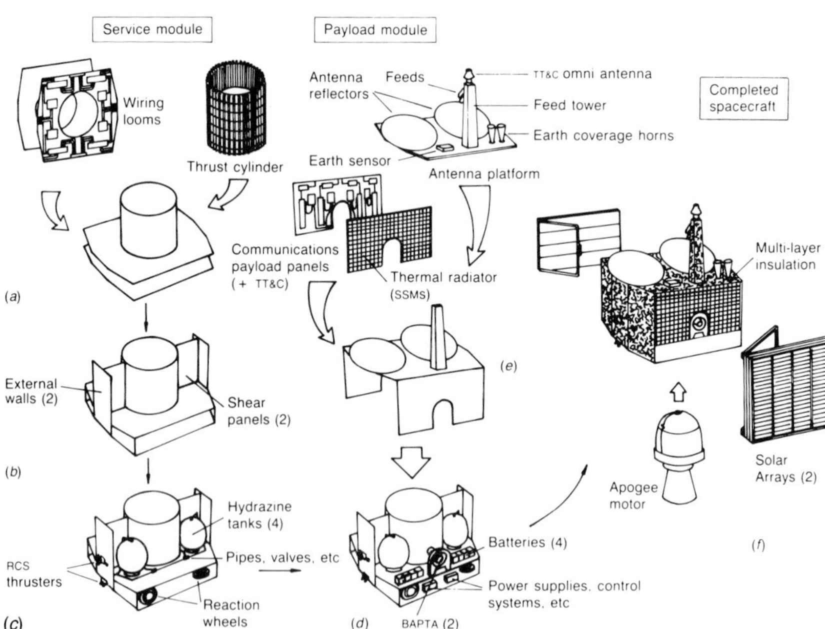

Figure A2.4 Satellite subsystem integration. (*a*) Primary structure integration; (*b*) additional structure; (*c*) propulsion and AOCS subsystems; (*d*) power subsystem; (*e*) payload module integration; (*f*) integrated modules.

Now that the service module structure is almost complete the majority of the subsystem equipment can be installed. In addition to the wiring harness, the service module holds the fuel lines, valves, thrusters and tanks of the propulsion system and the numerous boxes of electronics and control systems constituting the AOCS and telemetry, tracking and command (TT&C) subsystems (*c*).

The payload module, comprising two panels for the communications payload and an antenna platform, is constructed in parallel with the service module (*e*). The panels constitute the north and south faces of the satellite in its orbital configuration and it is no coincidence that these faces, which receive the lowest solar illumination, are the best for mounting heat-producing equipment such as TWTAs [see figure A7.7]. For the same reason, the north and south walls of the service module offer useful mounting positions for the spacecraft batteries which produce heat when discharging. The service module houses the rest of the power subsystem (*d*) and includes, on the north and south faces, the mounting points for the solar arrays, the bearing and power transfer adapters or BAPTAs.

Figure A2.5 Integration of ECS-1 payload and service modules. [ESA]

The fixing of two side-walls to the service module and two to the payload module is a convenient design for two separate modules. Apart, both modules are relatively open structures, allowing easy access for assembly and test, while together they form a rigid cube (*f*). Figure A2.5 shows the integration of the two modules of the ECS 1 spacecraft. Since the majority of spacecraft equipment, with the notable exception of propellant tanks, is housed in flat-sided boxes, the flat panels of the three-axis-stabilised satellite are particularly convenient. The spinner offers a number of segments between its concentric cylinders and equipment floors but is conceptually a more difficult volume to utilise.

As a result of its explosive potential, the apogee kick motor (AKM) is the last major item to be integrated with the spacecraft. Mounting a solid propellant AKM inside the thrust cylinder is not only convenient in terms of packaging, but also ensures that the loads imparted by the motor are coaxial with the launch vehicle loads, so the same basic load paths can be used in the design. For a liquid system the fuel tanks are balanced symmetrically in and around the cylinder for the same reason.

Figure A2.6 HS376 payload module. Note TWTs and EPCs on outer rim; filters, multiplexers and waveguide outputs on upper floor. This module is mounted on top of the service module depicted in figure A2.7. [Hughes Aircraft Company]

It should be mentioned, however, that the advent of combined liquid propulsion systems [see chapter A3] has offered the opportunity to design alternative primary structures without a central thrust tube. An example of an alternative design for a three-axis spacecraft is the Spacebus 300 (from Aerospatiale and MBB/ERNO) which has no thrust tube and features entirely separate service and propulsion modules in addition to the communications module/antenna platform.

A similar integration philosophy to that of the three-axis-stabilised satellite is followed in the construction of a spinner, in that the modules can be built and tested separately. The main difference is in the packaging: equipment is generally confined to a number of horizontal shelves, although it can be mounted on both upper and lower surfaces. Figures A2.6 and A2.7 illustrate respectively the payload and service modules of an HS376 spacecraft and figure A2.8 shows the modules combined. Heat is dissipated via heat sinks mounted on the equipment

Figure A2.7 HS376 service module. Note propellant tanks beneath shelf and batteries mounted around upper edges. The de-spun section (payload module and antennas) is mounted on the thrust cone at centre. [Hughes Aircraft Company]

platforms which radiate to the upper and lower end-shields and solar array, which in turn re-radiate to space. Larger spinners, such as this Hughes HS376 design, include a radiator panel around the periphery of the top section of the array through which the higher heat loads can be dissipated more directly to space [see figure A1.4].

Figure A2.8 Telstar 3 satellite showing integrated payload, service and antenna modules. Top: deployable communications antenna with deployable TT&C antenna folded on top. Centre: payload module showing SSPAs (left half) and TWTAs (right half). Bottom: service module showing propellant tanks. [Hughes Aircraft Company]

2.4 THE SPACE ENVIRONMENT

Having established a structural design, it is necessary to decide which materials will be best suited to the individual requirements. When called on to specify materials for spacecraft, the engineer has as a starting point those used in high-performance aircraft and missiles, where the low-weight/high-strength criterion is also important. As far as mechanical loads are concerned, the forces of acceleration for a launch vehicle are of the same order as those for a jet fighter, but this is where the similarity ends.

Even before a satellite leaves the ground, it is prone to a number of purely terrestrial problems such as oxidation and corrosion, water absorption or losses by evaporation, creep under load and biological attack. Some of these problems can be minimised by careful control of the satellite's immediate environment, using clean-room techniques, but making a satellite more fragile than it needs to be is potentially troublesome, and materials which offer some resistance to these effects must be chosen in preference.

The space environment places even more restrictions on the choice of materials. Perhaps the most obvious characteristic of space is the hard vacuum, especially in geostationary orbit, high above the outer reaches of the atmosphere. The problem here is that many materials (resins, adhesives and even some metals) evaporate in a vacuum, a process known as *outgassing*. Since there is no force available to carry the particles away from the spacecraft, they remain in a cloud around it or condense onto its colder surfaces. In this way, outgassing can degrade optical surfaces, radiators and solar arrays, and even cause shorting of electrical circuits and promote *corona* or electrical discharge.

As might be suspected, liquid lubricants also evaporate in a vacuum, and even the familiar solid graphite loses its moisture and becomes an abrasive. Although direct lubrication is not a primary consideration of the structural engineer, it represents a definite secondary skill. On Earth the atmosphere provides a number of natural lubricants, when in small quantities, such as oxide coatings and water vapour. In space there is no corrosion, and any water vapour absorbed on the ground and outgassed into the vacuum constitutes a nuisance. The lack of a lubricant or even an air gap between two metal surfaces in space causes them to diffuse into each other, a process known as *cold welding*. The solution to the problem of lubrication is to use soft metals or alloys, bearings made of ceramic materials or hybrid metal–ceramic compounds, or low-vapour-pressure grease.

A second characteristic of space is the temperature regime: the range is much larger than on Earth and the rate of change of temperature is more severe because of the lack of convection and conduction. The great extremes in temperature cause a number of problems. Most materials become brittle at low temperatures, and repeated changes in temperature, or *thermal cycling*, can produce fatigue, fracture and debonding through differential expansion and contraction. Zinc, cadmium, lead and magnesium sublimate at low temperatures and are therefore unsuitable for component surfaces. In general low temperatures promote condensation while high temperatures encourage outgassing.

An understanding of material properties enables a sensible choice to be made for a particular application. Some metals (e.g. titanium, beryllium and many steels) are largely insensitive to the duration of high-temperature exposure but have a critical temperature beyond which their desirable characteristics deteriorate rapidly. Most aluminium alloys cannot withstand high temperatures for long periods, but for low temperatures aluminium-, titanium- and nickel-based alloys are among the best metals. The majority of ceramics, however, can withstand a much wider temperature range than most metals and are little affected by time at either high or low temperature. Exposed structures such as antenna reflectors and solar arrays experience, and must be designed to withstand, significant temperature extremes. The design of the primary structure is not so constrained, because the body temperature has to be controlled to protect the payload and subsystem equipment.

The effect of external and internal sources of heat and its removal from the structure is an area where structural and thermal engineers work closely together. In fact the structural design contributes to the thermal design by providing conducting and insulating paths throughout the structure as well as suitable sites for the thermal subsystem equipment. A problem for the thermal engineer is that the low-density materials required to reduce the overall spacecraft mass are inherently poor conductors of heat [see chapter A6].

The space environment is also renowned for its exposure to the full solar spectrum and cosmic rays. To an extent a satellite must be designed to withstand bombardment by infrared (IR), ultraviolet (UV), x- and gamma radiation, alpha particles, protons and electrons, all of which, in sufficient quantities, can have an effect on the material structure. UV

BOX A2.2: PROBABILITY OF COLLISION WITH ORBITAL DEBRIS

Research has shown [Wolfe *et al* 1982] that in one year the probability of collision between a satellite in geostationary orbit (GEO) and a tracked object is between 10^{-6} and 10^{-5} (about a one-in-a-million chance). However, it is only possible to detect objects greater than 1 m in diameter at this altitude (36 000 km), so the probability of collision with much smaller objects must be appreciably greater. In low Earth orbits (LEO), where 10 cm objects are tracked and monitored by North American Air Defense (NORAD) computers, the probability of collision with a Space Shuttle orbiter on a typical 7 day mission is calculated to be 4×10^{-6}.

Extrapolating to a future space station of 50 m diameter in LEO, the probability in one year is 0.001 at an altitude of 500 km and 0.003 at 850 km, the height at which the maximum density of tracked material is observed. Although debris smaller than 10 cm across cannot be tracked directly, pieces as small as about 4 cm can be observed on re-entry: the probability of collision with objects greater than or equal to 4 cm is calculated at 0.007 at 500 km and 0.016 at 850 km. There are no data for objects less than 4 cm, but it is estimated that a 1 cm object in LEO could penetrate a 5 cm thickness of solid aluminium, which makes the apparently quite low risk seem more serious.

radiation can degrade polymeric materials, for example, and x- and gamma radiation scatter electrons in metals, which may eventually decrease their electrical conductivity, a particularly undesirable effect where low signal currents are concerned.

The more mechanical forms of bombardment from atomic particles and micrometeoroids must also be considered. Although only important for low Earth orbits, residual oxygen atoms have been shown to cause damage to polymeric materials on the Space Shuttle [Marsh 1986]. The probability of meteoroid impact on an object the size of a communications satellite is fairly low, but the increasing amount of man-made orbital debris may prove to be more of a problem in future [see box A2.2].

However, despite its vacuum, temperature and radiation characteristics, the space environment is not all bad: it presents no external source of vibration, next-to-no gravity, no corrosion, no precipitation and no wind for a spacecraft structure to withstand. In addition, space offers inherently good electrical insulation, meaning that components can be positioned closer together before arcing problems occur.

As far as materials are concerned, there is always the ideal material to aim for. It should have high dimensional stability under mechanical and thermal loads; a low susceptibility to fatigue, radiation damage and the influences of Earth's atmosphere; and, above all, a high strength, low weight and realistic cost!

The probability of collision of a 1 cm object with a communications satellite in GEO seems to be anybody's guess, especially since there have, as yet, been no known collisions. Bleazard [1985], for instance, states that the calculated probability of a 1 m^2 cross section of aluminium being penetrated to a depth of 5 mm and resulting in a hole 3 mm in diameter is once per century. We could play with the numbers for ever, but even if that once-a-century hit is scored, like a bullet passing through the human body, the precise point of entry is crucial: a direct hit need not necessarily destroy the satellite. The best way to reduce the probability of collision is of course to prevent the accumulation of debris, which for geostationary satellites is achieved by removing them from orbit at the end of their lives.

The NORAD catalogue for 30 July 1987 showed 6746 man-made objects in Earth orbit, 5108 of which were classed as debris [Schwetje 1987]. A principal source of debris is the deliberate or accidental destruction of space vehicles, several examples of which have been recorded [Lorenz 1988]. Owing to the limited traffic in manned space vehicles, 'eye-witness' accounts of collisions, with either natural or man-made objects, are hard to come by. However, at least two events have been documented: an object, believed to be a micrometeoroid, hit the window of a Salyut space station in July 1983, and another, thought to be a chip of paint, hit a window of the Space Shuttle *Challenger* in June 1983.

2.5 MATERIALS

Measured against terrestrial standards, the materials used to construct contemporary spacecraft range from the conventional to the highly exotic, depending on the specific application. Whatever their qualities, they must be capable of surviving about three years of manufacturing and processing, environmental testing, storage and transportation even before the spacecraft leaves the ground. Whereas the punishing launch environment exerts the maximum mechanical stress on materials, the extended period in Earth orbit, which for a successful mission may be up to 10 years, exposes materials to processes where time is the damaging factor. There are therefore many aspects to take into account when specifying materials for spacecraft.

Spacecraft materials can be divided broadly into two main categories: metals and their alloys, and composites, including fibre-reinforced plastics and metal-matrix composites.

2.5.1 Metals and their alloys

The aerospace metal with the longest pedigree is undoubtedly aluminium (Al). Its principal advantages are its relatively low density (half that of steel) and its resistance to corrosion, what there is taking the form of a thin transparent film which protects the deeper layers without affecting the material's appearance. Pure aluminium is a relatively soft metal but, as the German metallurgist Alfred Wilm discovered in 1906, it alloys easily with other metals to produce a more durable material. In fact aluminium alloyed with copper and magnesium, known as 'Duralumin' after the Durener Metal Works which bought the patent from Wilm, has a long history: it was used in the zeppelins in World War I.

Using standard methods, aluminium will alloy with only a handful of elements close to it in the periodic table, and it has been usual to limit this to only six: magnesium, zinc, copper, silicon, manganese and, most notably, lithium. However, new techniques, including rapid solidification technology, treble the number of elements which will alloy with aluminium, thereby expanding the available variety of materials. Lithium (Li) is notable amongst aluminium alloys since it combines a 10% increase in stiffness with a 10% decrease in density when compared with conventional non-lithium alloys. An additional useful attribute of an Al–Li plate is that it can be moulded by heat and pressure in a process called superplastic forming (SPF). Although not yet widely used for spacecraft manufacture, SPF has the potential to decrease structural mass by reducing the number of separate panels and obviating the need for the brackets, etc, formerly required to join them.

For many years aluminium was the predominant basic material in spacecraft structures, later joined by another metal widely used in the aerospace industry, titanium (Ti), which exhibits the useful properties of strength, low density and corrosion resistance, and the ability to alloy easily with other metals. In recent years, however, their virtual monopoly has been challenged by the increasing availability of more unusual metals, the result of constant advances in materials technology.

One example is beryllium (Be) which is two-thirds the weight of aluminium and little heavier than graphite–epoxy composite [see below]. It can tolerate high temperatures and has high specific stiffness, high specific heat capacity and good thermal conductivity. In powder-derived form it is ductile and machinable, and has the additional characteristic that it can raise the natural vibration frequency of a structure to avoid resonance [Dunn 1987]. There are, however, disadvantages. Owing to its hexagonal close-packed atomic structure, beryllium is relatively brittle and susceptible to surface damage from machining. Such damage can lead to *twinning*, where a change in the orientation of part of the crystal structure leads to a weakening non-uniformity and, unless the surface layer is etched away, to surface microcracks which can then lead to brittle fracture.

Beryllium has similar low-weight properties to lithium when alloyed with aluminium but has been used less, partly because of this tendency for brittle fracture, but also because of its toxicity. The special facilities required for both handling and machining make beryllium expensive. Nevertheless, it has been used extensively in US military communications satellites where, presumably, cost is less of a constraint. The use of beryllium for ESA or NASA spacecraft components must be 'approved', since brittle fracture could result in catastrophic failure. Magnesium alloys, on the other hand, are virtually excluded from current ESA spacecraft designs because of their low resistance to surface corrosion and so-called stress corrosion cracking (SCC) which results from a combination of corrosion and mechanical stress.

Many other metals and alloys are used, including the familiar stainless steel. Nickel and chromium form a series of alloys, known as the nimonic alloys, capable of withstanding high stress at high temperatures. They were developed during World War II for use in the jet engine. For even higher temperatures, refractory (heat-resistant) metals, defined as those with melting points above 1800 °C, are used (e.g. chromium 1850, niobium 2415, molybdenum 2610, tantalum 3000 and tungsten 3410 °C). Alloying titanium, cobalt and the refractory metals produces the 'superalloys' used in rocket motors.

The variety of metal alloys and their inherent properties is, no doubt, a metallurgist's paradise, but here it is chiefly their application that concerns us. One of the most notable structural design solutions using aluminium and its alloys is the honeycomb panel, comprising two aluminium alloy face skins bonded to an aluminium honeycombed core with epoxy film adhesive. Figure A2.9 shows that there are several variables with such a construction: skin thickness, core depth, coil thickness and cell size. By optimising the core depth and skin thickness a high bending stiffness can be achieved for low mass. Increased loads can be carried by a higher density core, and highly stressed areas can be locally reinforced using *doublers*, a term widely used for a structural element which increases the thickness of a panel. Where it is necessary to attach structural members or provide mounting points for equipment, threaded *inserts* are added to the panel. At these points stress pads, analogous to paper page reinforcers, can be bonded around the hole to provide additional strength. The honeycomb panel structure is one that is still widely used on all types of spacecraft. Indeed, it is hard to improve upon metal skins for equipment panels, where thermal and electrical conductivity are desirable. Despite this, in more recent years the aluminium facesheet has, in many cases, been replaced by skins composed of lighter materials, particularly composites.

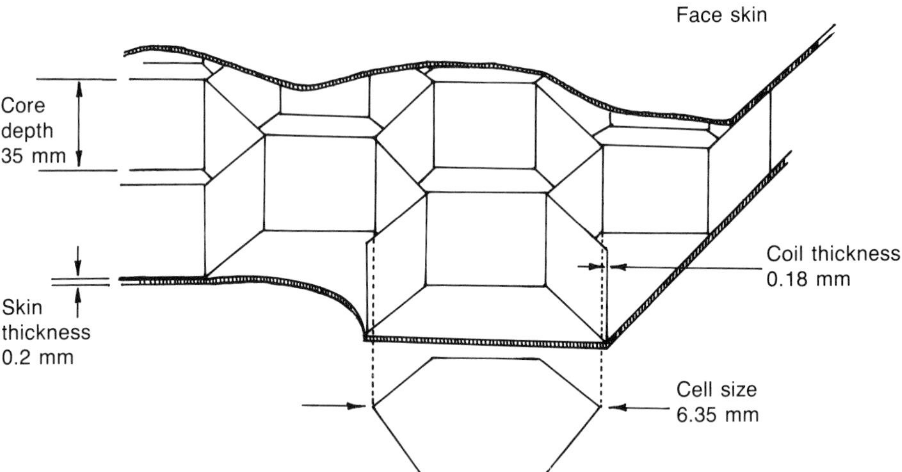

Figure A2.9 Structure of an aluminium honeycomb sandwich.

2.5.2 Composites

The 1960s marked another step in the evolution of materials with the development of fibre-reinforced polymer composites, in which long fibres of the reinforcing material are set in a *matrix* of thermosetting polymers such as the polyesters, phenols and epoxies.

The polyesters are a group of synthetic polymers, containing recurring –COO– groups, used as plastics and textiles (e.g. terylene). Setting fibres of glass in a polyester matrix gave rise to one widely used class of composites called glass-reinforced plastics (GRPs), more commonly known as fibreglass. The phenols (C_6H_5OH) are a class of organic compounds containing one or more hydroxyl (OH) groups bound directly to a carbon atom in an aromatic (unsaturated) ring. Phenolic resins are used in paints, adhesives and as thermosetting plastics. The epoxies are a set of synthetic thermosetting resins used mainly as adhesives, not least in the aerospace industries. Where once metals were exclusively joined by welding, there is now the option to glue them together.

Of the many materials used to reinforce polymer matrices, one of the best known is carbon fibre. It was developed in 1963 at the Royal Aircraft Establishment (RAE), Farnborough, where research into the effect of heat on materials produced long strongly bonded chains of pure carbon. When used to reinforce plastics, it was found to have high stiffness, four times the strength of an equivalent weight of high tensile steel and the ability to withstand high temperatures. Its high stiffness-to-weight ratio made it an instant candidate for aerospace, and particularly spacecraft, structures. Carbon fibres, with names like Fortafil, T-50 PAN, T-75 and GY-70, form composites with typically 65% of the density of aluminium. In the modern factory process carbon fibres are produced from specially treated fibres of polyacrylonitrile (PAN), a carbon-based polymer known as an acrylic precursor. The controlled oxidation of the PAN precursor converts the acrylic fibres to a heat-resistant polymer which retains its textile character, making it possible to wind, weave or knit the final product. The level of oxidation can be adjusted to determine the precise properties of the resultant carbon fibre, dependent upon its eventual application.

Many similar terms will be found in the literature apertaining to composite materials used in spacecraft, for instance carbon fibre, carbon composite, graphite composite, graphite epoxy and carbon-fibre-reinforced plastic (CFRP). In general terms these descriptions refer to the same material, that is threads or fibres of the element carbon bound up in a polymer matrix. In more technical circles, a distinction is drawn between carbon- and graphite-based products, in that the former tend to be based on the relatively untreated *amorphous carbon*, whereas the latter have been subjected to high-temperature and/or high-pressure processing to produce a more ordered structure, typically with the properties of higher electrical and thermal conductivity. The possible differences between particular products are many and often subtle, for instance the degree to which the precursor has been oxidised, the proportions of carbon and graphite, the constitution of the matrix, etc. From the same basic raw materials a multitude of different composites can be produced.

Fibres can be produced as continuous filaments or chopped into a variety of lengths. The chopped fibres are added to a matrix, forming a composite analogous to the GRP used for boat hulls, whereas the continuous filaments can be tailored in a more precise fashion to govern the strength of the resultant material.

Metals have a crystalline structure, the integrity of which provides their strength; composites, on the other hand, obtain their strength in part from a lack of crystal structure since paths for crack propagation are limited. The strength of a particular composite is mainly controlled by the orientation of its fibres, which means the structure of a fibre-based material can be designed for a specific application. Where vertical loading is known to be greater than horizontal loading, for instance, fibres can be laid proportionally to the axes of the expected loads. Additionally, when woven, different amounts of different types of fibre can be used for the warp and weft.

Fibre laying can be unidirectional or multidirectional. One of the simplest forms is the unidirectional woven tape, where the continuous carbon fibres form a warp which is effectively 'tied together' by a light-weight glass-fibre weft. As part of the production process the tape is pre-impregnated with a matrix material to form a *pre-preg*, which is available in roll form and can be cut to shapes defined by a computer-aided design (CAD) system. In a process known as 'laying up', the pre-preg is applied in layers to a former, which governs the final shape. The *lay-up* is then cured in an autoclave under conditions of high temperature and pressure. This process, whereby a material hardens permanently after an application of heat and pressure, is known as thermosetting.

Where hollow shapes are required a continuous length of pre-preg can be wrapped around a former. This *tape wrapping* is a coarser version of a method known as *filament winding*, which is used to produce items such as propellant tanks [see figure A2.10]. Another good example of filament-winding technology is given by the MAGE solid apogee boost motor (ABM) described in chapter A3, which is made from Kevlar-49 reinforced plastic (KRP) [see figure A3.3]. Kevlar, a form of the organic polymer polyparabenzamide, is an example of a proprietary material developed by Du Pont in the wake of carbon fibre to provide even greater weight savings. Kevlar exhibits a typical density of only 55% that of aluminium with the additional advantage of improved impact resistance, a characteristic used to effect by manufacturers of bullet-proof vests!

As might be expected from the low-mass high-strength requirements of spacecraft structures described earlier, composite products have found a multitude of applications on spacecraft [see box A2.3]. One of the first uses of CFRP was as a thin face skin bonded to an aluminium honeycomb to form an antenna reflector. It provided good electrical conductivity and a stable surface profile, even under severe thermal cycling *in vacuo*. An occasional delamination between lay-ups or fatigue of the epoxy – polyamide film bonding face skins to honeycomb observed in the late 1970s showed that this technology was not without its teething troubles, but holographic detection techniques now help to ensure that antennas can be made virtually defect free.

Figure 2.10 Filament-wound motor case for a solid propellant motor. [Mark Williamson]

Another component formerly made exclusively in metal, due to its requirement for electrical conductivity, is the waveguide, which can now be produced in CFRP with a performance up to the best metal equivalents and a 50% weight saving over the lightest metals available.

Fibre-reinforced composites are now used in a wide range of consumer items, from bicycle frames to tennis rackets, largely because of a reduction in cost: carbon fibres costing between $800 and $1300 per kilogram in the late 1960s cost only $40–$70 per kilogram by 1986 [Marsh 1986].

An alternative to the thermosetting polymer, mentioned above, is the thermoplastic, which becomes soft when heated and rehardens on cooling without an appreciable change of properties. Although generally used less than the thermosetting compounds because of their comparatively low temperature tolerance (400 – 500 °C [Kelly 1987]), thermoplastics present the engineer with more options since they can be moulded rather than laid-up and set. They are also amenable to metal-working processes such as roll forming, sheet stamping and machining with normal metalwork tooling. Continuous fibre-reinforced thermoplastics were first introduced by ICI in 1981 in the form of pre-preg tape based on carbon-fibre-reinforced polyether-etherketone (PEEK).

A further development is the metal-matrix composite (MMC) in which the fibres reinforce a metal instead of a plastic. Example matrices are aluminium, lead, copper, magnesium and titanium; prime candidate fibres are graphite, boron and silicon carbide, but tungsten, molybdenum and alumina are also usable. MMCs have high strength and stiffness at high temperatures, exceptional dimensional stability and high thermal and electrical conductivities. They have lower thermal expansion

BOX A2.3: USE OF MATERIALS—SPACECRAFT EXAMPLES

Example applications of materials on contemporary spacecraft.

Aluminium alloy

General: panel face skins; mounting brackets and fittings (machined); launch vehicle adapter rings (forged); cleats (to join panels; also glass fibre).

Aluminium honeycomb

General: structural body panels, antenna reflector cores; Olympus: all major structures except thrust cylinder; Palapa: solar panel drum.

Titanium alloy

General: highly stressed structural components, fasteners; Space Shuttle: thrust frame and parts of main engines.

Beryllium

General: solar array drive mechanisms, BAPTAs, APMs, casings and mirrors; Space Shuttle: window frames, brake and door components; Viking (Mars lander): RCS engines; Cassini/Huygens Titan entry probe (proposed): aeroshell nosecap.

Carbon composite

General: panels, face skins, structural components; Marecs: antenna reflectors; Telecom: antenna reflectors, feedhorn tower, solar array panels;

coefficients than the thermosetting polymers and exhibit good resistance to radiation damage, little outgassing and no low-temperature brittleness. Their main drawback has been their high cost of production, but recent advances with silicon carbide fibres in aluminium suggest they are becoming more economic [Heany 1987]. The material is 50 to 70% stiffer than aluminium and weighs 7% less than graphite epoxy, so the chances that it may be developed as a future material for spacecraft look good.

Another class of composites has been developed for high-temperature resistance. Carbon–carbon composites have been produced since the early 1960s and have found particular application in rocket nozzle throats and exit cones (e.g. the McDonnell Douglas payload assist module (PAM)). The carbon matrix is produced by *hydrocarbon cracking* or by pyrolysing a resin to leave a carbon residue into which the fibres are embedded. Such composites can tolerate temperatures in excess of 1400 °C in an inert (non-oxidising) atmosphere. In oxidising environments, up to and above 2000 °C, one must resort to the use of ceramic

TDF: antenna reflectors, tower, solar array panels; TV-SAT: antenna reflectors, tower, solar array panels; Tele-X: feedhorn tower, solar array panels; Arabsat: feedhorn tower, solar array panels; Olympus: feedhorn tower, antenna reflectors; ERS: payload support structure; SPAS: primary structure; Eureca: primary structure; DFS: CFRP corrugated thrust cylinder; Ariane SPELDA: CFRP face sheets on aluminium alloy honeycomb.

Graphite composite

General: panels, face skins, structural components; Palapa: antenna support structure, outer surface face sheets for radiator section of drum; Intelsat V: primary tower structure, antennas (Fiberite, ICI); Hubble Space Telescope: main support structure (Fiberite, ICI); Space Shuttle: engine pods, payload bay doors, RMS boom (3 mm thick Fiberite).

Kevlar composite

General: panels, face skins, wound motor cases (e.g. MAGE solid AKM); Intelsat VI: antenna cores and face skins, shear webs, booms; Palapa: antenna cores and face skins, single-ply face sheets for solar cell sections of drum; Giotto: rear shield of dust-protection system (KRP and polyurethane-foam sandwich).

Carbon–carbon composite

General: PKM nozzles; Space Shuttle: black TPS tiles.

Ceramics

General: bearings.

materials, which have even higher melting points (e.g. alumina, used in ablative shields, melts at 2054 °C). Ceramics are well known for their fragility, but fibre-reinforced ceramic-matrix composites (CMCs) have a greatly improved resistance to crack propagation.

Composite materials offer a mind-boggling variety of products for spacecraft and other structures both now and for the future. With fibres as varied as glass, graphite and silicon carbide, and matrices of plastics, metals and ceramics, the days are long gone when the best option was to corrugate a sheet of aluminium alloy!

2.6 ADVANCED SPACECRAFT APPLICATIONS

Apart from the range of applications mentioned already, a number of advanced developments in the use of materials are worthy of mention. These include a 1 m diameter dichroic subreflector comprising a honeycomb core bounded by two Kevlar face skins. What makes this antenna different is that the skins include copper dipole crosses oriented to provide sensitivity to opposite polarisations from each face skin. The result is a one-piece dual-polarisation reflector capable of replacing two separate antennas of equivalent size.

Extensive work has also been conducted on inflatable antennas from 2 to 12 m in diameter. Special resins impregnate the fibre matrix forming a pre-preg which can be inflated in space; the resin is cured either by exposure to UV radiation and solar heating or by including the curing agent in the nitrogen supply used for inflation [Dunn 1987].

One particularly interesting material is the Nitinol (55% nickel, 45% titanium) *shape-memory metal* or *memory alloy*, used as a release trigger for the solar array drive mechanism (SADM) of the Hubble Space Telescope. The memory alloy elements are mounted in a pre-bent form to depress a number of release latches. When the elements are heated to their 115 °C transition temperature, they become straight and lift the latches, releasing the mechanism. It has been demonstrated that Nitinol elements can be a reliable replacement for pyrotechnics, so the ubiquitous 'explosive bolt' release mechanism could be replaced by an element of Nitinol and an electrical resistance heater. It remains to be seen how widespread the application of this material will become.

2.7 CONCLUSION

To improve continually the design of a satellite structure, the structural engineer needs a knowledge of both advanced materials and analysis techniques. Before the advent of contemporary computer-modelling techniques, the best way to know whether a structural design would meet the specification was to build it and test it. Nowadays it is possible to predict the behaviour of a structure to a close approximation using *finite element analysis*, a computer-based technique which divides the structure into a finite number of separate parts. The elements are small enough to be assigned realistic values for mechanical loading, etc, but not too numerous that the integration process becomes unwieldy. The technique allows the engineer to assess accurately mechanical loads and the behaviour of a structure under stress at the design concept stage. Now, although spacecraft structural models are still built and tested, the function of the test has undergone a subtle change: instead of using the test results from an actual spacecraft to *reveal* the dynamic characteristics of its complex structure, they are used to *confirm* the predicted structural behaviour.

Greater knowledge of materials and structural behaviour leads to a change in test objectives, since over-large margins are generally a product of ignorance. Safety margins are still necessary, but over-conservative design, originating from a variability in material properties or a 'gloom factor' added out of ignorance of structural behaviour, should be a thing of the past. In the final analysis, this means that structures should account for a lower percentage of the mass budget while retaining their ability to meet the stringent requirements placed upon them.

Of course, the mass budget is not entirely the responsibility of the structural design engineer. Within the constraints of the payload envelope and the mass budget the structural engineer must provide accommodation, support and protection for the equipment of the various satellite subsystems such as propulsion, AOCS, power, thermal, TT&C and payload. In return, the onus is on the subsystem designers to make their respective units as diminutive and insubstantial as possible.

REFERENCES

Bleazard G B 1985 *Introducing Satellite Communications* (Manchester: NCC) ch.4

Dunn B D 1987 *ESA J.* **11** 153–66

Heany J E 1987 *Aviat. Week Space Technol.* 12 Oct. 71–106

Kalweit C 1981 *ESA J.* **5** 57–64

Kelly A 1987 *Phys. Bull.* **38** 337–9

Lorenz R D 1988 *Spaceflight* **30** 4–7

Marsh G 1986 *Space* **2** no. 3 10–15

Schwetje F K 1987 *Int. Astronautical Fed. Congr.* IISL-87-37

Wolfe M, Chobotov V, Kessler D and Reynolds R 1982 *Earth-Oriented Appl. Space Technol.* **2** 161–6

A3 Propulsion

3.1 INTRODUCTION

Once the long process of design, manufacture and testing is complete, the satellite can be launched. This chapter is intended to provide a general grounding in propulsion systems, including descriptions of hardware and the types of propellant used in connection with satellites. Greater detail of the specific applications to attitude and orbital control systems (AOCS) and launch vehicles can be found in chapters A4 and A10 respectively. Propulsion systems for launch vehicles and spacecraft are different in their requirements but are based on the same physical principles, which will be discussed below.

The purpose of propulsion, as far as a communications satellite is concerned, can be summed up as delivering the satellite to orbit and keeping it there. The delivery system comprises a launch vehicle and a number of smaller devices, known as perigee and apogee motors, and station keeping is performed using reaction control thrusters which are smaller still.

The principles discussed in this chapter can be applied equally to spacecraft on deep-space missions following trajectories which take them out of the solar system, remote sensing spacecraft in polar orbit, and satellites and manned spacecraft in a multitude of different orbits around the Earth, Moon and other planetary bodies. Here, however, we shall confine ourselves to the communications satellite in geostationary orbit.

3.2 GEOSTATIONARY ORBIT—DELIVERY SYSTEMS

Geostationary orbit, commonly abbreviated to GEO, is nominally circular, coplanar with the Earth's equator and has an average height of 35 786 km (22 237 miles). According to the laws of orbital dynamics, the orbital period of a satellite in GEO is equal to the Earth's rotation period (23 h 56 min 4 s), which means that a spacecraft will appear to be approximately stationary with respect to the Earth—hence 'geostationary'. Since this eliminates the need for small earth station antennas to track the satellite, it is the orbit utilised for most communications satellites. The unique attributes of geostationary orbit were first brought to public attention by the science fiction writer Arthur C Clarke in the October 1945 issue of *Wireless World*. It is therefore sometimes referred to as a Clarke orbit.

It is advisable at this point to eradicate a widespread misuse of terms regarding this orbit. The term geosynchronous is often used in place of geostationary; in fact any orbit whose period of rotation is some multiple or submultiple of the Earth's can be termed *geosynchronous*. Only the orbit whose period is exactly the same as the Earth's and whose plane lies in the plane of the equator is *geostationary*. The geostationary orbit is a special case in the family of 24 h geosynchronous orbits [see appendix B]. A simple derivation of the orbit's radius is given in appendix C3.

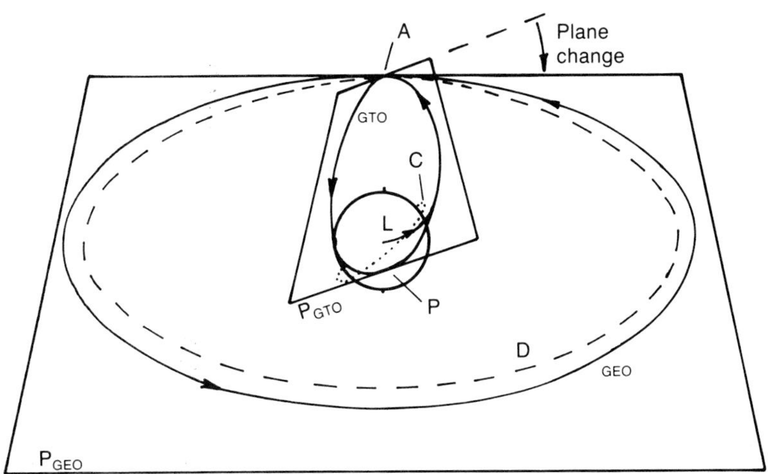

Figure A3.1 The route to geostationary orbit (Earth and orbits approximately to scale). GEO, geostationary orbit; GTO, geostationary transfer orbit; P_{GEO}, plane of GEO; P_{GTO}, plane of GTO; A, apogee; P, perigee; D, drift orbit; C, circular low Earth orbit (e.g. as attained by the Space Shuttle); L, launch trajectory to GTO.

A conventional multi-stage launcher like Ariane delivers a spacecraft into an elliptical transfer orbit with its apogee (or highest point) at geostationary height. The lowest point of this geostationary transfer orbit (GTO), as it is known, is called the perigee [see figure A3.1]. Although most expendable launch vehicles (ELVs) inject their payload directly into

GTO, satellites launched by the reusable Space Shuttle require an extra rocket motor, called a perigee kick motor (PKM), to attain this orbit. This is because the Shuttle operates from a circular low Earth orbit (LEO) or parking orbit [orbit C in figure A3.1].

Geostationary transfer orbit is an example of a Hohmann transfer orbit, named after the German physicist, Walter Hohmann, who first described the trajectory in 1925 as the most economical path between two circular orbits. He was particularly interested in the minimum energy trajectories between the orbits of the planets, but a transfer from LEO to GEO follows the same physical laws: a maximum of payload can be transferred by a spacecraft using a minimum amount of propellant. In propulsion terms, the transfer requires two separate velocity increments known as *delta-Vs* (ΔV): one to inject the satellite into the transfer orbit and another, at apogee, to place it in GEO. In practice the first is supplied either by the third stage of an ELV or a Shuttle PKM, and the second by the satellite's integral apogee kick motor (AKM). Apogee and perigee motors are basically the same, but the perigee motor is the larger of the two since it has to carry the satellite, including its apogee motor, from a low orbit to geostationary height, and since the change in velocity (ΔV) at perigee is greater than that required at apogee.

Most rocket launches are made in the same direction as the Earth's rotation, thus saving energy by allowing the Earth to impart a velocity of almost 1700 km h^{-1} (about 1000 mph) when launched from the equator [see appendix C1]. The Ariane launch site at Kourou in French Guiana, at 5.23°N, is thus in a favourable position to launch geostationary satellites. However, a vehicle launched from a non-equatorial launch site places its payload in a transfer orbit whose plane is inclined to the plane of geostationary orbit by an angle dependent on the latitude of the launch site. Of course this means that a change of plane must be made at some point. The change in orbital inclination can be made at apogee or perigee or both, but in practice most are made at apogee since the velocity is lower and less propellant is required.

Once in transfer orbit the satellite makes a number of revolutions, depending on the time it takes to make sufficiently accurate orbit and attitude measurements and on the desired orbital position. Then, at the chosen apogee (typically between the second and the eleventh), the AKM is fired. This injects the satellite into a good approximation to geostationary orbit, called a *drift orbit*. Its height is arranged to be marginally higher or lower than GEO so that the satellite's orbital period is, respectively, slightly greater or less than 24 h. This allows the spacecraft to drift slowly westwards or eastwards towards its final orbital position, the direction being chosen to minimise the drift period. When the satellite reaches this position, its thrusters are fired to arrest the drift and render it geostationary. The sequence of events which follows, including its stabilisation and the deployment of its solar arrays, is described in §A4.6.1.

The satellite is tracked by a controlling ground station throughout its time in drift orbit, which may be 2 or 3 weeks, and telemetry is available from the spacecraft so that preliminary in-orbit checks can be undertaken. Once on-station, the transponders are activated and checked out in preparation for the start of the satellite's service life. This in-orbit commissioning and acceptance testing phase, which confirms whether the spacecraft is operating according to the design specifications and prepares it for handing over to the customer, can last a month or more depending on the complexity of the satellite.

Once in orbit, the propulsion requirements of the satellite are limited to attitude control, which maintains the orientation of its antenna platform towards the Earth, and orbital control or *station keeping*, the maintenance of its orbital position [see chapter A4]. The reaction thrusters which provide these functions and the much larger apogee and perigee motors operate according to the same physical principles, which are discussed below.

3.3 PROPULSION PRINCIPLES

Spacecraft propulsion offers the perfect demonstration of Newton's third law—every action produces an equal and opposite reaction. Figure A3.2 illustrates the principle of the rocket. In a gas-filled container the pressure acts with equal force on all surfaces and there is no resultant motion. If one end of the container is removed, the gas flows out; the pressure is no longer balanced at the other end, so the container moves. The higher the gas pressure, the greater the speed. This result is of course analogous to what happens to an inflated balloon held by the neck and then released. In terms of spacecraft propulsion, the 'action' of high-velocity gas molecules expelled from a rocket nozzle causes a 'reaction' of the spacecraft in the opposite direction.

If a specially shaped exit cone, or nozzle, is added to the container, the performance will be improved. Figure A3.2 shows that a rocket needs no surface to bear against and can operate in a vacuum. Indeed performance is improved *in vacuo* since there is no atmospheric resistance to impede forward motion and no atmospheric backpressure (i.e. the pressure of the air restricting the exit of the propulsive gases).

3.4 PROPELLANT TYPES

Almost without exception, the propellants used to produce this thrust are supplied in either solid or liquid form and are based on well

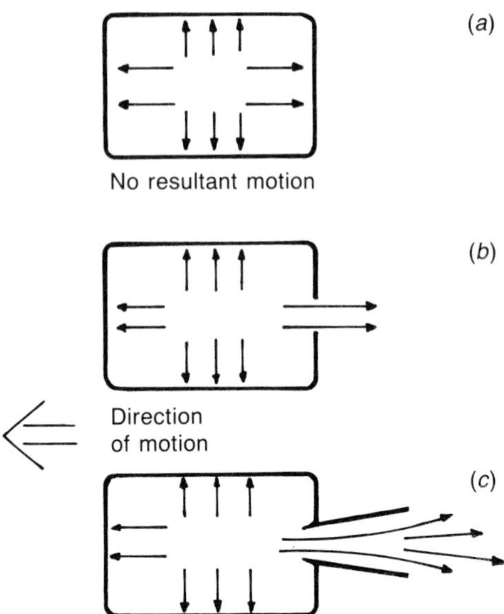

Figure A3.2 Principle of the rocket. (*a*) Sealed container. Gas pressure produces equal force on all surfaces. (*b*) Remove one end. Forces no longer balanced so container moves according to Newton's third law: the 'action' of the jet of gas produces an equal and opposite 'reaction' of the container. (*c*) Add nozzle to improve performance.

understood chemistry. The physical state of the propellant is mirrored in the names given to the hardware which uses it (e.g. solid rocket booster, liquid apogee engine). Although the terms engine and motor are generally interchangeable, it is customary for propulsion devices using liquid propellant to be called *engines* and those using solids to be called *motors*.

Solid propellant motors are the simpler of the two, comprising only a few major components: a motor case which contains the *propellant grain*, a surrounding insulating blanket, an exhaust nozzle and an ignition system. With solid motors, the motor case is both the propellant tank and the combustion chamber, consisting of a pressure vessel and a skirt to enable it to be mated to the spacecraft. Figure A3.3 compares a cutaway and a photograph of the European MAGE-2 motor, the case of which is filament wound from Kevlar fibres in an epoxy matrix; the skirt is a wound composite of Kevlar fibres and carbon-fibre cloth. The insulating blanket or *propellant liner*, an asbestos-filled polymer rubber compound, acts as both a thermal insulator and a flame inhibitor. This thermal blanket also supports the propellant grain which is mixed and poured into the open-ended motor case.

Liquid engines, on the other hand, require a network of tanks, pipes, valves and (in some cases) pumps to deliver the propellant to the combustion chamber, a far more complicated system as shown in figure A3.4.

(*a*)

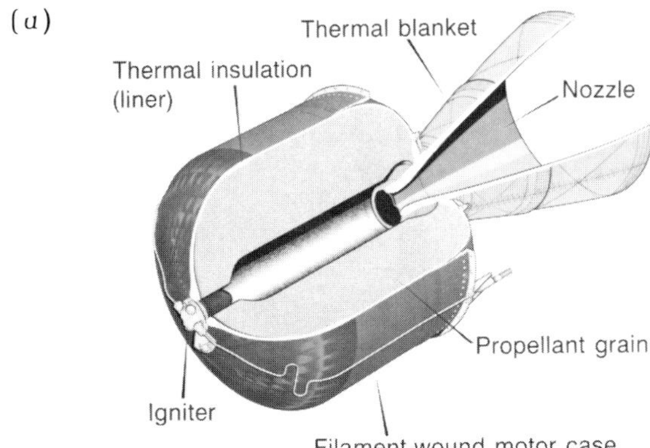

(*b*)

Figure A3.3 (*a*) Cutaway of a MAGE 2 apogee motor. [SEP]; (*b*) The MAGE 2 AKM used on the Giotto Comet Halley interceptor. [SEP]

Most of the world's major launchers use liquid propellants for the main rocket stages, although solid propellant *strap-on* boosters are quite

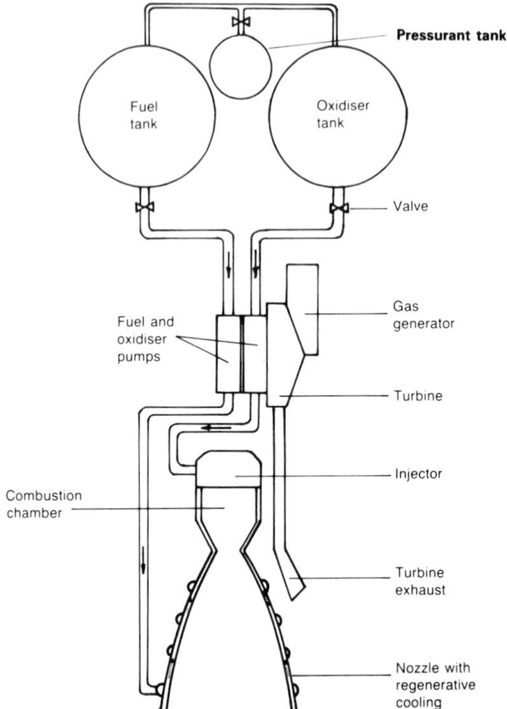

Figure A3.4 A liquid propellant rocket engine.

common. Perigee and apogee motors have been predominantly solid, but liquid apogee engines are now becoming more common. Various types of liquid propellant thrusters have been widely used for satellite reaction control systems.

One fundamental difference between rocket propulsion and aircraft propulsion is that the rocket system must carry its own source of oxygen in which to burn the fuel. Rocket propellant is therefore a term which covers both fuels and oxidisers.

In addition to the distinction between the solid and liquid state, propellants are divided into two classes: monopropellants and bipropellants. A monopropellant system is one which uses a single chemical component, but solid propellants which comprise a stable mixture of both oxidiser and fuel can also be regarded as monopropellants, since they are essentially self-contained and used as a single-component propellant. A bipropellant system uses two components, a fuel and an oxidiser, stored in separate locations and brought together only at the moment of ignition.

If ignition is provided spontaneously, by introducing one component to the other, the propellants are termed *hypergolic*. Since it is impractical to mix two solids in a combustion chamber, bipropellant systems use predominantly liquid propellants. However, it is possible to combine a liquid oxidiser with a solid fuel (more practicable than vice versa) in a *hybrid* propulsion system.

Early liquid monopropellants such as hydrogen peroxide are now very rarely used, but monopropellant hydrazine, stored as a liquid and decomposed to form a gas, is widely used in reaction control systems. Solid propellants can be regarded as monopropellants.

3.4.1 Solid propellant

Two main types of solid propellant can be identified: the homogeneous or double-base propellants and the heterogeneous or composite type. The double-base propellants consist of nitrocellulose plasticised with nitroglycerine with an admixture of stabilising products but are limited in power. The solids generally used for space applications are the composites, consisting of a mixture of fuel and oxidiser. An oxidiser in common use is crystalline ammonium perchlorate (NH_4ClO_4) and is mixed with an organic fuel like polyurethane or polybutadiene, which also binds the two components together. The propellant used in the MAGE (moteur apogée géostationaire) is known as CTPB-1612 (carboxyl terminated polybutadiene 1612) and contains 72% ammonium perchlorate, 12% CTPB binder and 16% aluminium powder [Asad 1983]. The inclusion of a plastic binder like polybutadiene reproduces the consistency of a pencil rubber, making the propellant relatively easy to handle.

Performance is enhanced by the addition of finely ground (10μm) metal particles of aluminium, for instance. Apart from their advantage of high density, they increase the heat of the reaction due to the formation of metal oxides. For example, the solid propellant of the Shuttle SRBs comprises 14% polybutadiene acrylic acid acrylonitrile (binder/fuel), 16% aluminium powder (fuel), 69.93% NH_4ClO_4 (oxidiser) and 0.07% iron oxide powder (catalyst).

3.4.2 Liquid propellant

There are also two main types of liquid propellants: *storable* and *non-storable*. The distinction is one of temperature, the former being a storable liquid at terrestrial environmental temperatures and the latter needing various degrees of cooling below these temperatures to render them liquid. The non-storable propellants are otherwise termed cryogenic after the branch of physics concerned with very low temperatures. The best known cryogenic oxidiser is liquid oxygen (LO_2 or 'LOX'), which boils at −183 °C. A possible alternative is fluorine (F_2), which is potentially a more potent oxidiser than LOX, but it is extremely corrosive and toxic and therefore difficult to work with. If fluorine is to be used in the future, it would most likely be employed in compound forms such as fluorine chlorate ($FClO_3$), chlorine trifluoride (ClF_3) or fluorine oxide (F_2O). The potential oxidiser ozone (O_3) produces the maximum energy release with any fuel but is highly combustible and decomposes explosively in the presence of trace impurities [Frisbee 1985a].

The cryogenic *fuel* with the most promise and the most significant track record is liquid hydrogen (LH_2) which boils at −253 °C. One of the most common propellant combinations for launch vehicles is LH_2 and LOX, used in the second and third stages of the Saturn V and the third stage of Ariane. The disadvantages associated with hydrogen are due mainly to its low density, which means that the tanks to hold it have to be large, and its cryogenic nature, which necessitates thicker insulation which adds to the structural weight of the vehicle.

As for storable liquid fuels, the methyl and ethyl alcohols, used in early rockets such as the V2 with liquid oxygen as the oxidiser, have since been replaced by hydrocarbons such as kerosene which provides a more efficient mixture. Kerosene was used with liquid oxygen in the first stage of the Saturn V. Examples of storable liquid oxidisers widely used for rocket propulsion in the past are red fuming nitric acid (RFNA) and hydrogen peroxide (H_2O_2); currently more common is nitrogen tetroxide (N_2O_4). Two fuels which have been used with nitric acid are hydrazine (N_2H_4) and unsymmetrical dimethylhydrazine (UDMH, $(CH_3)_2N.NH_2$). Hydrazine is now more commonly used with N_2O_4, with which it is hypergolic. UDMH is also used with N_2O_4—in the first two stages of Ariane for instance.

As mentioned above, hydrazine is widely used in reaction control thrusters without an oxidiser. It is now equally common to design liquid bipropellant systems which typically use the hypergolic combination of nitrogen tetroxide and monomethyl hydrazine (MMH, $CH_3NH.NH_2$) [see §A3.9].

3.5 SOLID MOTOR OPERATION

Current spacecraft propulsion technology is founded on relatively basic chemical and physical principles: a chemical mixture undergoes combustion to produce a gas, then the flow of the exhaust gases is controlled so that their rearward momentum causes the vehicle to move forwards. Despite the many differences between solid propellant motors and their liquid counterparts, this simplistic view of spacecraft propulsion is essentially correct for all chemical propulsion systems.

The solid rocket motor can be divided into two distinct sections according to function: the propellant container where combustion occurs, and the nozzle which controls the ejection of the exhaust gases. As mentioned above, it is the propellant grain itself which forms the 'combustion chamber'. Once the propellant has been ignited, it will burn until the propellant liner is reached, when combustion will be *inhibited*.

In the MAGE-2 motor shown in figure A3.3, the igniter is located on the main axis at the forward end. It is operated by a current pulse from the spacecraft power supply to a pyrotechnic cartridge which ignites a small propellant grain in a steel or glass-fibre housing in the combustion chamber. This produces a controlled amount of hot gas and, in a similar way to a detonator in an explosive device, ignites the main motor. The thrust produced throughout the *burn* is controlled by the rate of combustion of the propellant grain. Perhaps the most obvious method of burning the propellant is by *cigarette combustion* where one end of the block of propellant is ignited and then burns to the other end like a cigarette. In this case the cross section of the block gives the surface area of active combustion and therefore limits the thrust.

One way to increase the active surface area is to have a cylindrical hole through the centre of the block, but as the propellant burns the hole enlarges radially and the thrust increases. Most engines, and certainly launcher boosters, require the maximum thrust at the moment of ignition to overcome the inertia of the vehicle. The central hole is therefore star shaped in cross section so that the combustion area remains more nearly constant in the initial phases and then, if required, decreases towards the end of the burning time [figure A3.5]. The grain used in the Shuttle solid rocket boosters has the cross section of an 11-pointed star, for example. It is designed to decrease thrust by about a third at 55s into flight to avoid overstressing the vehicle during the period of maximum dynamic pressure (known as 'Max-Q') when the combination of vehicle speed and atmospheric resistance reaches a peak.

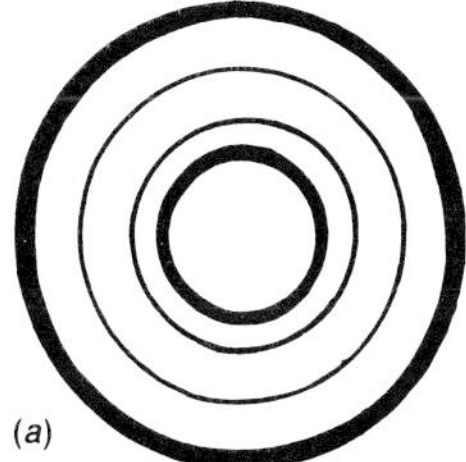

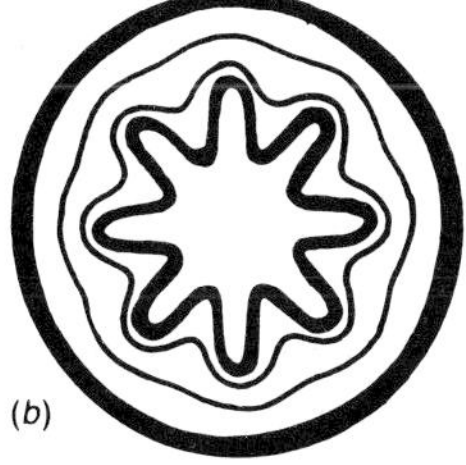

Figure A3.5 Solid propellant grain cross sections. (*a*) With a circular-section central hole the active surface area of combustion enlarges with time and thrust increases. (*b*) A star-shaped section provides a large initial surface area giving greater thrust. Surface area remains approximately constant as the hole becomes circular.

It is important that the propellant grain should not be cracked or contain voids. A larger than average local surface area provides a greater area for combustion, leading to a local increase in pressure which can cause the motor to explode. Quality control is therefore critical. As part of the quality control process the propellant undergoes examination by x-rays and CAT scan (computer-aided tomography).

Since it is the gas products of the chemical reaction that produce the thrust, the nozzle which controls the ejection of these exhaust gases is a critical component in producing the optimum thrust: it converts the thermal energy of the combustion gases to the kinetic energy of the exhaust plume. A typical nozzle comprises a convergent section and a divergent exit cone, linked by a narrow throat. The gases move naturally from the high pressure of the combustion chamber to the vacuum of space, expanding and accelerating rapidly as they leave the chamber. The throat, as the narrowest part of the nozzle, is designed to maintain the required pressure within the chamber and regulate the outflow of combustion gases. The convergent–divergent shape of the nozzle is due to the variation in the rates of change of the velocity and specific volume (volume per unit mass) of the combustion gases: since the rate of increase of specific volume is at first less than, but finally greater than, the rate of increase of velocity, the nozzle must first converge and then diverge. The throat is the region of transition from subsonic to supersonic flow and is very prone to erosion. By way of protection, all MAGE motors, for example, have a throat section made from a high-density erosion-resistant carbon–carbon composite (Sepcarb 4 D). The exit cone of the MAGE-2 uses a low-density carbon–carbon composite (Sepcarb 2 D) allowing a thickness of only about 2 mm. The exit cones of the MAGE-1 and 1S motors are made from a thicker carbon–phenolic material [Asad 1983]. The exit cone controls the expansion of the exhaust plume which attains its final exhaust velocity at the outlet.

The degree to which the exhaust gases are expanded is given by the *expansion ratio*, the ratio of the area of the widest part of the exit cone to that of the throat. Expansion ratios range typically between about 50 and 150 depending on the individual design and restriction on the available space within the launch vehicle. In addition to this *area expansion ratio*, propulsion engineers use a *pressure expansion ratio*, defined as the ratio of the pressure in the combustion chamber to that at the exit of the exhaust nozzle; the two are related by a complex expression.

Exhaust velocity is important because it is one of the quantities which defines the *thrust* of a rocket engine, such that thrust is the product of the mass of gas ejected per unit time and the exhaust velocity:

$$t = \dot{m}v_0 \tag{A3.1}$$

where t is the thrust (N), $\dot{m}$ is the mass flow rate (kg s^{-1}) and v_0 is the exhaust velocity (m s^{-1}). For more precise determinations a pressure term may be added such that $t = \dot{m}v_0 + A_e(P_e - P_a)$, where A_e is the area of the nozzle exit, P_e is the exit pressure and P_a is the ambient pressure.

Exhaust velocity is fundamentally limited by the chemical energy of the propellants which governs the heat and pressure of the reaction. It is also limited by the temperatures which the walls of combustion chambers (for liquids) and exhaust nozzles can withstand. Apart from the inherent heat of the reaction, particles moving at speeds of around 3 to 4 km s^{-1} produce significant frictional heating and eventual erosion of nozzle components, and it is for this reason that carbon–carbon and graphite–epoxy nozzles have been developed.

Another quantity, which is related to the exhaust velocity, is *specific impulse* (I_{sp}) given by the expression

$$I_{sp} = I/m \tag{A3.2}$$

where I is the impulse (N s) and m is the mass (kg). The units in which specific impulse is measured can cause confusion. Historically the unit used was the second (s), but this is gradually giving way to m s^{-1} [see box A3.1]. However, since the majority of sources, both published papers and manufacturers' promotional material, quote I_{sp} in seconds, it has been decided to use the historical standard in this volume.

BOX A3.1: THE UNITS OF SPECIFIC IMPULSE

The units of specific impulse as given by equation (A3.2) are N s kg^{-1}, which can be reduced to m s^{-1} (since N = kg m s^{-2}). An alternative, but equivalent, definition is the ratio of thrust (N) to the rate of propellant consumption (kg s^{-1}), which gives the same units for I_{sp} (m s^{-1}). Specific impulse is therefore equivalent to exhaust velocity (m s^{-1}).

Originally I_{sp} was defined as impulse per unit weight of propellant consumed, I/Mg, with I in newton seconds and Mg in newtons. This gave the second (s) as the unit of specific impulse. In terms of the alternative definition, when thrust was measured in pounds force and rate of consumption in pounds per second, the units lb/(lb/s) also gave the second as the unit for I_{sp}. This was a convenient concept, since the specific impulse could be said to be the time (in seconds) over which the combustion of 1 kg of propellant produced a thrust of 1 kg: the longer the time, the less fuel used to produce an equivalent thrust and the more efficient the engine; the longer the time, the greater the specific impulse.

Specific impulse is still often quoted in seconds but can be converted to metres per second by multiplying by g, the acceleration due to gravity (9.81 m s^{-2}).

3.6 LIQUID ENGINE OPERATION

Engines using liquid propellant are far more varied in their design, depending primarily on the type of propellant used. In the case of the storable monopropellant, the simplest liquid system, the propellant can be stored in a single tank at normal environmental temperatures. It is fed to a combustion chamber where it is decomposed, either by the application of heat or in reaction with a catalyst, to form a gas which is then managed by an exhaust nozzle in the same way as with the solid propellants discussed above. Storable bipropellants are, by definition, stored at normal temperatures in separate tanks. They are fed to a combustion chamber, mixed and ignited and then handled in the usual way. The only major difference with cryogenic bipropellant engines is the temperature at which the propellants must be kept.

There are two methods of feeding propellant to a combustion chamber: by gas pressure or mechanical pumps. In the pressure-fed engine an inert gas, such as helium or nitrogen, pressurises both fuel and oxidiser tanks to provide the force to feed the propellants to the combustion chamber. Since all the tanks are effectively pressure vessels, they must have thicker walls. A standard tank may consist of two machined, forged titanium domes welded together. If a large tank is required, a cylindrical section can be interposed between the two end-domes, and for increased operating pressures the titanium tanks can be overwound with Kevlar.

As a consequence of the relatively heavy tanks, pressure-fed systems are typically used for the smaller engines which fire for relatively short durations (e.g. reaction control thrusters). Where the weight of the large thick pressure vessels would outweigh that of a system of pumps—in large long-burning engines—a pump-fed system is preferred. However, this is by no means a simple trade-off; pressure-fed systems are less complex and more reliable, so for certain applications the mass burden may be of secondary importance.

Although cryogenic propellants have distinct disadvantages in terms of handling, they have many advantages. One of these is the common practice of vaporising some of the propellant in a heat exchanger on the engine and using it to pressurise the tank. Prior to the launch of any cryogenically propelled rocket, clouds of this evaporant can be seen venting from the tanks as the internal pressure is adjusted. A second advantage of cryogenic propellants is their efficiency in cooling the exhaust nozzle, around which they can be pumped before entering the combustion chamber. The absorption of heat from the engine serves to increase the energy of the propellant. This *regenerative cooling* of the nozzle means that higher exhaust gas temperatures, leading to an increase in specific impulse, can be entertained without recourse to increasingly exotic heat-resistant materials.

Once the fuel and oxidiser are injected into the combustion chamber, atomised and mixed, they must be ignited. The simplest solution is of course to use hypergolic propellants, which ignite on contact. The fuel MMH, for instance, is hypergolic with most oxidisers, while the oxidiser fluorine is hypergolic with most fuels. However, when used with cryogenic hydrogen, the H_2–F_2 mixture produces the highly unpalatable exhaust gas hydrofluoric acid (HF), whereas the product of the common H_2–O_2 reaction produces nothing more harmful than water vapour (H_2O).

If the propellants are not self-igniting, a pyrotechnic system or an electrical resistance may produce the heat required for evaporation, ignition and final combustion. Initially a small amount of fuel meets the oxidiser and a *pilot flame*, produced by the ignition system, heats the liquids to their ignition point. The supply valves are then opened fully and combustion temperatures and pressures rapidly increase to produce a high-velocity exhaust.

This gradual build-up in thrust is evident from observation of the manner in which launch vehicles leave the ground: liquid propellant vehicles are released some seconds after ignition when the thrust is sufficient to support and lift the vehicle, whereas solid boosters have a relatively instantaneous effect. For example, the Space Shuttle leaves the launch pad when the solid rocket boosters (SRBs) are ignited, which is several seconds after ignition of the liquid-fuelled main engines (SSMEs). This procedure is, moreover, also followed for reasons other than the thrust characteristics of the different types of motor. If a problem is evident with the SSMEs prior to release they can be shut down and the launch can be aborted; once the SRBs are alight nothing can stop them.

This example highlights one of the fundamental differences between solid and liquid propellants. In choosing which one to use for a particular application, there are many such factors to take into account.

3.7 SOLIDS VERSUS LIQUIDS

In historical and contemporary descriptions of propulsion systems, there is often the implicit assumption that solid propellant somehow represents old-style technology and that liquids are 'the key to the future'. The fact that both solid and liquid propellants are still used in the most advanced propulsion systems, especially for launchers where they are often used together, indicates the simplistic nature of the assumption. It is not, however, difficult to see why this view persists: solids are simpler, less controllable and less efficient than liquids; the structure of the liquid engine with its branching tree of pipes, valves and turbo pumps seems a generation ahead of the 'overgrown firework' solid motor. However, an examination of the pros and cons shows the need for both types of propellant.

One of the main advantages of liquids is that they produce higher specific impulses; solids typically achieve between about 230 and 300 s, whereas liquids range between about 260 and 450 s. The different types of propellant used in the Space Shuttle system are indicative of contemporary achievable I_{sp} values: the SRBs develop on average 250 s dependent upon the altitude; the SSMEs, which use LOX and LH_2, are rated at 453 s, currently the highest I_{sp} of any western-built engine; the orbital manoeuvring system (OMS) and reaction control system (RCS) give around 315 and 270 s respectively. Although both the OMS and the RCS use the common bipropellant combination of MMH and nitrogen tetroxide, the specific impulse is higher in the OMS because of higher expansion ratios in the nozzles and higher pressures in the combustion chambers.

To add a little historical perspective, it is worth noting that the J-2 engine which powered the second and third stages of the Saturn V (burning LOX and LH_2) had an I_{sp} of 425 s, somewhat less than the more advanced SSMEs. The first-stage F-1 engines, still the most powerful known rocket motor ever produced, generating over 1.5 million pounds of thrust at sea level, burned LOX and kerosene with an I_{sp} of 265 s [Fuller and Minami 1987].

Perhaps equally important is the greater degree of control afforded by liquid engines. The flow of propellant can be varied, stopped and restarted, and the combustion chamber/nozzle assembly can be gimballed to vary the direction of the thrust vector, thereby steering the vehicle. Solids have to be designed to give precisely the required amount of thrust: they cannot be *throttled* and they cannot be restarted. The only way to control the magnitude of the thrust is by designing the propellant charge to offer varying surface areas for combustion throughout the burn. This is, however, the ultimate in pre-programming and can only offer relatively coarse thrust control. Some early solid motors were equipped with an explosive venting device, which involved blowing the top off the booster to allow a cancellation of the upward thrust by the application of an equivalent downward thrust. This method could prove disastrous and contemporary solid rocket motors used as launch vehicle strap-ons are now carefully matched to provide equivalent thrust magnitude and duration. Thrust vectoring is more difficult with a solid motor since only the nozzle can be gimballed, but systems incorporating the injection of inert gas into the nozzle to deflect the plume have been demonstrated.

In space, the virtual 'unstoppability' of the solid motor, once burning has commenced, presents on the one hand a simplicity which should make it more reliable, but on the other can result in a runaway. The failure of the perigee motors which should have transferred the Westar VI and Palapa B2 satellites from LEO to GEO in 1984 showed, in the most publicised satellite failures ever, that this could indeed happen.

The inherent simplicity of the solid motor should not be underrated, however. The ignition system of the liquid engine is relatively elaborate and therefore more prone to failure. This is an undesirable characteristic particularly for the upper stages of a multi-stage launcher since, once the vehicle has left the pad, there is no turning back. The failure of the Ariane V18 mission was a case in point: following perfect operation of the first two stages, the cryogenic third stage failed to ignite and the vehicle had to be destroyed, along with its payload, the Intelsat VA-F14 communications satellite.

One major disadvantage of cryogenic propellants is that they are difficult to handle, since they evaporate at environmental temperatures. Even the high-performance storable liquids are, to some extent, hazardous, being highly reactive and toxic. In contrast, solids are generally not too unpleasant, are easily handled and stored, and do not corrode their containers. The relative ease of storage is akin to leaving a pencil rubber and an ice cube on a plate: 2 or 3 days later the rubber will still be there; the ice will have become part of the atmosphere. In operation, the solid propellant motor has no need of complex piping, pumps and pressurisation systems: it is rather like comparing a safety match with a gas lighter. Liquid propellants are also more difficult to handle in space because of the inherent microgravity conditions [see box A3.2].

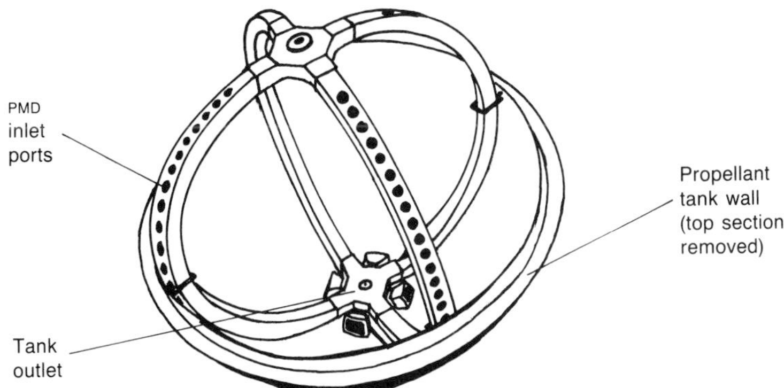

Figure A3.6 Propellant management device (PMD) used to ensure that propellant can be drawn from a tank in weightless conditions.

In addition to the major factors there are a number of more minor considerations which affect the design of propulsion systems. Since the solid propellant container is also the combustion chamber, it must be

BOX A3.2: LIQUID PROPELLANTS IN THE WEIGHTLESS ENVIRONMENT

When liquid propellants experience the force of gravity or the force provided by the thrust of a rocket engine, they will remain in the bottom of their tank in contact with the outlet. When this force is removed they become 'weightless' and float freely within the volume of the tank. If they are fluids which cannot 'wet' the inner surface of the tank, they will form one or more globules uselessly positioned away from the tank outlet.

Fortunately, cryogenic fluids behave as *totally wetting liquids* and their contact angle with metals is typically zero (compare this with the familiar substance mercury (Hg) which has a high surface tension, a contact angle greater than 90° and is therefore *non-wetting*). Although other liquid propellants are *wetting* to an extent, they have to be forced to remain near the outlet. One method is to spin the whole spacecraft; another is to include within the tank a flexible diaphragm or bladder which separates the propellant from a pressurising gas. Any pressurising system requires tanks of sufficient thickness to contain the pressure, which increases their weight. Depending on the material used, bladders and diaphragms may have limited lifetimes before they begin to contaminate the propellant. The slow chemical attack by the propellant on the elastomer of the diaphragm releases *filler compounds* which contaminate the fuel and become deposited within the engine. The effect was diagnosed as the cause of a progressive decrease in thrust after the Orbital Test Satellite (OTS) had spent 5 years in orbit. The solution is to use an all-metal surface tension device.

The need for lightweight, non-pressurising propellant management devices, or PMDs, has led to a multitude of complicated-looking structures being inserted into propellant tanks. Using a combination of perforated metal sheets, screens, traps, troughs and vanes, PMDs have proved themselves capable of holding sufficient propellant over the tank outlet to allow propulsion at any time. Using surface tension and capillary forces, PMDs can be designed to hold any proportion of the propellant (free of pressurising gas) within reach of the outlet. If a continuous thrust is to produce the retaining force over a period of time, the PMD can be of the simplest partial control type, holding just enough propellant to get the engine started. If in the case of short-period usage, say in a spacecraft attitude control system, a continuing thrust cannot be expected, and a total control PMD is required to keep the outlet covered.

Figure A3.6 shows an extension of the total control PMD, called a total communication PMD, which ensures a flow path to the outlet from wherever liquid may be located. It comprises a number of curved 'passageways' with circular openings, known as galleries, which transfer the propellant to the outlet by means of capillary action.

heavier to withstand the pressure. Liquid propellant containers are separate from the combustion chamber and can be very light in a pump-fed system, although the low density of liquid hydrogen means that its tanks have to be large. Overall, the liquid system is the heavier because of the additional hardware and, by and large, launch vehicles using liquid propulsion are biased towards slower lift-offs than the simpler solids.

Whereas regenerative cooling of the nozzle and combustion chamber walls is possible with liquids, solids do not have this advantage and have a limited burn time dependent on the materials used in the nozzle. Liquids in contrast have a theoretically unlimited burn time.

The same can be said for the size of the respective engines. Large solid propellant grains can give problems with homogeneity and inspection. Finally, solids are relatively cheap to manufacture and maintain, whereas liquid engines are expensive because of their complexity, and the propellants can be expensive to obtain and, if cryogenic, maintain at their operating temperature.

The trade-offs between solid and liquid propellants are many and complex. In choosing which one to use, any of these many factors may completely outweigh all of the disadvantages: for example, the use of solid propellants for missiles carrying nuclear warheads where the overriding requirements are long-term storage and instant readiness.

Little mention has been made of solid/liquid hybrid motors, because in modern times their use has not been so energetically pursued as with the pure solids or liquids. At the time of writing, however, the American Rocket Company (Amroc) are proposing to use a solid polybutadiene fuel with liquid oxygen for their Industrial Launch Vehicle, a commercial endeavour designed to carry small payloads to low Earth orbit. In addition to being more efficient than a pure solid, the composite engine is less complex than a pure liquid system although it retains the cut-off/restart capability. Only time will tell whether hybrid motors are to have a commercial future. It is, however, virtually certain that both solid motors and liquid engines will be widely used well into the next century.

3.8 SATELLITE PROPULSION SYSTEMS

So far, propulsion has been dealt with in a fairly general sense, with the occasional allusion to the propulsive devices used on board a satellite. Figure A3.7 is a schematic diagram of a typical propulsion system for a communications satellite. It consists of a solid propellant apogee motor, like those discussed above, and an independent reaction control system comprising a number of tanks for the liquid propellant, a network of pipes and valves, and a number of thrusters. The majority of communications satellites built in the last decade or so have used this type of propulsion system.

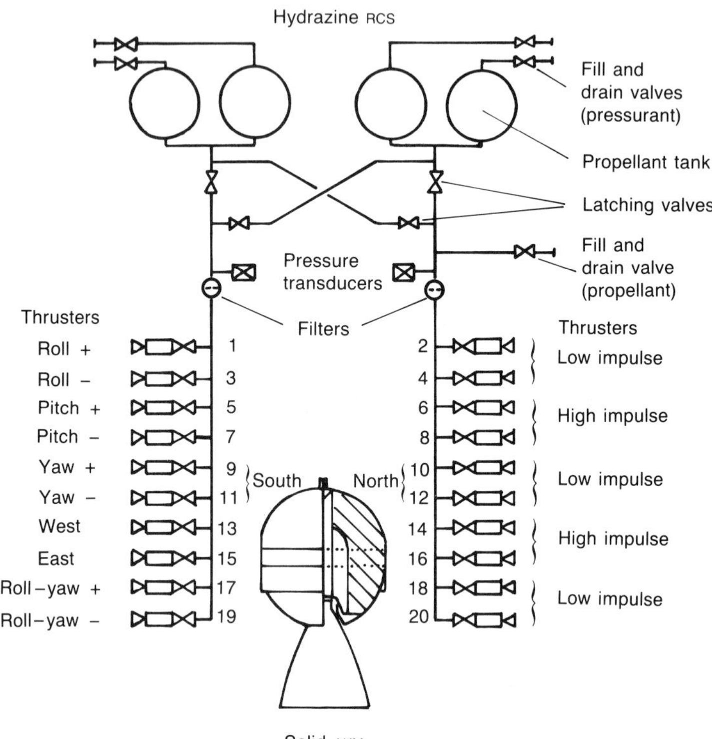

Figure A3.7 Typical communications satellite propulsion system with a hydrazine reaction control system and a solid propellant apogee motor (thruster labelling refers to both odd- and even-numbered thrusters).

3.8.1 Reaction thrusters

The historical development of spacecraft reaction control systems began with *cold-gas* thrusters which used a gas (e.g. nitrogen, argon, freon, propane) stored at high pressure and fed to an expansion nozzle via a pressure reducer. The term 'cold' is used because there is no combustion or heating of the propellant. The ejection of the gas in a tiny jet led to the colloquial term *gas jet*. The early cold-gas thrusters delivered a typical I_{sp} of only about 65 s [Berry 1984], but much greater I_{sp} values are now possible. A problem is the greater volume required for the storage of the lighter gases, but they are still used for space vehicles requiring very fine attitude control. Contemporary examples include Exosat, Hipparcos and Eureca-1, the last one having a very low acceleration limit of 1×10^{-5}g due to the sensitive experimental payloads carried. For this application the cold-gas thruster is ideal.

Communications satellites and geostationary meteorological satellites, however, require a higher performance than that available from the cold-gas thruster. There are three main varieties of thruster in common use for geostationary satellites: the monopropellant hydrazine thruster, the *hiphet* and the bipropellant thruster.

The shape of the thrusters is similar to their larger brothers, the apogee and the perigee motors, in that they require a combustion chamber with a narrow opening leading to a divergent exit cone. The physics is the same, but the scale is vastly different: thruster dimensions are measured in millimetres as opposed to the motors which may be nearly 2 m in length [compare figures A3.8 and A10.14], and the thrust produced can be as low as 0.5 N in one case and 50 kN in the other. Typical thrusts produced by RCS thrusters are 2, 5 or 10 kN, although thrusts as low as 0.5 N are possible using a device about the size of the word THRUSTER, as printed here.

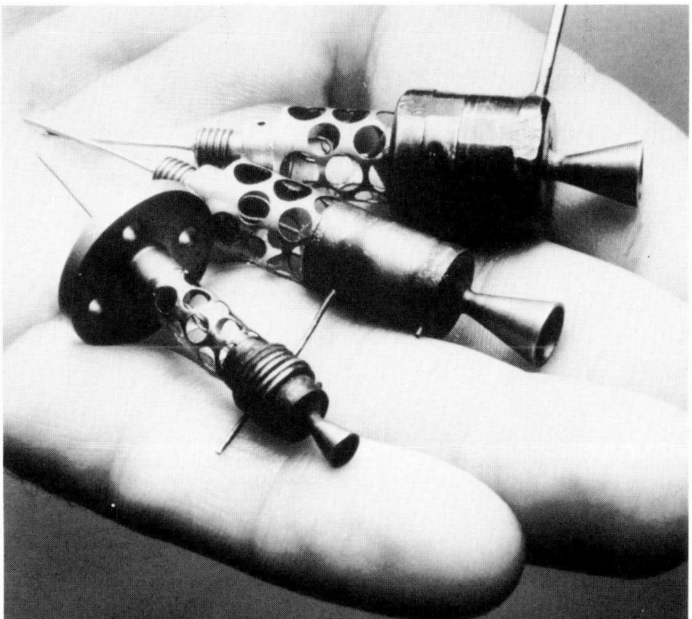

Figure A3.8 A handful of thrusters: tiny rocket engines that produce only 1 or 2 N of thrust. [British Aerospace]

3.8.2 The hydrazine thruster

Hydrazine has been the standard on-board propellant since the late 1960s, largely because of its technical simplicity, high reliability, predictable performance and low cost. It is stored as a liquid and, in the three-axis-stabilised satellite, fed to the thrusters by the action of a pressurant such as gaseous nitrogen. In the spin-stabilised satellite, propellant delivery to the thrusters also benefits from the inherent centrifugal force.

The hydrazine (N_2H_4) is decomposed to ammonia and nitrogen with the aid of a catalyst in the combustion chamber. The propellant is sprayed from an injector onto a bed of the catalyst [see figure A3.9]. The catalyst used almost exclusively in monopropellant hydrazine systems is a platinum/iridium compound finely dispersed by impregnation into a porous aluminium oxide (Al_2O_3) substrate to provide a large surface area [Berry 1984]. Thrust is available in short or long bursts and even in pulses as short as 12.5 ms. The minimum duration is limited by the time taken for the valves to open and close—about 5 ms in each case—in that the repeatability of thruster firings becomes difficult when the valve actuation time is a significant proportion of the thrust duration. Continuous thrust mode tends to be used for orbital control and pulsed mode for attitude control, since the former requires the most energy [see chapter A4]. The specific impulse of the hydrazine thruster is typically about 235 s in continuous mode and about 180 s in pulsed mode [Hudson and Gartrell 1987].

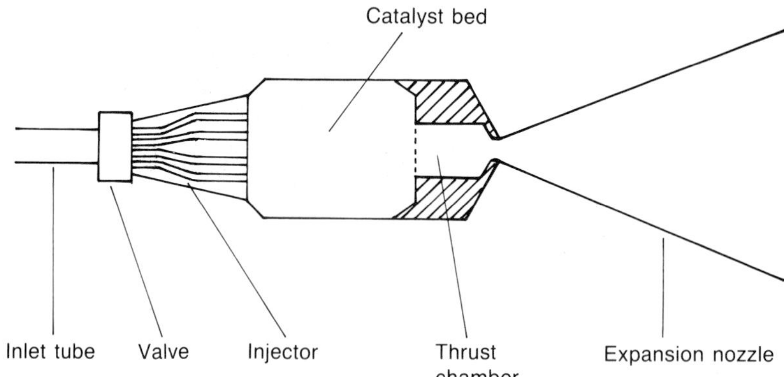

Figure A3.9 Structure of a hydrazine thruster.

3.8.3 The hiphet thruster

To increase the characteristic exhaust velocity of the standard hydrazine thruster the combustion gases can be heated, which typically improves the I_{sp} to about 300 s. This type of device is known variously as the electrically heated thruster (EHT), the power-augmented hydrazine thruster (PAHT) or the high-performance electrothermal hydrazine thruster (HiPEHT), colloquially termed the *hiphet* because it is easier to pronounce.

Between the combustion chamber and the exit cone, the hot gases pass through an electrically heated tube, which naturally increases the temperature of the gases and in turn increases their velocity. The resultant hot gases, at around 1800 °C, require the use of refractory metals such as rhenium, molybdenum and their alloys. Although 10 to 15% more efficient in their use of propellant, the hiphets consume large amounts of electrical power, approximately 1.4 W mN^{-1} [Hudson and Gartrell 1987]. Hiphets cannot be used in pulse mode, and so tend to be restricted to north–south station keeping where relatively long-duration burns are required. This type of thruster has been used on the RCA Satcom series of satellites, all of RCA's series 3000 and 4000 spacecraft and Ford's Intelsat V, amongst others.

The third type of thruster, that using bipropellants, is best described as part of the *unified* or *combined propulsion system.*

3.9 BIPROPELLANT COMBINED PROPULSION SYSTEMS

It has been standard practice for some years to divide satellite on-board propulsion into two separate systems: an apogee boost motor, usually a solid, and a reaction control system, commonly using hydrazine. It is now becoming more common to combine the two functions in one liquid bipropellant system known as a combined or unified propulsion system. Use of bipropellants began in the late 1970s with the Symphonie spacecraft, which used a liquid apogee engine, followed by Insat 1 with its combined system [Fritz *et al* 1983]. Current examples of satellites using this system are Intelsat VI, JC Sat [shown in figure A3.10], Olympus and the Eurostar and Spacebus platforms (e.g. TV-SAT and Arabsat).

In the solid AKM/hydrazine RCS system, the apogee motor is sized, as accurately as possible, to inject the spacecraft from its elliptical transfer orbit to the circular geostationary orbit, an operation it has only one chance to perform. Once the satellite is approximately on-station, its position has to be adjusted using the reaction thrusters of the orbital control system. The apogee motor is no longer of any use. The combined system offers several advantages, not least of which is its operational flexibility. Instead of one pre-ordained engine burn to circularise the orbit, a liquid system is capable of several separate burns, each providing a more accurate approximation to geostationary orbit [see figure A3.11].

Figure A3.10 JC Sat, an HS393-series spin-stabilised satellite. One of two apogee engines of the bipropellant propulsion system is visible below the row of propellant tanks; two RCS thrusters can also be seen. The photograph also shows the folded antenna, the TWTAs of the communications payload and the honeycomb construction of the spacecraft 'floors' or 'shelves'. [Hughes Aircraft Company]

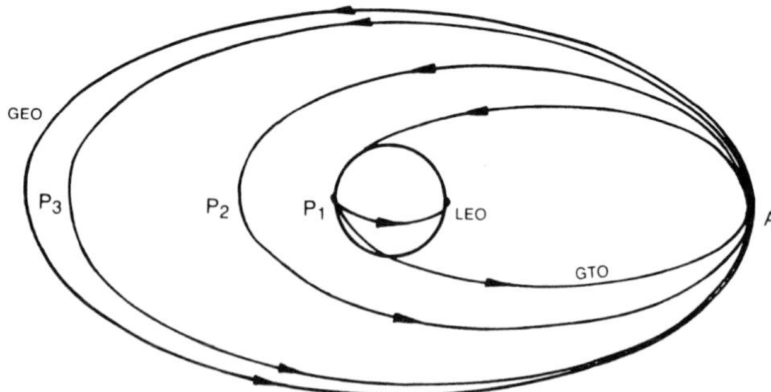

Figure A3.11 Three-burn apogee manoeuvre with liquid apogee engine (LAE). GEO, geostationary orbit; GTO, geostationary transfer orbit; LEO, low Earth orbit; A, apogee; P, perigee. First burn raises perigee from P_1 to P_2; second burn from P_2 to P_3; third burn injects satellite into GEO.

The multiple-burn philosophy is more complex, but it allows greater overall manoeuvring efficiency since the engine can be calibrated on the first burn and subsequent burns made more accurate. By varying the timing and magnitude of the burns to adjust the periods of intermediate transfer orbits, the final burn can be performed at the desired on-station longitude, injecting the satellite directly into its final orbit at the correct position. This technique, known as parasitic station acquisition, can save a significant amount of station-keeping fuel and increase the satellite's lifetime [Pocha 1984].

Once in GEO, the liquid apogee engine (LAE) can be isolated from the rest of the system and all propellant not used for the apogee burns can be made available to the attitude and orbital control system, thereby increasing the life of the spacecraft. On the other hand, if the launch has been slightly inaccurate and placed the satellite in a lower energy orbit than desirable, the liquid combined system can make up the difference, something impossible with the fixed-charge solid motor. It also offers greater flexibility in the payload capacity of the satellite. An increase in the payload mass, relatively late in the design process, could be accommodated without major redesign of the platform. The extra mass would require a greater consumption of propellant on the apogee burn(s), but the system makes it possible to trade payload mass against lifetime.

As far as engine performance is concerned, the liquid system provides higher specific impulses than the solid, perhaps 310 s compared with 290 s for the apogee boost and 285 s compared with 220 s for the other functions. The lower thrust levels available during the apogee boost phase and the attendant lower accelerations offer further flexibility in terms of the deployment of antennas and solar arrays in the transfer orbit. With a solid AKM system, these items can only be deployed in geostationary orbit where mechanical loads are low. The kick from a solid motor would, likely as not, tear the solar panels from their mountings. Similarly, antennas rely on precise alignment for their efficient operation and could easily be misaligned by an AKM firing.

Naturally there are also a number of disadvantages with the combined liquid bipropellant system. For instance, due to its inherent mobility, a large mass of fluid can make stability more of a problem than it is with a large mass of solid propellant. Secondly, two-liquid propellants need to be mixed in the correct ratio to avoid wastage, a problem not encountered with a single liquid. Finally, multiple apogee burns provide a more complex scenario for orbital injection, especially if one of them has to be aborted. The advantages do outweigh the disadvantages, however, which is why the combined system is finding favour with more and more spacecraft designers.

The operation of the combined propulsion system is as follows. At launch the propellant tanks are partially pressurised and isolated from the rest of the system by pyrotechnic valves. After separation from the launch vehicle the valves are fired to pressurise the tanks fully. The apogee boost manoeuvre typically requires more propellant than all the other manoeuvres performed throughout the spacecraft's lifetime (about 80%), so it is most efficient, in terms of providing a uniform thrust and high I_{sp}, to feed the propellant to the liquid apogee engine under constant pressure. After the LAE firing the pressurant tanks are permanently isolated by firing a normally open pyrotechnic valve; from then on the system operates in *blow-down* mode, which is simple and reliable. In this mode, pressure is derived from the remaining pressurant in the tanks and gradually decreases as the tanks are emptied.

Communications satellites are traditionally spin stabilised in the transfer orbit because it is simple. The spin speed is selected to ensure that the spin axis maintains a predictable orientation throughout the apogee boost phase, which in turn ensures an accurate placement. The low-thrust long-burn-duration nature of the LAE makes three-axis stabilisation feasible for the transfer orbit, thus offering a choice.

The choice for spacecraft with combined propulsion systems is by no means clear cut and depends on the particular spacecraft; for instance Olympus is three-axis stabilised in the transfer orbit whereas Eurostar is spin stabilised. Both spacecraft use combined systems and both are three-axis stabilised in GEO. Put simply, the Eurostar points itself in the right direction before each apogee motor firing and relies on its spin for direction control, whereas the Olympus uses small thrusters for attitude control (steering) throughout the transfer phase. One reason for the choice of three-axis stabilisation for Olympus is the potential instability caused by the large mass of liquid propellant it carries.

Bipropellant engines are the subject of continual development. For instance, storable liquid bipropellants are being considered for integral spacecraft perigee motors to replace the separate PKM used with the Space Shuttle with its duplications of TT&C, power and guidance systems [Fritz *et al* 1983]. Effectively combining the perigee and apogee motors into a single system should save considerable mass but, like the combined propulsion systems discussed above, it adds complexity to the system and completely does away with the tried and tested concept of *staging*, where the spent PKM is discarded. The size of the tanks required to hold the extra propellant may be the limiting factor.

Another proposal is a dual-mode system which utilises a bipropellant MON3†/hydrazine engine for orbital manoeuvring combined with mono-propellant hydrazine thrusters for station keeping and attitude control. The hydrazine is fed either to the apogee engine to be combined with

†MON is mixed oxides of nitrogen, i.e. N_2O_4 with added NO (3% by weight for MON3).

the MON3 or alone to the thrusters. The system combines the flexibility of bipropellant combined propulsion systems and the simplicity and reliability of the monopropellant hydrazine thrusters.

3.10 ALTERNATIVE PROPULSION SYSTEMS

Spacecraft propulsion is an ever-widening and continually developing subject, and more methods seem to have been suggested for propelling spacecraft through the cosmos than for powering the humble motor car, which has itself had a fair amount of effort devoted to it. The list includes nuclear, electric, electromagnetic and antimatter propulsion, as well as solar sailing and the laser-heated hydrogen rocket [Frisbee 1985b]. So far, contemporary spacecraft, including planetary probes, have used conventional chemical propulsion systems since they represent current reasonably reliable technology. One particular alternative, that of *electric* or *ion propulsion*, is at an advanced stage of development, however, and seems likely to find application on production spacecraft by the mid 1990s. The form of electric propulsion currently under discussion should not be confused with electrothermal propulsion in which chemical combustion gases are heated electrically to increase the efficiency of the device, as with the hiphet. It concerns the ionisation of a liquid or gas to produce ions which are accelerated rearwards by an electrostatic or electromagnetic field to produce a thrust.

The Kaufman electron bombardment ion thruster typifies the technology in its mode of operation [Fearn 1982]. Electrons strafe the atoms of the propellant vapour forming a cloud of positive ions, which drifts towards a pair of perforated electrode plates known as the screen grid and accelerator grid, which together form the *exit port* of the discharge chamber. The ions are accelerated by the high potential gradient across the grids and ejected as a high-velocity exhaust, thereby producing the thrust. As an alternative to electron bombardment, an RF (typically 1 MHz) generator coil around a quartz discharge chamber can be used to induce ionisation [Berry 1986]. This method is used by the radio-frequency ionisation thruster assembly (RITA), developed by MBB/ERNO in Germany. The RITA 10 version, which has a 10 cm discharge chamber, is expected to have its first in-orbit test on board the Eureca-1 spacecraft.

Proponents of electric propulsion are encouraged largely by the promise of specific impulses an order of magnitude greater than those available from chemical propulsion systems. The RITA 10 provides a thrust of about 10 mN with an I_{sp} of around 3000 s for mercury and 3500 s using xenon as the propellant. Mercury was commonly used in the research stages because of its high density and ease of storage, but fears of contamination and disruption of spacecraft equipment by its deposition have forced a change to rare gases such as xenon. Apart from the I_{sp} advantage, there is, of course, no need to vaporise xenon before injection, which simplifies the engine slightly.

The small thrusts available from the ion engine are indicative of the novelty of the concept as compared with chemical propulsion: chemicals give high thrusts for short periods; ions provide low thrusts for long periods. A major advantage of the ion engine is that it uses about a tenth of the propellant [Martin 1986], which for the right-sized spacecraft can allow an increase of 20–25% in payload mass. This proviso explains why ion engines have not replaced, or even augmented, current satellite reaction control systems. Although the ion engine uses less propellant, the engine itself and its associated power conditioning unit (PCU) weigh more than conventional systems and by their very nature consume more power. Satellites of less than a tonne and only a kilowatt of DC power are not considered capable of generating sufficient extra revenue were the ion system to be installed, but spacecraft weighing several tonnes and providing several kilowatts, which are now becoming more prevalent, make the mass fraction of the propulsion system less significant.

Although the application of the low-thrust concept to interplanetary and interstellar spacecraft has featured large in journals of astronautics for decades, it is only in more recent years that its utility for the common or garden communications satellite has been recognised. In future it seems likely to be used for north–south station keeping and fine attitude control of geostationary satellites and to compensate for the drag of the Earth's atmosphere in the case of low-orbit spacecraft. Other applications include the propellant-efficient position control of satellites 'stored' in orbit for long periods, relocation from one orbital position to another and the transfer of satellites from GEO at the end of their useful lives.

Little progress has so far been made in nuclear propulsion. The development of the NERVA rocket, designed to use the heat from a nuclear reactor to vaporise liquid hydrogen, was recommended by President Kennedy in 1961, but by the end of the decade the programme had been cancelled, partly because of its high cost but also because of the political impact of the Vietnam war. Although the propellant load of a nuclear rocket would tend to make it lighter than the equivalent chemical booster, the weight of shielding, necessary even for unmanned payloads, may be a limiting factor. Despite this, nuclear propulsion remains an attractive prospect for the future, since it produces very high specific impulses (around 1000 s).

3.11 CONCLUSION

This review of spacecraft propulsion systems has covered the most common types of chemical propellant and the propulsive devices that use them. Future propulsion research will probably include the physics of plasmas and the behaviour of matter in electromagnetic fields, but, in the meantime, we can expect to see a continuing development of chemical propellants for use in Earth orbit and beyond.

REFERENCES

Asad W 1983 *ESA Bull.* no. 33 6–11
Berry W 1984 *ESA Bull.* no. 40 6–24
—— 1986 *ESA Bull.* no. 45 24–33
Fearn D G 1982 *J. Br. Interplanet. Soc.* **35** 156–66
Frisbee R H 1985a *Space Educ.* **1** 394–400
—— 1985b *Space Educ.* **1** 448–59
Fritz D E, Sackheim R L and Macklis H 1983 *Space Commun. Broadcast.* **1** 173–88
Fuller P N and Minami H M 1987 *Space* **3** no. 4 55–9
Hudson W R and Gartrell C F 1987 *Int. Astronautical Fed. Congr.* IAF-87-07
Martin A 1986 *Space* **2** no. 2 12–16
Pocha J J 1984 *Space Commun. Broadcast.* **2** 59–70

A4 Attitude and Orbital Control

4.1 INTRODUCTION

Whether the satellite is spin or three-axis stabilised, the attitude and orbital control system can be considered a 'lead' subsystem, since virtually all major design decisions are related to the choice of attitude control. The discussion of constraints in chapter A1 and the basic structural design and layout covered in chapter A2 lend support to this statement. This chapter discusses the need for attitude and orbital control, the hardware required to provide it and some of the methods for its realisation.

It is easy to understand why certain spacecraft might need an attitude and orbital control system. Manned vehicles, for instance, are often required to change orbits, and astronomical satellites need to point in a different direction for each observation. Although perhaps less obvious, the needs of a communications satellite are equally well defined.

Satellites in geostationary orbit are allocated a position in orbit measured in degrees of longitude [see chapter B2]. This *orbital position* is referred to the Earth's longitude at the sub-satellite point—the point of intersection, with the equator, of a line drawn between the satellite and the Earth's centre. The onus is on the operators of the satellite to ensure that it remains within a reasonable distance of the *nominal orbital position* to limit the possibility of interference with other services using the geostationary arc, to allow the use of fixed antennas on the ground and, indeed, to avoid collisions [see box A2.2]. The spacecraft system that performs this function is termed the orbital control system.

The spacecraft's attitude is a description of the orientation of its axes with reference to a number of known points, either on Earth or in space. The required orientation is, of course, for the antenna platform to point towards the Earth while the solar arrays point towards the Sun. Although some satellites have antennas with individual pointing mechanisms, it is the responsibility of the attitude control system to provide overall pointing control of the antenna platform. The system is also required to control the spacecraft's attitude during the transfer orbit.

In terms of motion, attitude control is characterised by rotation about the centre of mass and orbital control by translation. The two functions are usually combined in a common subsystem, the *attitude and orbital control system* (AOCS), the need for which is bound up in the word *perturbations*.

4.2 ORBITAL PERTURBATIONS

If the Earth were a perfect homogeneous sphere, or better still a point mass, *and* the only celestial body in the Universe, a spacecraft in geostationary orbit would be stationary with respect to its associated ground stations. Unfortunately none of this is true, which means that 'geostationary' is very much an approximation. There are three main mechanisms that result in the perturbation of a spacecraft's orbit, as described below.

4.2.1 Triaxiality

The first mechanism is due to the shape of the Earth. The rotation of the Earth has caused a bulging of the equator and a flattening of the poles, producing a polar radius some 21 km shorter than the average equatorial radius. On its own this oblateness would not cause a problem for equatorial orbits, but the Earth's southern hemisphere is larger than the northern, which places its centre of mass south of the equator. Moreover, the equatorial radius varies, so that a cross section through the equator approximates an ellipse, meaning that the Earth also lacks gravitational symmetry in the equatorial plane. The difference between the minimum and maximum radii is only about 70 m [Perek 1987], but this is sufficient to affect a satellite in equatorial orbit [see §A4.4.2]. In fact, whereas the equatorial bulge has been known since Newton's time, the ellipticity of the equator was only detected through its effect on the orbital paths of the early artificial satellites. The attribute of being an ellipsoid with three dissimilar axes is known as *triaxiality*.

4.2.2 Luni-solar gravity

The second source of perturbation is the gravitational pull of the Moon and the Sun, called collectively luni-solar gravity. Because of the Moon's proximity, and because it orbits the Earth, its gravitational influence is dominant, but the Sun has a contribution which can be classed as a second-order effect. The result is analogous to the Earth's tides which, seen on a planetary scale, distort the seas into an ellipsoid with its major axis oriented towards the Moon. The Earth rotates once a day within a 'bubble' of water which rotates with the Moon once a month, producing two tides per day. Likewise, a satellite in GEO tends to 'rise and fall' with the tides.

However, the situation is complicated by the relative orientations of the planetary bodies and their orbital planes. The Earth and the Sun lie in a plane known as the *ecliptic*, but the Earth's spin axis, and hence the geostationary orbital plane, is inclined at 23.5° to the ecliptic. This is actually quite fortuitous for satellite communications since, if the two orbits were coplanar, the satellite would be eclipsed by the Earth every day [see chapter A5]. The Moon's orbit is inclined at about 5° to the ecliptic, which means that the direction of its gravitational pull on a satellite varies constantly with their relative positions. The result of this varying attraction is described in §A4.4.3.

4.2.3 Solar wind

The third perturbation mechanism is that of solar radiation pressure,

BOX A4.1: ATTITUDE CONTROL ACTUATORS

The actuators discussed in the text are reaction control thrusters, reaction wheels and momentum wheels. Another actuator is the nutation damper, a device designed to reduce the 'nodding' motion of a spinning spacecraft [see figure A4.9]. This is obviously of prime importance for the spin-stabilised satellite but is also used for three-axis satellites which are spin stabilised during their transfer orbits. Active nutation damping is realised using thrusters; passive damping involves a device containing a viscous liquid in a sealed circuit or a steel ball in a gas-filled tube. The energy of nutation is converted to heat in the working fluid, which is typically mercury in the first instance and neon in the second, and the motion is damped out.

An increasingly common actuator is the solar array panel, used in the manner of a solar sail. On the Eurostar satellites, for instance, roll and yaw control is currently provided by solar sailing, following tests made on the OTS satellite which used it as the baseline for its attitude control during the later part of its life [Howle *et al* 1985]. The angle of the arrays with respect to the Sun can be adjusted to control orbital position and attitude. The arrays can be arranged to generate a windmill torque around the Sun–satellite axis, giving control in roll or yaw dependent upon the satellite's orientation relative to the Sun. Flaps attached to the arrays allow array angles to be minimised, which keeps the power loss due to array mispointing to about 1%.

otherwise known as the *solar wind*, a continuous stream of atomic particles ejected from the Sun. Depending on the state of solar activity, the velocity of the plasma varies between about 300 and 500 km s^{-1}, a veritable solar hurricane! However, since the mass of individual particles is vanishingly small, the effective force per unit area is correspondingly small despite the high velocity—about 9.1 N km^{-2}. Nevertheless, the force of the solar wind is appreciable when it acts for long periods on the large areas of satellite solar array panels. As power requirements for satellites and space stations increase, their solar arrays become more like the sails of an ocean-going yacht, a concept which has already been proposed for space travel in the inner solar system. The exploitation of the solar wind for satellite attitude and orbital control is discussed in box A4.1.

Having identified the three perturbing influences on a satellite in geostationary orbit, we shall now consider the control systems which have been devised to counter them.

Two other passive methods of attitude stabilisation, used predominantly in low Earth orbits, are worth mentioning briefly. One makes use of the decrease in the Earth's gravitational field with distance from its centre: it is called gravity gradient stabilisation. Its use requires that the spacecraft body should be as long as possible with its mass concentrated about an axis aligned with the Earth's centre (the yaw axis). The difference in gravitational attraction between the two ends will tend to maintain the body's orientation, since in the near vacuum the disturbing torques are small. Although the body can rotate about the yaw axis its orientation is restricted, and this method of attitude control has been useful for only the simplest satellites in LEO. However, it is feasible that future space platforms could use this method. This use of gravity in orbit, incidentally, explains why the term *microgravity* should strictly be used in place of zero gravity for low Earth orbit applications.

The other method makes use of the Earth's magnetic field. It is based on the principle that a freely suspended current-carrying coil will align itself with the local magnetic field. Again the satellite must be axially stabilised with one axis pointing towards the Earth. A battery-fed coil mounted in the structure of the satellite is aligned with the Earth's magnetic field. Small disturbances in the satellite's attitude will be damped out as the coil acts to realign itself with the Earth. Some degree of attitude control is available about the two axes perpendicular to the Earth's field, but the effect varies with the strength of the field and is best suited to applications in LEO.

4.3 CONTROL SYSTEMS

The primary effect of the Earth's triaxiality, luni-solar gravity and the solar wind is the perturbation of the satellite's orbit. Instead of following a perfectly circular path in the plane of the Earth's equator, the satellite is pulled and pushed into a gyrating ellipse which is anything but geostationary. This, of course, affects the satellite's orbital position and produces the requirement for an orbital control system.

The attitude of the satellite is affected in a more round about way, since the gravitational perturbations act on the satellite as if it were a point mass. The spacecraft's attitude is therefore affected *as a consequence* of the perturbation of its orbit: in other words, if it is no longer where it should be in orbit, it cannot be expected to be pointing at the right earth station. The only one of the three mechanisms that can have a direct, rather than secondary, effect on the spacecraft's attitude is the solar wind, in the case when the spacecraft presents different areas on either side of its centre of mass. An asymmetry in solar array panels, for instance, would cause a moment about the centre of mass and the spacecraft would rotate.

Why then not ensure the spacecraft's symmetry and do away with the attitude control system? Unfortunately the degree of accuracy with which the antenna beams have to be pointed militates against this. To maintain *pointing* by orbital control alone would require more accurate tracking equipment than is practicable and more station-keeping propellant than can be economically carried. Apart from this, orbital control manoeuvres themselves can cause *mispointing* [see §A4.4.3].

The mention of tracking equipment highlights the difference in the methods of feedback for the two control systems. Although the orbital position of a spacecraft can be determined using star sensors and an onboard computer, the most common solution for communications satellites is orbital control by ground tracking, the so-called *open loop* philosophy. There are, however, two schools of thought amongst control system designers. One opts for a maximum of onboard control (so-called *closed loop* systems), whereas the other argues against it. The input for this loop—the measurement of distance and direction from the ground station—is obtained by the station itself, since onboard position measurement is not sufficiently accurate. Since the bulk of the processing is still ground based, there is said to be no advantage in putting the control loop on the spacecraft [Soop 1987]. Conversely, although a satellite's attitude can be determined by ground-based analysis of sensor outputs, this is relatively slow. Thus the requirement for closed loop attitude control systems has been led by the required speed of response to the depointing mechanisms.

Before describing the hardware concerned with attitude control, we shall investigate in greater detail the need for an orbital control system.

4.4 ORBITAL CONTROL

4.4.1 Station-keeping components

There are two main components of orbital control: east–west station keeping and north–south station keeping. The former relates to a drift in longitude along the geostationary arc away from the nominal orbital position and the latter to a drift north or south of the equatorial plane.

A third orthogonal 'axis of freedom' does exist—that passing through the Earth's centre and the satellite—but motion along this axis manifests itself as an east–west drift in the following manner. Consider a satellite orbiting from west to east in an equatorial orbit. Since the Earth's mass is not distributed symmetrically around the equator, the satellite experiences a variation in the acceleration due to gravity throughout its orbit. If the force of attraction towards the Earth increases, the satellite's orbital height decreases. Then, according to the laws of orbital mechanics, the orbital period also decreases and the satellite's ground speed increases—it appears to move eastwards. Conversely, if the orbital height increases, the orbital period increases and the satellite's ground speed decreases—it appears to move westwards. The change in the direction of the gravitational attraction has caused an apparent acceleration or deceleration along the direction of orbital motion. The remedy is, clearly, to instigate *east–west* station keeping. So, although the spacecraft may move in a very complex fashion in all three dimensions, its effective motion as seen from the Earth can be described in two dimensions.

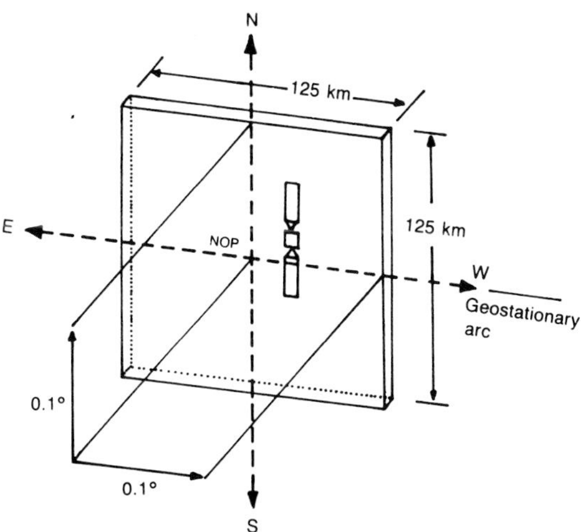

Figure A4.1 Tolerances on a satellite's orbital position; a ± 0.1° 'box' (NOP, nominal orbital position).

The tolerances placed on this two-dimensional motion are illustrated in figure A4.1 as a 'box' with N–S and E–W dimensions centred on the spacecraft's nominal orbital position (NOP). A common specification requires the communications satellite to remain within a box subtending an angle of ±0.1° from the Earth's centre, which is equivalent to about 125 km (77.5 miles) at geostationary height.

4.4.2 East–west station keeping

If we consider the effects of triaxiality in isolation the aspect most important to E–W station keeping is the elliptical cross section of the Earth's equator. This produces four *equilibrium points* on the geostationary orbit: two *stable equilibrium* points aligned with the minor axis of the Earth's equatorial ellipse, and two *unstable equilibrium* points aligned with the major axis.

The ends of the minor axis mark the stable points where, if there were no other perturbations, no station-keeping manoeuvres would be required: the ends of the major axis mark the unstable points (which, by analogy, offer the sort of stability possessed by a coin balanced on its edge). The approximate longitudes of the stable points are 75°E and 255°E, while the unstable points are at 165°E and 345°E, giving symmetrical 90° spacing between stable and unstable points. Following the recognised terminology of GEO position allocations and terrestrial cartography, which measure longitude both east and west from the Greenwich meridian, the same points are, respectively, 75°E, 105°W, 165°E and 15°W [see figure A4.2].

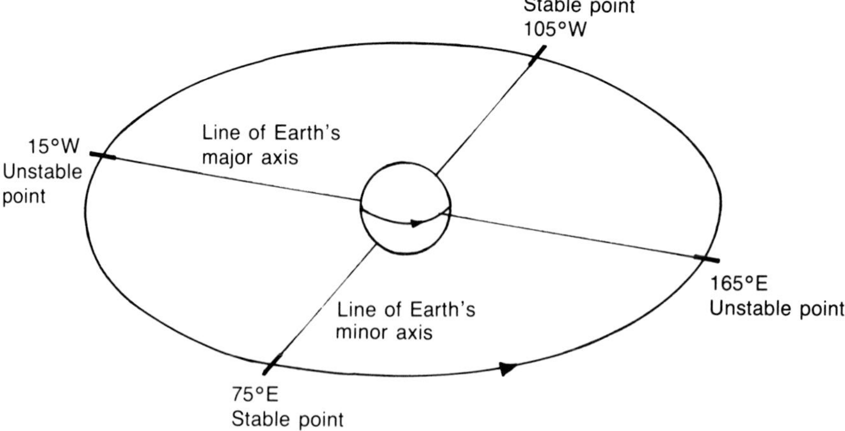

Figure A4.2 Geostationary orbit equilibrium points.

For a satellite stationed at an arbitrary position between the equilibrium points, the triaxiality mechanism operates in the following manner. If the effective component of the gravitational force is in the same direction as the satellite's motion (west to east), energy is added to the system and the orbital height increases. The satellite decelerates, its orbital period increases and it drifts back towards the equilibrium point on the minor axis. Passing the equilibrium point, the satellite experiences a force in opposition to its direction of motion, which reduces its energy. The orbital height decreases and the satellite accelerates, reducing its orbital period. It moves back towards the equilibrium point and the cycle continues. During its eastward drift, the satellite is up to 34 km below the nominal height of geostationary orbit, and during the westward drift it is above it to the same extent [Perek 1987].

Analysis of satellite motions has shown that, if the motion remains uncorrected, the spacecraft will tend to oscillate about a stable point at a maximum drift rate of 0.4° per day, with a period of about 840 days. This is known as the *libration period*. Of course, an operational satellite never completes its libration period, because it is required to remain within its 'box'. This is achieved using the reaction thrusters of the orbital control system. The necessary velocity increment (or change in velocity, ΔV) required to correct the motion is a maximum of about 2 m s^{-1} per year depending on the nominal orbital position [Pocha 1987]. One effective orbital control strategy involves establishing the satellite at one side of its box and giving it an initial drift rate which is just reduced to zero (by the action of the perturbing force) by the time it reaches the opposite side. The satellite then moves back across the box to the starting point, whereupon the thrusters are fired once again to repeat the process. This strategy makes the maximum use of the allowable box and maximises the time between corrections.

So far we have assumed that triaxiality is the only perturbation mechanism affecting east–west station keeping. However, the solar wind also has an effect. It tends to increase the eccentricity of the orbit by lowering one side to a perigee and raising the other to an apogee, while leaving the semi-major axis (half the perigee–apogee distance) approximately constant [see figure A4.3]. To correct for this, thruster burns have to be made at particular times, dependent upon the orbital orientation with respect to the Sun. Triaxiality correction manoevures can, however, be made at any convenient time, as long as two burns are made 12 h (180°) apart to avoid changing the orbit's eccentricity. This means that the two corrections can be combined, in two thruster burns, to correct for both triaxiality and the solar wind.

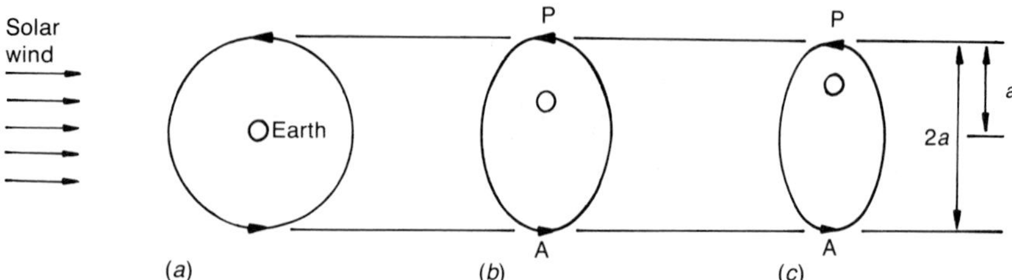

Figure A4.3 Effect of the solar wind on geostationary orbit: eccentricity increases from (*a*) to (*c*). A, apogee; P, perigee; *a*, semi-major axis.

4.4.3 North–south station keeping

As we have seen, the principal effects of triaxiality and the solar wind are manifested by satellite motion *in the plane of the orbit* which, to a ground-based observer, translates as a movement in longitude requiring east–west station keeping. The perturbations due to luni-solar gravity, on the other hand, cause an apparent drift in latitude, which requires north–south station keeping.

Since the Sun and the Moon are not coplanar with GEO, there is a component of their gravitational attraction, at right angles to the plane of geostationary orbit, which acts so as to increase the inclination of the satellite's orbital plane. The outcome is that the satellite appears to drift in latitude with a period equal to the orbital period (about 24 h for GEO) and exhibits a maximum excursion in latitude equal to the orbital inclination. Figure A4.4 shows how a satellite in an orbit inclined to the geostationary plane appears to move to a ground-based observer: it describes a figure of eight in the sky every 24 h.

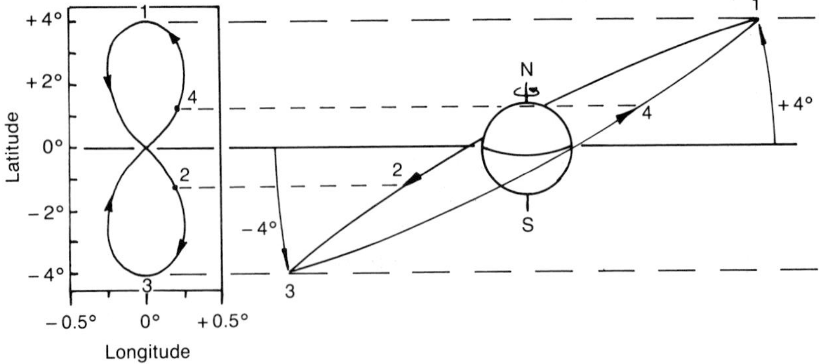

Figure A4.4 Apparent daily motion of a satellite in a 4° inclination 'geostationary' orbit as seen by a ground-based observer.

If this mechanism is taken in isolation, a satellite initially in GEO would have its inclination increased to a maximum of about 15° in 27.5 years [Pocha 1987]. As far as corrective measures are concerned, plane changes require far more fuel than manoeuvres *in* the plane, so north–south station keeping is relatively expensive in terms of propellant consumption. The necessary velocity increment, which varies from year to year, is around 50 m s^{-1} per year, and 1 or 2 min burns are therefore not unusual (compare this with the 2 m s^{-1} per year for east–west station keeping). The higher burn magnitude makes it even more important to align the thrust vector correctly, since any misalignment will produce a thrust component in the east–west direction which will need to be corrected by an east–west station-keeping manoeuvre. Station-keeping strategies therefore dictate that north–south adjustments are made before east–west adjustments, so that error compensation can be included in the east–west manoeuvre.

The thrust misalignment mentioned above is due largely to errors in the mechanical alignment of the thrusters on the spacecraft body, although close control of the spacecraft integration process helps to minimise mounting errors. Another source of error in thruster firings is the thrust level produced by individual devices, which can be divided into mean and random components. The mean error is usually derived by analysis of several thruster operations during the early part of the mission and then allowed for in subsequent manoeuvres. The random error can be assessed by an exhaustive ground test programme prior to launch. An error band can then be incorporated within the station-keeping box to ensure against additional corrective manoeuvres and extra fuel consumption. The 'random on-station' entry in the spacecraft propellant budget [table A4.1] is an additional fuel allocation to account for the variation in the specific impulse of the thrusters [see chapter A3]. Propellant consumption is usually monitored from readings of tank and line pressures and measurements of temperature, from which the remaining propellant mass is computed. Calculating consumption from individual thruster firings, based on nominal performance, produces unacceptable errors.

In fact, north–south station-keeping fuel accounts for a relatively large proportion of the propellant budget. In the example of table A4.1, the satellite has a mass of 1200 kg in geostationary transfer orbit and the total propellant budget exceeds 600 kg. Half of the mass launched into GTO is that of propellant, and typically 20% of this is used for north–south station keeping. It is for this reason that the *free-drift strategy* has been developed. Whereas east–west station keeping confines the satellite to a box of about ±0.1° to obviate interference with other satellite systems, the north–south *dead-band*, as it is known, can be much greater. The capability of the earth segment is the deciding factor: if the earth stations can accommodate a large excursion in latitude (a tall figure 8) by tracking the satellite, the north–south station-keeping requirement can be reduced. The Marecs maritime communications satellite, operated by Inmarsat, is a good example: it is allowed a north–south dead-band of ±3.0° throughout its 7 year life. Since the predominantly ship-based terminals have to be steered anyway, to take account of the sea's motion, a gradual movement in latitude presents no problems.

Table A4.1 Propellant budget for a communications satellite with a design life of 10 years.

Function	Mass (kg)	Mass (%)
Transfer orbit	0.2	0.03
Apogee motor	466.8	76.3
Drift orbit	1.2	0.2
Launch vehicle placement variations	4.9	0.8
In GEO:		
Orbital control: N–S station keeping	119.3	19.5
Orbital control: E–W station keeping	4.9	0.8
Attitude control	0.9	0.15
Relocation	1.5	0.25
Random on-station	1.5	0.25
De-orbiting	0.7	0.12
Residuals	9.8	1.6
Total:	611.7	100
Example satellite:		
Mass into GTO	1200.0 kg	
Mass of propellant used in GTO	473.1 kg	
Mass into GEO	726.9 kg	
Mass of propellant used in GEO	138.6 kg	(including residuals)
Mass of pressurant	2.5 kg	
Dry mass of spacecraft	585.8 kg	

Under the right conditions, a satellite can be given an orbital inclination at the outset which causes it to describe a figure of eight completely filling its box. The effects of the luni-solar perturbations decrease the inclination, reducing the size of the figure of eight to a minimum about halfway through the satellite's lifetime, and then increase it until a maximum is reached at the end. The operational variations between satellites, which may call for smaller boxes or longer lifetimes, often mean that a combination of free drift and north–south station keeping is required.

In the main, the aspects of orbital control as discussed apply equally to spin-stabilised spacecraft, as far as triaxiality and luni-solar gravity are concerned, since they act upon the satellite as if it were a point mass. The solar arrays of a three-axis-stabilised spacecraft, however, present a much larger area for the solar wind to act upon, so the effect of this perturbation would be somewhat less for a spinner. Despite the different layout of reaction control thrusters the station-keeping strategies for both types of spacecraft are essentially the same.

4.5 ATTITUDE CONTROL

4.5.1 Satellite axes and orientation

Once a satellite is released from the launch vehicle its attitude is monitored and controlled from the ground until it is safely *on-station*. In the transfer orbit a spinner is spin stabilised, as it will be when it reaches geostationary orbit. A three-axis-stabilised spacecraft may be either spin or three-axis stabilised in the transfer orbit. The spin rate, or the maintenance of a stabilised attitude, is effected using reaction control thrusters, as in the case of on-station orbital control. Once the three-axis-stabilised satellite reaches GEO its attitude control is largely the responsibility of a system of rotating flywheels, as we shall see. First, however, the satellite must be given a spin rate equal to the rotation period of the Earth, which is of course the same as the satellite's orbital period. This spin rate of 15° per hour ensures that the spacecraft will remain pointing towards the Earth at all times [see figure A4.5].

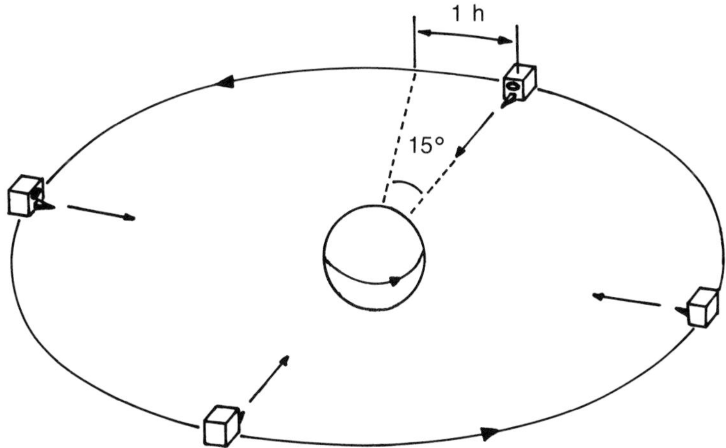

Figure A4.5 A spin rate of $15°\ h^{-1}$ ensures that a satellite will remain pointing towards Earth.

Easily 99% of all promotional material from manufacturers of communications satellites gives a completely misleading idea of a satellite's orientation in geostationary orbit. This seems to be due to a doctrine which states that a satellite only looks attractive when angled at 45° across the face of a brochure cover, which is akin to picturing a car balanced on its front bumper!

Figure A4.6 depicts two satellites and a rowing boat in geostationary orbit. The rowing boat, considered to be circumnavigating the equator from west to east, is there to act as a mnemonic for the roll, pitch and yaw axes of a satellite. The satellite's roll axis is aligned with its direction of travel along the geostationary arc. Rotation about this axis results, therefore, in a north–south movement of an antenna beam projected onto the Earth's surface. Similarly, rotation about the pitch axis, aligned approximately with the Earth's spin axis, produces an east–west movement of the beam. The yaw axis passes through the satellite's thrust cylinder and, if the attitude control system is operating correctly, close to the centre of the Earth. Rotation about the yaw axis causes an equivalent rotation of the antenna beam.

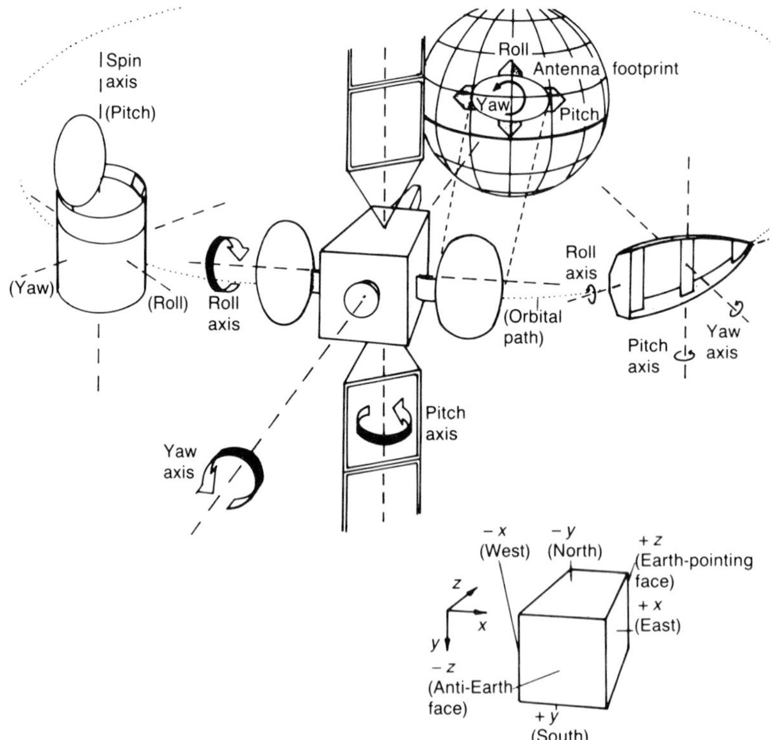

Figure A4.6 Satellite axes and orientation.

For convenience in the compilation of engineering drawings and for subsequent technical discussion, the three orthogonal axes can be labelled in cartesian fashion as x = roll, y = pitch, z = yaw. For the uninitiated, mention of the 'plus z' face can be confusing: the extension of figure A4.6 illustrates the convention which, for instance, makes the 'plus z face' the Earth-pointing face.

Independent of the orientation of the spacecraft body, the solar arrays must always face the Sun [see figure A4.7]. This is arranged by rotating the arrays, once every 24 h, in a direction opposite to that of the 15° per hour spin rate mentioned above. In other words, the orientation of the arrays remains constant with respect to the Sun. It can be seen from figure A4.6 that the axis of rotation of the satellite *and* its arrays is the pitch axis.

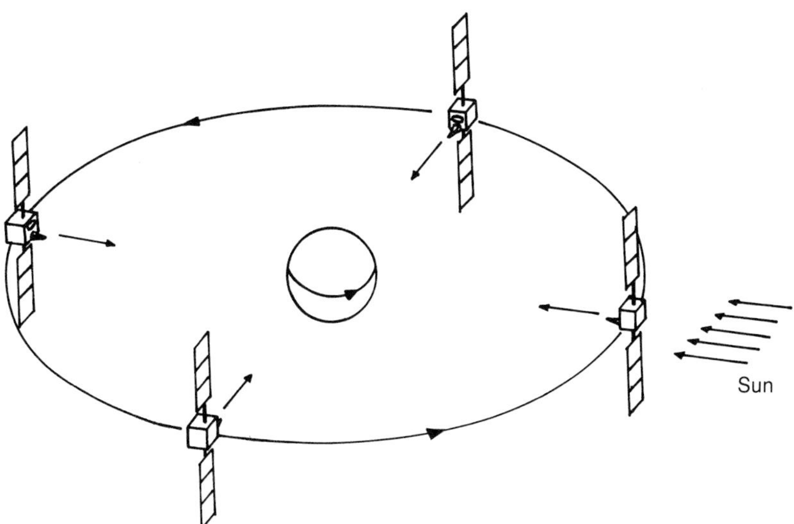

Figure A4.7 Orientation of the solar arrays of a three-axis-stabilised satellite throughout a 24 h orbit.

The pitch axis is also the spin axis for the spin-stabilised satellite. An east–west adjustment of the antenna beam can be obtained by adjusting the spin rate of the de-spun antenna platform; a north–south adjustment is usually engineered by means of an antenna pointing mechanism. Whole-spacecraft adjustments in roll or yaw can be made using reaction control thrusters to provide a moment about the respective axis, as with the three-axis satellite.

4.5.2 Reaction control thrusters

Now that the axes have been defined, it is possible to see how the thrusters should be oriented to enable them to correct for the perturbations described earlier in this chapter. Figure A4.8(*a*) shows a stylised rendition of a reaction control system capable of providing rotation about and translation along all three axes in either the positive or negative direction. Figure A4.8(*b*) shows a more practical arrangement which groups the thrusters more closely together, particularly by removing them from the Earth-pointing face where they could interfere with the antenna platform.

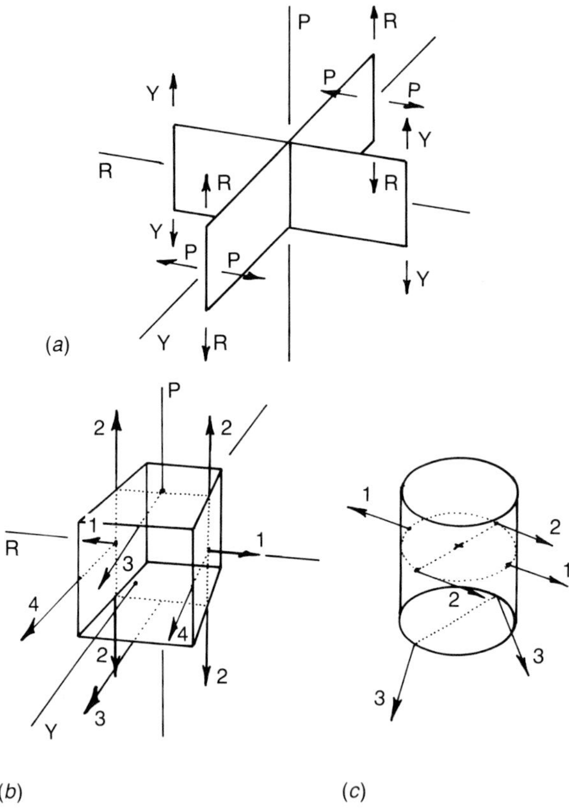

Figure A4.8 Reaction control thruster layout: (*a*) schematic representation; (*b*) practical arrangement for a three-axis satellite; (*c*) practical arrangement for a spinner. P, pitch; R, roll; Y, yaw.

Thruster groups labelled 1 and 2 are used for orbital control: 1 for east–west station keeping and orbital repositioning; 2 for north–south station keeping (using thrusters on the same face) and spin/de-spin operations in the transfer orbit (using diagonally opposed thrusters). By the same token, thruster group 2 can be used for yaw control. Groups 3 and 4 rotate the spacecraft about the centre of mass, group 3 in roll and group 4 in pitch. Roll, pitch and yaw adjustments come under the heading of attitude control which, on a three-axis-stabilised satellite, is the province of the reaction wheel as much as the reaction thruster.

Relating figure 4.8(*a*) to 4.8(*b*), Y corresponds to 2, R to 3 and P to 4, but thruster layouts vary from satellite to satellite. The stylised diagram can therefore be a useful reminder of the basic requirements before a specific layout is considered. Moreover, where one thruster has been shown, it is likely that two will be mounted to provide redundancy.

Figure A4.8(*c*) shows a typical thruster layout for a spin-stabilised satellite, which has thruster groups known as *radials* (1 and 2) and *axials* (3), dependent upon their orientation to the spin axis. Group 1 thrusters are used for east–west station keeping and orbital repositioning, group 2 for spin/de-spin, and group 3 for north–south station keeping and precession control. A loss of stability for this type of spacecraft is manifested mainly by the precession of its spin axis; that is the axis rotates, describing a circle, and the spacecraft 'wobbles' [see figure A4.9 and box A4.1]. By controlling precession, thruster group 3 effectively cancels out roll and yaw. Pitch control is performed by the de-spun platform. Since the spacecraft is spinning, most thruster firings have to be made in pulse mode, pulsing each time the spacecraft's rotation brings the thruster into alignment with the desired thrust vector.

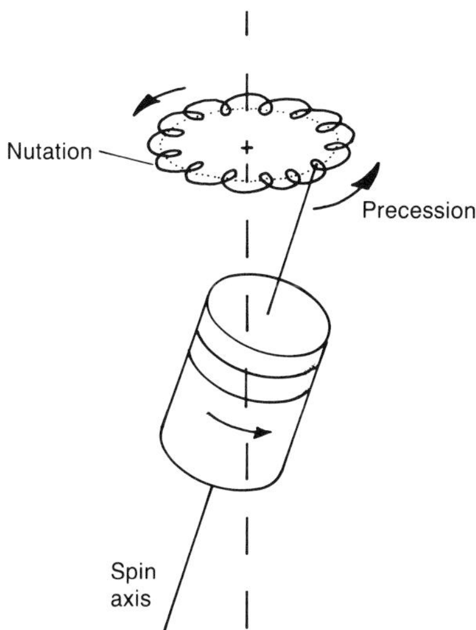

Figure A4.9 Spin axis precession and nutation.

One aspect of thruster firings which may not be immediately obvious is the need to cancel the induced motion of the spacecraft once the desired orientation is acquired. This is particularly true for rotational adjustments made by the attitude control thrusters, since these adjustments are not designed to provide cancellation forces for gravitational perturbations, etc, in the way that many orbital control manoeuvres are.

Spacecraft attitude control provides an ideal demonstration of Newton's first law of motion: a body remains in a state of rest or constant velocity when no external forces act upon it (i.e. no force, no change). A firing of the attitude control thrusters imparts a constant angular velocity to the spacecraft about the desired axis. Since there is no available external force (such as friction, air resistance, etc) to resist the motion, it will continue for an exceedingly long time. The force to halt the rotation is derived from a pair of thrusters directly opposing those that initiated the motion: the same value of thrust that began the rotation will cancel it out. In this way, quite fine adjustments can be made to the spacecraft's attitude.

The siting of the thrusters themselves is important, because the impingement of a thruster plume on a part of the spacecraft can produce an unwanted torque. Despite what the engineer may desire, the exhaust gases from a thruster are not emitted in a narrow pencil beam: about 90% of the gas flow is released in a cone 60° wide. One example of the effect has been documented for the Orbital Test Satellite (OTS), whose yaw thrusters produced these so-called *windmill torques* by their impingement on its solar panels [Lechte *et al* 1985]. As a solution, thruster burns can be timed for favourable array positions, or a system of automatic correction can be instigated (as with the European Communications Satellite, ECS). In addition to such undesired forces, thruster plumes can produce heating effects, erosion due to high-speed droplets of unburnt propellant, corrosion due to the propellant and its combustion products, and the contamination of optical surfaces.

4.5.3 Momentum wheels and reaction wheels

An arguably more elegant means of attitude control can be provided by a number of wheels within the body of the satellite. In controlling the attitude of a spacecraft, the physical law followed by the wheels is the conservation of angular momentum. For an understanding of the application of this law, it is necessary to picture the satellite as a 'physical system' in complete isolation from other physical systems, which is a fair approximation for a body in geostationary orbit!

A wheel spinning inside the satellite has a certain amount of angular momentum, hence the term *momentum wheel* [see figure A4.10]. If a power input from its electric motor causes it to spin faster, its angular momentum will increase. This fact alone would infringe the law of conservation of momentum, since the total angular momentum of the system has increased. However, the law is satisfied by a transfer of angular momentum from the wheel to the satellite, which rotates about the wheel axis in the opposite direction, thereby conserving the total angular momentum of the system. Naturally the satellite rotates much more slowly than the wheel, in proportion to their relative moments of inertia. If it was required to rotate the satellite in the opposite sense, the wheel would be decelerated: it would lose angular momentum and the satellite would gain it by rotating in the opposite direction. In both cases, in the same way that attitude control thruster firings require a cancellation force, a further change of wheel momentum would be required to halt the rotation of the satellite. It is this 'action and reaction' philosophy that gives rise to the term *reaction wheel*.

Figure A4.10 A spinning momentum wheel (top cover removed). [Teldix]

Although both momentum wheels and reaction wheels have this ability, they differ in their operational characteristics. A spin-stabilised satellite maintains its stability in the same way as a gyroscope or a spinning top. A three-axis-stabilised satellite can mimic this approach by incorporating a momentum wheel, aligned with its north–south axis, to become a sort of 'inside-out spinner'. When the wheel is spinning it provides a resistance to perturbing forces in roll and yaw while allowing the body to rotate once every 24 h about the pitch axis. An alternative system uses two wheels set at a slight angle to each other but symmetrically on either side of the north–south axis. When run together the wheels will act as a double-sized single wheel, but altering the speed of one will upset the balance (albeit in a controllable way) to provide a degree of attitude control. These arrangements are examples of the *momentum bias* system.

Whereas a momentum wheel is normally spinning at a high rate to give stability to the spacecraft, the nominal speed of a reaction wheel is zero and it can be rotated in either direction. A system using reaction wheels is known as a *zero momentum* system. Reaction wheels are typically fitted in sets of three—one for each of the orthogonal axes—although it is quite common to install a fourth wheel, mounted at an angle to all three axes, as a spare [see figure A4.11]. If one of the first three wheels fails, the fourth can be used in conjunction with the remaining wheels to provide full three-axis attitude control.

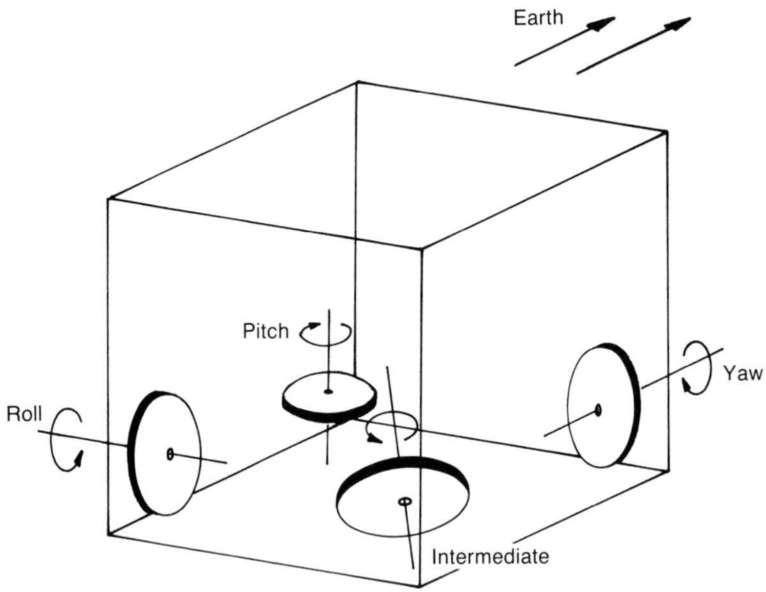

Figure A4.11 Typical reaction wheel layout [compare axes with figure A4.6].

A typical spin speed for a reaction wheel might be ±2500 rpm; that for a momentum wheel might range between 6000 and 12 000 rpm, depending on the type of wheel, with a variation of ±10% of the nominal value for control adjustments. Momentum wheels are typically larger, heavier and, in providing a greater 'momentum storage capacity', consume more power. Wheels between 0.2 and 0.6 m in diameter with masses between about 2 and 40 kg can provide momentum ranges between 1 and 100 N ms, although 25–50 N ms is most common for contemporary communications satellites.

The suspension systems used vary from a conventional design using ball bearings to one using samarium–cobalt magnets to suspend the wheel. In the former, the ball races are lubricated typically by a highly refined mineral oil, of which a few milligrams per year are delivered automatically by centrifugal force. The wheel is sealed inside an evacuated housing to protect it, and to eliminate contamination of other devices and the space environment generally. The obvious advantage of magnetic suspension is the lack of mechanical contact, and since the wheel requires no lubrication there is also no risk of contamination and no need for an evacuated housing.

There comes a point, after several weeks or even days of operation, when a particular momentum wheel has reached its maximum allowable spin speed. It is then necessary to decelerate the wheel while, at the same time, firing an opposing thruster to maintain the spacecraft's attitude. This is described as offloading momentum or *momentum dumping*. Wheels can also be used in conjunction with thrusters to provide a more flexible attitude control system. Other devices used to correct spacecraft attitude are considered in box A4.1.

4.6 SENSORS FOR ATTITUDE CONTROL

4.6.1 Roles and requirements

Correcting the spacecraft's attitude is only one part of the problem: in order to detect an error in the first place, the spacecraft's attitude must be determined and compared with a reference. Luckily there are several reference sources available in space, namely the Sun, the stars and the Earth itself. The onboard sensors designed to detect these reference sources begin their work in the geostationary transfer orbit (GTO), where sun sensors and Earth horizon sensors determine the satellite's orientation. In drift orbit, the approximation to the circular geostationary orbit, infrared earth sensors and gyroscopes help to maintain the spacecraft's attitude.

Before we describe these sensors, it is useful to follow the sequence of events concerned with the stabilisation of a three-axis spacecraft [see figure A4.12]. During the GTO and throughout the firing of the apogee kick motor, the satellite is spin stabilised (*a*). Once in GEO the spacecraft is de-spun by ground command (*b*). Then it is rotated to bring the Sun into the field of view of the sun sensor; the roll axis is steered towards the Sun using the pitch and yaw thrusters (*c*). When this position has been acquired, the solar arrays are deployed and locked onto the

Sun using their individual sun sensors (*d*). The Earth is acquired in two stages: first, with the roll axis still pointing towards the Sun, the yaw axis is steered (rotating about the roll axis) towards the Earth using infrared earth sensors mounted on the Earth-facing antenna platform (*e*); then the yaw axis is steered about the pitch axis with the Sun maintained in the roll–yaw plane (*f*). The yaw axis, and therefore the antenna platform, should then be pointing directly towards the centre of the Earth. It remains only to erect the pitch axis so that it is normal to the orbit plane (g); up to this point it has been normal to the ecliptic. At this point the antennas can be deployed and the wheels can be 'spun up'. The satellite is then stationed at the correct orbital position.

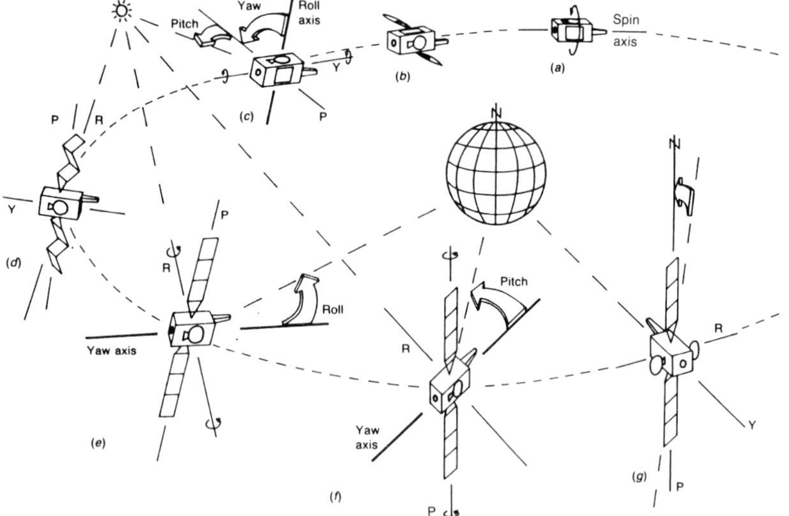

Figure A4.12 Stabilisation sequence for a three-axis-stabilised satellite: (*a*) spinning, (*b*) de-spun, (*c*) roll axis steered to Sun in yaw and pitch, (*d*) arrays deployed and orientated to Sun, (*e*) yaw axis steered to Earth in roll, (*f*) yaw axis steered to Earth in pitch, (g) pitch axis erected to orbit normal, antennas deployed, etc.

4.6.2 Sun sensors

A common direction-sensing technique involves an arrangement of sensors grouped around the pointing axis so that they each receive the same illumination when the arrangement is exactly aligned with the target source. A simple sun sensor, for instance, might comprise four identical solar cells with an opaque screen mounted centrally above them [see figure A4.13]. When the sensor is directly in line with the Sun the output from each of the four cells is the same, since the screen masks an equal amount of each cell. If the sensor is tilted with respect to the

Sun, the shadow cast by the screen masks different areas of cell and the output changes accordingly. This type of differential output can be used to control a satellite's reaction wheels or thrusters to correct the mispointing.

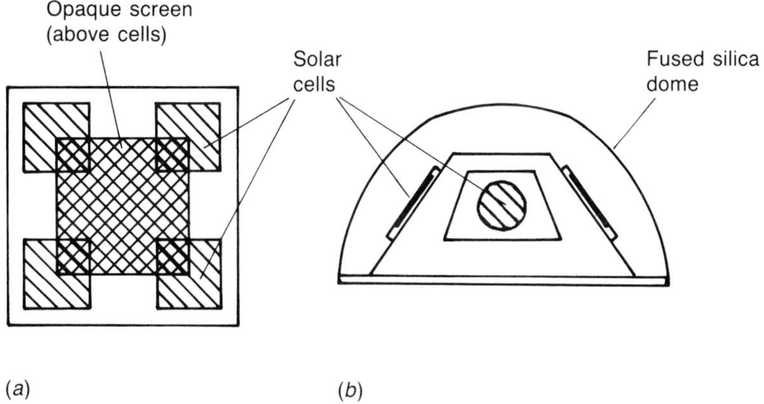

Figure A4.13 sun sensor using null-seeking technique: (*a*) simple illustrative design; (*b*) proprietary design.

Proprietary sun sensors utilising this *null-seeking* technique typically feature a number of cells mounted on the four faces of a pyramid, sealed inside a fused silica dome for protection. Null-seeking sensors often have a wide field of view (FOV), perhaps ±20°, and accuracies of about 0.1°. This wide FOV type is particularly useful for the acquisition phase, and subsequently should a pointing problem occur, but narrow FOV sensors can be used for normal in-orbit operations.

4.6.3 Earth sensors

The same is true for earth sensors. Spin-stabilised satellites make use of two types of earth sensor, both of which use the satellite's spin to scan across the Earth's disk. They contain infrared detectors which sense the difference between 'cold space' and the relatively warm Earth as the Earth's disk crosses into and out of the sensor's FOV [see figure A4.14]. Two types of sensor are in common use: the telescope type has a limited FOV and relatively high accuracy; the fan-beam type senses the Earth for a longer period but is less accurate. Scanning beams can also be used on three-axis satellites, the scanning motion being provided by a rotating mirror; *horizon scanners*, as they are called, are usually employed in low Earth orbit.

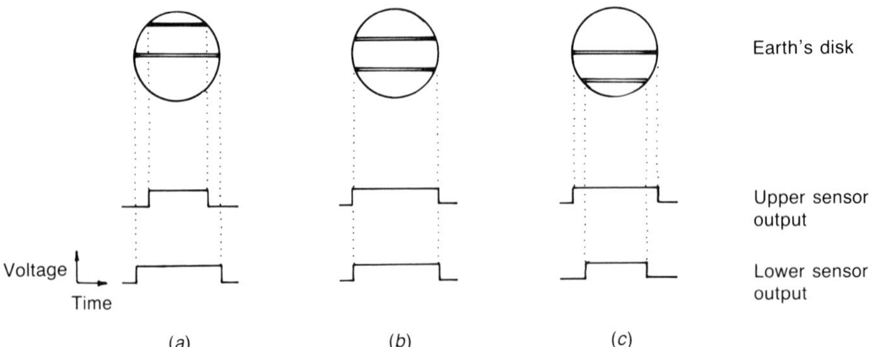

Figure A4.14 Operation of a scanning beam earth sensor: (*a*) satellite pointing error to north; (*b*) no pointing error; (*c*) pointing error to south. Bands on earth disks show fields of view of upper and lower sensors.

Static sensing, as opposed to the above dynamic sensing, is more prevalent in geostationary orbit. The static sensor follows the null-seeking philosophy described above for the sun sensor. In this case an infrared image of the Earth is projected onto an array of thermopiles which convert the incident infrared radiation to an electric current. Since the unit has fixed optics, it can only be used in an orbit in which the Earth's apparent size remains constant (the Earth subtends an angle of about 17.4° from geostationary orbit). This allows the Earth's disk to cover the sensing elements in identical proportions when the sensor axis is pointing towards the centre of the Earth, and to provide difference signals when it moves off axis. Sensor accuracies between 0.05 and 0.25° are common. Since the temperature of the Earth's atmosphere varies with the time of day, a filter limits the incoming radiation to the CO_2 absorption band (around 15 μm) where the radiance is nearly constant. As with any thermal detection system, the temperature of the device itself must be kept constant, so it is well insulated and may feature an active thermal control system.

4.6.4 Other sensors

Other common reference sources for spacecraft sensors are the stars. Star sensors are particularly useful for astronomical satellites which use several of the brightest stars as reference points. Charge coupled devices (CCDs) are finding increasing application in this field.

The *celestial sphere* is also the frame of reference for a completely different type of sensor—the gyroscope, an instrument for the detection of rotation relative to inertial space, which for all practical purposes means *the stars*.

A gyroscope or 'gyro' is a spinning flywheel which possesses inherent stability about its axis of spin. Although gyroscopes are better known for their role in aircraft and rocket guidance, they are also used in satellites. The concept of the *inertial platform* is, however, easier to understand with reference to a rocket moving on a trajectory relative to the Earth. An inertial platform typically comprises three gyros and three accelerometers mounted on gimbals which isolate them from the vehicle. The gyros are set spinning with their axes in the three orthogonal directions which, however the vehicle moves, will remain fixed in inertial space; that is if gyro 1 is pointing towards Polaris it will remain so throughout the vehicle's flight. Any movement of the gimbal frames around the gyros will be detected and the signal thus generated used to drive the gimbal motors to hold the gyro cluster in a fixed orientation. The three gyros thus measure the vehicle's attitude. Any acceleration imposed upon the platform is detected by the accelerometers, the outputs from which are processed to calculate the vehicle's velocity and position. The stable gyro platform provides the accelerometers with a reference. The actual trajectory is compared with a theoretical trajectory and the vehicle autopilot corrects any errors. The problem with a gyroscope/accelerometer system is that it depends upon the integration of small changes in attitude to derive an orientation with respect to an initial known value. Errors are cumulative, and periodic updates based on an external frame of reference (e.g. the fixed stars) are therefore required. Despite this, the accuracy of inertial pointing control systems (at about 0.001°) is impressive compared with Earth-pointing systems (at about 0.1°) [Hudson and Gartrell 1987].

Where greater accuracy is required for Earth pointing, which is often the case for direct broadcasting satellites where the service area may be relatively small, a radio-frequency sensor is used. The sensor, often referred to as a monopulse RF sensor, is tuned to detect an RF signal at a particular frequency transmitted from a known point in the service area. Monopulse techniques are described in §A9.7.

4.7 ATTITUDE CONTROL ELECTRONICS (ACE)

The devices in a spacecraft's attitude control system can be divided into two categories: sensors and actuators. Having described the sensors that detect the change in a spacecraft's attitude and the actuators that correct it, it remains only to discuss the controller which marries them into a system. All attitude control systems operate in accordance with the simple block diagram [figure A4.15]: the sensors compare their input with a reference, commands are sent to the actuators and the attitude is corrected. The brain of the system, the control electronics unit (CEU) accepts inputs from the sensors and directs outputs to the actuators as well as accepting commands from, and returning telemetry to, the Earth.

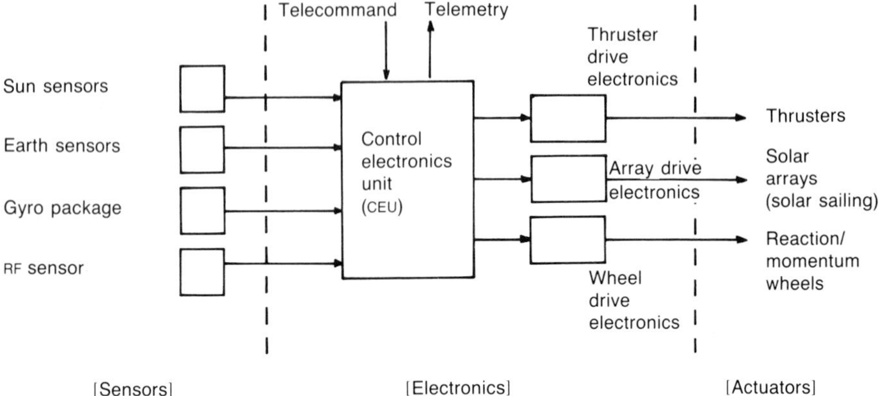

Figure A4.15 Typical satellite attitude control system.

With the increasing use of microprocessors, satellite onboard control systems are becoming ever more complex. A typical modern system continually checks both input and output data and its own operation, so that it can correct errors and switch to a redundant unit in cases of failure. Should a major pointing error occur, for instance if a thruster valve should become stuck in the open position, the CEU can effect a recovery to the normal on-station configuration by switching to automatic reconfiguration mode (ARM). If the unit is unable to maintain even coarse Earth pointing, for whatever reason, it is programmed to put the spacecraft into a 'safe' configuration, typically with the roll axis oriented towards the Sun as it is during the initial stabilisation phase [see figure A4.12]. This is termed ESR, for emergency Sun reacquisition.

There is every reason to believe that the complexity and capability of control systems will continue to rise, following the general rule of increasing in-orbit complexity. The in-orbit control of a number of satellites clustered about the same nominal orbital position has been mooted as the next major challenge in satellite orbital control [see chapter A13]. This and the requirements for ever-increasing pointing accuracy suggest that improvements in both sensors and actuators are also needed for the satellite systems of the next century.

4.8 CONCLUSION

This chapter has attempted to show that *geostationary* is a misnomer: the shape of the Earth, the gravitational attraction of other celestial bodies and the force of the solar wind combine to make this so. Indeed the forces acting on a satellite in geostationary orbit are sufficient to require the development of a complete spacecraft subsystem designed almost entirely with a view to their cancellation. Since all satellites are subject to these forces and all require an attitude and orbital control system of some description, the subsystem holds a position in satellite technology as fundamental as the forces themselves.

REFERENCES

Howle D, van Holtz L, Leggett P, Strijk S and Ashford E 1985 *OTS: Seventh Year in Orbit* (ESA)

Hudson W R and Gartrell C F 1987 *Int. Astronautical Fed. Congr.* IAF-87-07

Lechte H, Hughes C and Ashford E 1985 *ESA Bull.* no. 43 58–63

Perek L 1987 *Int. Astronautical Fed. Congr.* IAF-87-635

Pocha J J 1987 *An Introduction to Mission Design for Geostationary Satellites* (Dordrecht: Reidel)

Soop E M 1987 *ESA Bull.* no. 52 42–6

A5 Power

5.1 INTRODUCTION

One of the most difficult engineering problems of the modern world in general is the storage of electrical power†. On the scale of national power generation, the best solution is the conversion of electrical energy to some other form of energy, such as the potential energy of the head of water in a pumped-storage scheme. As for portable electrical storage media, the only practical solution is the electrical cell or battery. The difficulties involved are well illustrated by the limited availability of electrically powered road vehicles and the poor performance of those that *are* used. The mass of the batteries accounts for a large proportion of the total and they require regular recharging.

The same is true for spacecraft. Only the very earliest spacecraft relied upon batteries as their sole source of electrical power, and only then because the durations of their missions were measured in hours rather than days. It comes as no surprise, therefore, that the solution involved the direct conversion of solar energy to electrical power via the solar cell. Indeed, the power of the Sun has been the mainstay of spacecraft electrical systems for the past three decades.

This chapter concentrates on the methods of power provision for communications satellites which, without exception, use the solar cell. It should, however, be mentioned that contemporary spacecraft of different kinds utilise other forms of power generation. Planetary exploration probes, destined for Jupiter and the other planets of the outer solar system, cannot benefit from solar power because of its diminished flux at these great distances. They therefore use a form of nuclear power

†To the physicist, power is the *rate of doing work* (measured in W or J s^{-1}) and energy is the *capacity for doing work* (measured in J). However, in engineering and in everyday use the word power generally describes a *commodity* that can be generated, stored and used—in the same way as heat, another form of energy. It is used in this way here.

in the guise of the radioisotope thermoelectric generator (RTG), which derives its power from an array of thermocouples heated by the decay of a radioactive source. The RTG offers few advantages in the inner solar system, however, due to the mass of shielding required and the potential danger in manned vehicles.

Since the Gemini capsule, the solution for manned spacecraft has been the fuel cell, a device which generates power by the electrochemical combination of hydrogen and oxygen. A by-product of the reaction is water which, while useful in manned spacecraft, would be a nuisance in a satellite. The requirement for consumables naturally limits the duration of the mission and makes the fuel cell highly unsuitable for satellites expected to operate for up to 10 years unattended.

5.2 POWER REQUIREMENTS

The average power demand of satellites in the early 1960s was a few hundred watts, but by the mid 1980s this had risen to about a kilowatt. Communications satellites now being launched have power demands of several kilowatts and contemporary satellite buses are capable of producing power up to 7 or 8 kilowatts. It is interesting to note, in this context, that a satellite providing several thousand intercontinental communications channels has the same power rating as the average domestic fan heater or electric kettle! The specific power requirements of a communications satellite are many and varied: from power for reaction wheels, solar array drives and antenna-pointing mechanisms, to power for the communications payload itself.

The spacecraft power subsystem can be divided into three main parts concerned respectively with energy collection, energy storage and energy management. Thus subsequent sections of this chapter will deal with the collection and conversion of solar energy by the solar arrays (the so-called primary power supply); interim storage by the batteries (the secondary power supply); and the power distribution and management system which provides a standard voltage to subsystem equipment. When the solar arrays, batteries and other equipment are taken into account, the power subsystem represents one of the largest fractions of the total spacecraft dry mass [see table A2.1]. As with every other subsystem there is a continual desire to increase the efficiency of the various components to reduce the mass.

5.3 THE SOLAR ARRAY

Nowhere is this desire for mass reduction better illustrated than by the solar array, or solar generator, the mass efficiency of which is measured by a quantity called *specific power* (W kg^{-1}), the ratio of array power output to array mass. From the early 1960s, this factor has improved from about 7 W kg^{-1} to better than 20 W kg^{-1} for arrays using silicon cells [Hudson and Gartrell 1987]. The overall power subsystem can also be assigned a value of specific power which, over the same period, has improved from 1 to 6 W kg^{-1}. This indicates the mass burden of other power subsystem equipment, particularly the batteries. Improvements are in hand, however: a solar array, potentially capable of some 60 W kg^{-1}, has been demonstrated aboard the Space Shuttle [see §A5.3.2] and changes to the design, incorporating large-area cells and thin-cell technology, may improve this to around 120 W kg^{-1}. Apart from the improvements in the solar cells themselves, the mass efficiency of arrays is improving as a result of the specification of low-density materials [see chapter A2].

Mechanically speaking, arrays are of two different types: rigid and flexible. Currently most common for communications satellites is the rigid panel type, depicted in many of the illustrations in this volume [e.g. figure A5.1].

Figure A5.1 Arabsat solar array during deployment tests. The four panels are suspended from a deployment rig since they are unable to support themselves in the 1 g environment. Note the array hold-down clamps, or 'stirrups', on the top and bottom of the end panel. [Aerospatiale]

5.3.1 Rigid arrays

In the three-axis-stabilised design, a hinged array of rigid panels is connected to the satellite by a lightweight yoke structure attached to the solar array drive mechanism. Array frames and ribs are typically made of carbon-fibre-reinforced plastic (CFRP) to reduce their mass. A sandwich 'filling' of aluminium honeycomb improves stability and tends to decrease the thermal gradient between the front (sunlit) and rear (shadowed) frames.

The arrays on spin-stabilised spacecraft are, of course, rigid since they form the outer skin of the spacecraft body. Spinners with higher power capabilities have an additional array, in the shape of a rigid cylindrical *skirt*, mounted over the prime array [see figure A1.4]. Once the spacecraft is in orbit, a set of motors deploy the outer array from the main body in a telescopic fashion.

The panel arrays for a three-axis spacecraft require a more complex set of mechanisms to ensure a controlled concertina-style deployment. These devices have three main requirements: to hold and protect the array from damage during launch, to release it when required and to deploy it in a controlled manner. 'Hold-down' is accomplished by clamping the array against protective pads on the north and south walls of the spacecraft. The clamps are released simultaneously by a pyrotechnic cable cutter, and a spring-loaded hinge initiates deployment. An alternative hold-down and release system uses heated wires at each hold-down point to cut a Kevlar release cable. Some arrays allow initial deployment of the outboard panels only, to provide power in the transfer orbit, with final full deployment following once the satellite is on-station.

In the modern array, the deployment mechanism is a complex amalgam of gears, cable drives, centrifugal brakes and speed governors, all of which are designed to ensure predictable constant-velocity deployment—it takes about 18 s to deploy a 6 m array wing. The interpanel hinges are spring actuated and latch when straight to make sure that the panels remain coplanar (one could imagine the panels bouncing back into the closed position if this were not the case). Not all mechanisms are this complex, but the ability to predict deployment kinematics and latching shocks has been shown to increase the reliability of this stage of a satellite's life. This is important since, without a fully deployed array, the satellite cannot hope to provide full payload power for the full design lifetime.

In addition to the constraints of the deployment mechanisms and the reliability problems associated with such structures, rigid panel arrays are limited in size by simple physical constraints. The number of panels for three-axis spacecraft arrays is typically between two and six, individual panels tending to be about the same size as the north–south faces of the spacecraft upon which they are mounted. The thickness of the panel stack is limited by the launch vehicle envelope, however, and this ultimately limits the total area of the array. One way to increase the area is to pack the array into a smaller volume, an improvement offered by advances in flexible array design.

5.3.2 Flexible arrays

Flexible arrays are not yet widely used but offer the possibility of larger, lighter-weight arrays for three-axis-stabilised spacecraft. The array designed for the Olympus satellite is an example [see figures A5.2 and A5.3]: it can be sized to accommodate missions with power requirements, after 10 years in geostationary orbit, of between 2 and 7.8 kW.

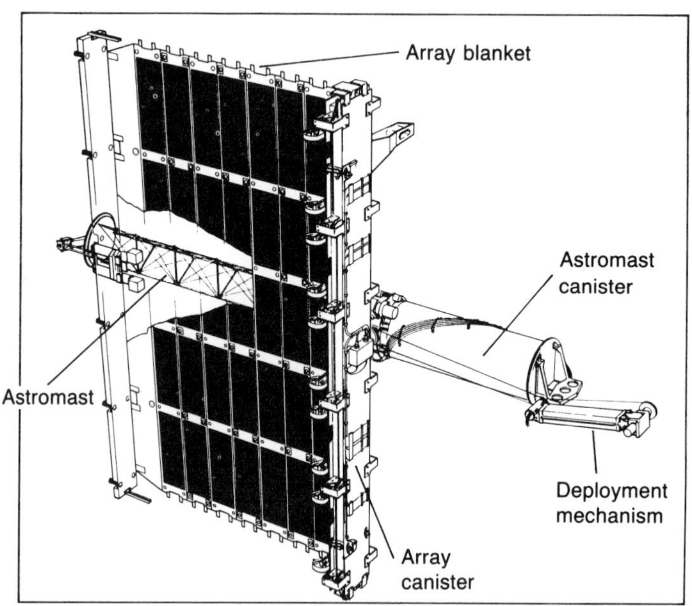

Figure A5.2 Olympus solar array in partially deployed configuration. [Fokker]

The array is stored folded in a box which is swung out from the side of the spacecraft on a hinged arm. An extendable truss structure, called an astromast, pulls the array blanket out of its box at a rate of 2 cm s^{-1}, at first to an extension of 8 m to provide power in the transfer orbit and, in GEO, to a maximum length of 24 m. It takes little arithmetic ability to confirm that the Olympus spacecraft will, in its largest version, be a significant structure—over 50 m from wing tip to wing tip.

Figure A5.3 Artist's impression of the Olympus 1 communications satellite in its on-station configuration. [British Aerospace]

At the time of writing, the largest array of this type ever deployed in space was the NASA Solar Array Experiment (SAE) flown on the STS 41-D mission [see figure A5.4]. It was 32 m long, 4 m wide and comprised 84 panels together producing some 12.5 kW. The array was extended and retracted several times during the mission using a mast similar to that for Olympus and folded concertina fashion into a stack only 19 cm thick, 0.6% of its extended length.

It is evident that rigid arrays are ideal for low-to-medium-power satellites but that flexible arrays are preferable, if not indispensible, for higher power satellites and larger vehicles such as space stations. As far as future spacecraft are concerned, there may be some advantage in an array which can be easily retracted during service. Although modern switching techniques allow the active area of the array to be adapted to suit specific electrical load conditions, the partial retraction of an array would reduce the area exposed to the degrading effects of solar radiation [see §A5.4.3], thereby increasing its potential lifetime. This may be useful for spacecraft, such as man-tended platforms, which require peak power only when inhabited. In addition, the decreased moment of inertia of a retracted array could allow more economic use of the attitude control system by making rotation of the spacecraft easier.

Figure A5.4 NASA Solar Array Experiment (flown on STS 41-D) in cleanroom. The white circles are targets for a stereo TV system which recorded the motion of the array when the orbiter's attitude control thrusters were fired. [NASA]

5.3.3 The BAPTA

Whatever the type of satellite, and whether the array is rigid or flexible, a rotating interface between the satellite body and the array is required to keep the array pointing at the Sun for the lifetime of the satellite and to reliably transfer the power generated to the spacecraft equipment. The device is known either as a bearing and power transfer assembly (BAPTA), or a solar array drive assembly (SADA), a combination of solar array drive mechanism and electronics (SADM and SADE). Whatever the name, its requirements make the device a crucial, but largely unsung, piece of equipment.

Apart from the electronics, a typical BAPTA for a three-axis-stabilised satellite consists of a bearing, a drive motor and a slip-ring unit. The bearing unit is similar to that of some momentum wheels, comprising an alloy–steel ball race specially coated to eliminate cold welding. The array is rotated by electric stepper motors equipped with sensors which indicate their position with respect to the spacecraft bus. Their rotation is accomplished in a sequence of equal steps (about 10 000 for 360°) at a nominal rotation rate of 15° per hour, but faster speeds of perhaps 15°

per minute are possible for Sun acquisition or solar sailing [see box A4.1]. Finally an electrical slip-ring unit, similar to that in any rotating electrical device, couples the power from the array to the spacecraft bus. The commands for the positioning of the BAPTA are derived from the array-mounted sun sensors described in chapter A4.

One obvious difference between a BAPTA used for a spinner and one used on a three-axis-stabilised satellite is the nominal speed of rotation, that is about one revolution per second in the former case and once every 24 h in the latter. Spinners have only one BAPTA, whereas three-axis satellites have one for each array wing. Another difference is that the spinner's BAPTA has only to transfer power for the communications system located in the de-spun section and not that for the *housekeeping* functions such as thermal balance, AOCS and telemetry transmission. In terms of power transferred this is more significant for the smaller spacecraft, since on satellites providing several kilowatts the housekeeping load is only about 10–20% of the total load, most of the power being used by the payload.

From what has already been said it is apparent that a BAPTA, in common with other devices on board the average communications satellite, has to work flawlessly for up to 10 years without maintenance. It is worth making special note that to expect such performance from a mechanical device subject to the space environment is no small matter, but then this is the way things are in satellite engineering.

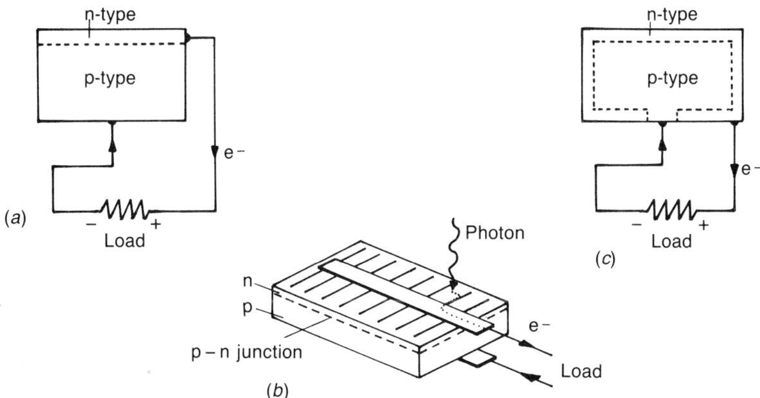

Figure A5.5 Structure of a typical solar cell: (*a*) n-on-p type in circuit; (*b*) a photon strike creates a voltage across the p–n junction; (*c*) 'wrap-around' type (e^- denotes electron flow).

5.4 THE SOLAR CELL

The first practical solar cell was developed at Bell Telephone Laboratories (USA) in 1954 and the first satellite to make use of it was

Vanguard 1, launched by the USA as long ago as March 1958. The device was then as likely to be called a solar battery as a solar cell, but the principle of its operation was precisely the same [see box A5.1].

5.4.1 Construction

Figure A5.6 is a schematic representation of a pair of interconnected solar cells mounted on a backing layer or substrate. The solar cells are assembled into modules and bonded onto the substrate, which is commonly made of Kapton and carbon-fibre cloth. Kapton is also a typical substrate for the flexible array due to its inherent pliability. Interposed between the cells and the substrate, there may also be a back-surface reflector (BSR) to reflect the long-wavelength radiation not absorbed by the cell. This reduces the heating of the substrate and minimises the operating temperature of the array, which enhances cell efficiency. The rear side of the substrate carries wiring connections

BOX A5.1: THE PHOTOVOLTAIC CELL

Theory

The solar cell operates in accordance with the photovoltaic effect. Solar photons cause the ejection of electrons from one part of the cell, creating a potential difference between the two dissimilar parts [see figure A5.5(*a*)]. The upper part of the cell is composed of n-type semiconductor material, the lower part a p-type semiconductor, where n stands for negative and p for positive. The n-type semiconductor has atoms in its crystal lattice with one more electron than the host material and the p-type has atoms with one less electron. Silicon, the predominant semiconductor used in solar cells, has four electrons in its outer (valence) shell for instance. If phosphorus, which has five valence electrons, is added (a process known as *doping*), the result is an n-type semiconductor. Silicon doped with boron, which has three valance electrons, produces p-type material. A silicon photocell doped in this manner therefore contains an electron-rich upper layer from which electrons are more likely to be ejected. The n-on-p cell has been found to be more resistant to radiation than the p-on-n cell [Stark 1986].

Figure A5.5(*a*) indicates how the cell's construction is conducive to electron flow, in other words a current. For a current to flow in the device, electrons must be removed from the *valence band* to the *conduction band* (valence electrons occupy lower bands of energy than conduction electrons). The electrons must therefore be given an energy greater than or equal to a certain value, known as the *band gap* of the semiconductor device. Silicon has a band gap of 1.12 eV. Cadmium sulphide (CdS) and gallium arsenide (GaAs), two other materials used in photocells, have band gaps of 1.2 eV and 1.35 eV respectively. These energy levels are provided which are often applied in a mesh form to decrease mass and increase reliability. Individual cells are connected in series to provide the voltage required for a particular power subsystem design [see §A5.7] and the series strings are connected in parallel to form an *array section*. This construction, with the addition of a silicon diode in the positive end of each string, ensures that single-cell failures do not short out whole sections of the array.

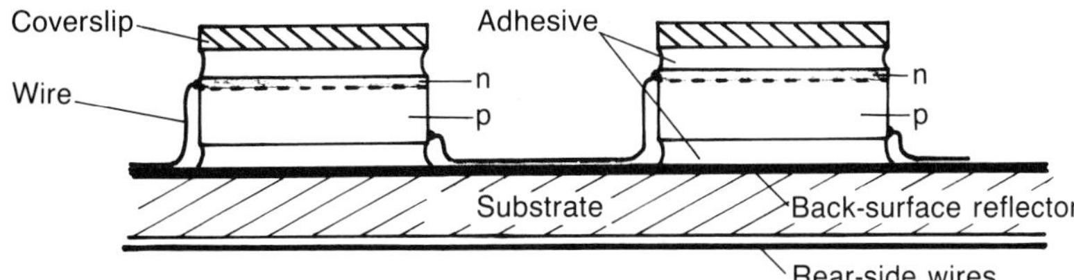

Figure A5.6 A pair of interconnected solar cells (schematic).

by photons with wavelengths less than 1.107 μm [Stark 1986], that is those in the near-infrared, visible and ultraviolet parts of the spectrum, the last having the highest energy.

The excess photon energy over and above that required to lift the electron into the conduction band is absorbed by the semiconductor crystal lattice where it acts so as to heat the device. The result is a reduction in conversion efficiency: a 1 °C increase in temperature results in a 2 mV drop in output voltage [Pratt and Bostian 1986].

Manufacture

Silicon cells are typically made from ingots (up to 15 cm in diameter and 45 cm long) of pure silicon doped with boron. The ingot is cut into disks and phosphorus is diffused into the top surface to make a semiconductor (p–n) junction across the surface of the cell. For satellite applications the disks are then cut into rectilinear shapes to increase the packing density of the resultant solar array. Metal is vacuum deposited on both sides of the cell to provide electrical contacts. A photon hit creates a voltage across the p–n junction and electrons flow along the contact wires [see figure A5.5(*b*)].

Rather than deposit contacts on the front face of the cell, where they only serve to decrease the effective cell area, the phosphorus diffusion can be extended around the cell in a manner described as 'wrap-around'. Wiring interconnections can then be made via the bottom surface alone [see figure A5.5(*c*)]. Although this can make for easier assembly, the gain in performance is less significant than expected because the increase in effective area is almost cancelled out by internal losses [Rauschenbach 1980].

Although other sizes are possible, cells used for spacecraft typically measure 2 × 2 cm or 2 × 4 cm; their thickness, commonly about 0.2 mm, is likely to decrease to about 0.05 mm in future [Stark 1986]. The cells are protected by individual coverslips, or cover glasses, bonded to their top surfaces. These vary in thickness between 0.05 and 0.5 mm.

Solar cells are extremely susceptible to damage from the Sun's high-energy radiation and particle flux. The coverslips offer some protection from this bombardment. In one design the borosilicate-type glass is doped with cerium dioxide (5%) to stabilise the glass and prevent discoloration due to electron and proton irradiation. It also filters out short-wavelength UV radiation to prevent darkening of the cover-glass adhesive. Coverslips may also be coated with antireflection agents such as magnesium fluoride or tantalum oxide, which help to ensure the maximum transmission of solar energy at the peak response of the cell. Other alternative coatings include blue or blue/red filters to screen out UV and IR wavelengths, the latter of which give rise to undesired heating of the cell. In addition, the application of a conductive coating, such as indium tin oxide, can help to prevent the build-up of static electricity,

BOX A5.2: SIZING SOLAR ARRAYS

At a distance of 1 astronomical unit (AU) (the mean distance between the Sun and the Earth) the Sun delivers an energy flux of about 1370 W m^{-2} (the solar constant). If the commonly quoted cell conversion efficiency of 10% is assumed, we might expect 1 m^2 of solar cells to produce 137 W of power. However, most practical arrays, using cells of this efficiency, produce somewhat less than 100 W m^{-2}. Several factors must be taken into account to explain this discrepancy.

They can be placed in two groups. The first concerns a practicality of array construction. Although the cells are cut into rectilinear shapes to increase the packing density, the gaps between the cells produce no power. There is therefore a difference between the physical area of the array and the effective cell area. This factor (known as the *packing factor*) and any occasional shadowing of the array by the spacecraft, at certain angles of the Sun, reduce the effective cell area by about 20%. The second group is composed mainly of electrical losses in the panel wiring and transmission losses in the cell cover glasses. This factor is also taken as a 20% reduction.

The area of array required for a given satellite can now be calculated from the rule-of-thumb expression

$$A = P/Snae \tag{A5.1}$$

where P is the required power (W), S is the solar constant, n is the cell conversion efficiency, a is the loss factor for unused array area and e is the factor for electrical and other losses.

the undesirable result of which can be electrostatic discharge [see box A2.1]. More generally, coverslips offer protection against micrometeoroids and, in LEO, the impact of residual atmospheric oxygen.

5.4.2 Efficiency

The efficiency of the solar cell in its conversion of solar photons to electrical current is a topic of interest in both terrestrial and extraterrestrial applications. The polycrystalline silicon used for many years has an efficiency of about 10% and this manifestly low figure has been widely quoted. In fact, used in its monocrystalline form, the maximum theoretical conversion efficiency for silicon is about 23%; that for gallium arsenide (GaAs), a semiconductor that has found increasing use throughout the 1980s, is about 26%. Solar cells can also be made from indium phosphide, though to date they are less common.

Typical practical efficiencies for silicon cells range between 13 and 16%, while GaAs delivers 16–19% [Hudson and Gartrell 1987]. GaAs cells are made in a slightly thicker form at present which affects their

Array degradation: As far as P is concerned, the power may be the beginning of life (BOL) or the end of life (EOL) power depending on whether the array degradation factor is to be taken into account (see text). For example, an ECS-class satellite such as Eutelsat 1 has EOL power requirements of about 1000 W: a 20% degradation over its 7 year life indicates the need for a BOL power of about 1250 W. Substituting the figure for P in equation (A5.1) gives an estimated array area of 14.25 m^2 (if a cell efficiency of 10% is assumed) and an effective power per unit area of 87.7 W m^{-2}. (As a check on the accuracy of this expression, the real Eutelsat 1 spacecraft has an array area of 13.26 m^2—six 1.3 × 1.7 m panels—which is fairly close to the estimated area.)

Cell efficiency: Substitution of a 1980s technology figure for silicon solar cell efficiency (see §A5.4.2) indicates the importance of improvements and their effects on the required size of solar array and the spacecraft mass budget itself. For a BOL power of 1250 W and a cell efficiency of 15%, the area of array would be 9.5 m^2, a one-third reduction. The relatively low conversion efficiency of the solar cell explains why solar arrays are so large.

Spinners: Everything else being equal, the array size for a spinner should theoretically be larger by a factor of π, since only a narrow strip of cells is orthogonal to the Sun's rays at any instant. The practically accepted factor, however, is taken to be about 2.5, because the array degrades less than a three-axis array throughout the satellite's lifetime (since its cells are in shadow or obliquely illuminated for most of the time).

efficiency advantage and has been partly responsible for their scarcity in space applications. They are, however, more resistant to radiation damage than silicon cells, an important factor for future spacecraft arrays.

5.4.3 Degradation

Despite the protection offered by coverslips and coatings, the solar radiation environment causes damage to the solar cell and results in an unavoidable degradation of the solar array. Degradation due to solar protons and electrons, energetic radiation and micrometeoroid impact is observed to decrease the power output of solar arrays by 2 or 3% per year. Over the 7 to 10 year life of current communications satellites the cumulative effect of this array degradation factor is considerable. It is therefore common to quote two different values for spacecraft solar array output: one for beginning of life (BOL) and another for end of life (EOL). As the above percentages show, the difference between the two can amount to 20 or 30%. What this means in practice is that arrays have to be oversized by a considerable amount, based on BOL power, to ensure that the satellite will be fully operational at EOL. The alternative would be to gradually shut down perfectly good transponders as the satellite grew old, an economic nonsense. The effect of cell efficiency, degradation and EOL power requirements on the size of the solar array is discussed in box A5.2.

5.5 ECLIPSES

5.5.1 Astronomical mechanisms

So far in this chapter we have assumed that power is continuously available via the solar array, which is not the case. It is evident that spacecraft in all low Earth orbits enter the Earth's shadow for a similar time on every orbit: although it depends on their precise orbital period, they spend about 35 min of a 90 min orbit in eclipse. The situation for a spacecraft in geostationary orbit is not so readily apparent, since the orbital plane is not in the plane of the ecliptic, as explained in chapter A4.

The fact that the Earth's axis is tilted at 23.5° to the plane of its orbit about the Sun brings about the situation depicted in figure A5.7. For most of the year the entire geostationary orbit is sunlit, since the plane of the orbit is tilted out of the Earth's shadow. As an example, figure A5.8(*a*) shows the relative orientation of geostationary orbit and the ecliptic when the Earth is at the winter solstice: a satellite in GEO can never be eclipsed by the Earth at this time of year. The diagram also shows, however, that a satellite orbiting at the height of an average low Earth orbit is always eclipsed.

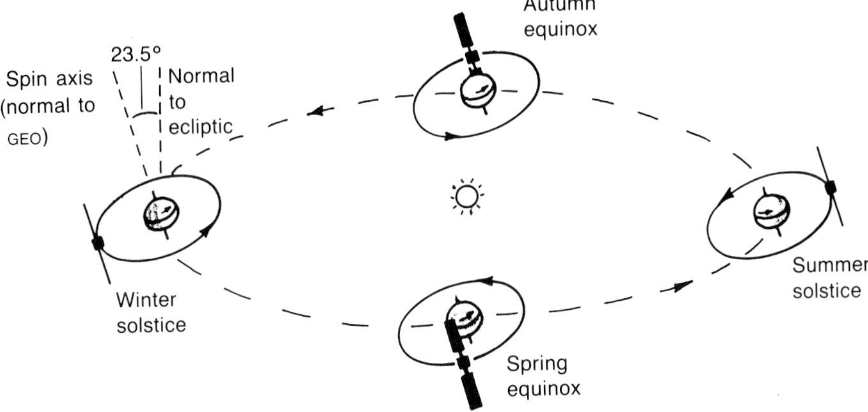

Figure A5.7 Eclipse geometry for a satellite in geostationary orbit. At the solstices the satellite is either above or below the ecliptic and does not pass through the Earth's shadow. At the equinoxes it crosses the ecliptic in the Earth's shadow (and passes into eclipse).

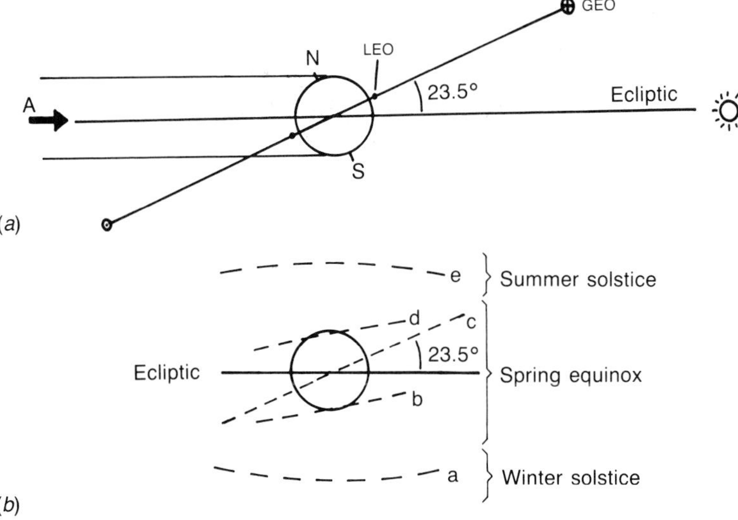

Figure A5.8 (*a*) Earth at winter solstice. A satellite in GEO is always sunlit; a satellite in an equatorial low Earth orbit is eclipsed on every orbit (orbital heights are to scale). (*b*) View along arrow A in A5.8(*a*) from a point in the plane of the ecliptic within the Earth's shadow (over a period of half a year): lines a–e show successive orbital tracks from winter solstice, a, to summer solstice, e; c is the track at the mid-point of the spring equinox 'eclipse season' [compare with figure A5.7].

For a satellite in GEO, figure A5.7 indicates how the situation changes. At and around the winter solstice (21 December) the satellite 'sees' the Sun under the south pole, while at the summer solstice (21 June) it sees it over the north pole. As the Earth approaches the vernal (spring) equinox (21 March) and the autumnal equinox (23 September), part of the orbit moves into the shadow cast by the Earth and the satellite becomes eclipsed for a small part of its circuit.

For an observer looking along the ecliptic from point A in figure A5.8(*a*) successive orbits of a satellite appear as in figure A5.8(*b*). At the winter solstice the satellite orbits well clear of the Earth's shadow as we have ascertained (a). At about 22 days before the spring equinox the satellite begins to clip the Earth's shadow (b) and is eclipsed for a short time. The next day the eclipse is longer and its duration increases to a maximum at equinox (c) when it peaks at about 72 min. The daily eclipse period decreases towards the end of the *eclipse season*, as it is called, about 22 days after equinox (d). The pattern is repeated, albeit in reverse, around the time of the autumn equinox. Figure A5.9 shows how the eclipse period varies according to the number of days from equinox.

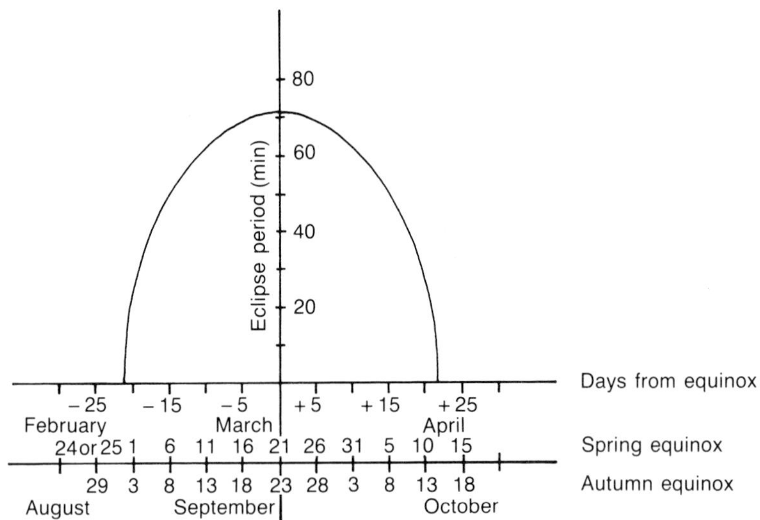

Figure A5.9 Eclipse duration for a satellite in geostationary orbit.

Students of astronomy will know that the shadow cast by an eclipsing astronomical body is composed of a central region of complete shadow, the umbra, surrounded by a fringe region of semi-shadow, the penumbra. This is due to the fact that the Sun is not a point source. From the point of view of the satellite, the solar illumination reduces steadily throughout the penumbra until the Sun is entirely eclipsed. The duration of the penumbral eclipse is 71.8 min and that of the umbra 67.5 min. In addition to the regular eclipse by the Earth, the Sun is eclipsed by the Moon, but on a much less regular basis.

5.5.2 Engineering consequences

The eclipse seasons have many effects on spacecraft design. For a start it is clear that the solar array sun sensors, which ordinarily provide the signals to keep the arrays perpendicular to the Sun direction, become inoperative during an eclipse. The problem is usually solved by disabling the sensors and commanding the arrays to turn at a constant rate (about 15° per hour; see chapter A4) until the Sun is reacquired at the end of the eclipse. More important is the effect on the thermal subsystem which has to cope with, at worst, a 72 min interruption of the 'solar heat wave'. How it does this is discussed in the next chapter.

Most fundamental of all to a communications satellite, however, is the cessation of power required for the payload, the subsystem which pays for the satellite's very existence. Of course the satellite cannot afford to be without power entirely and for this reason carries a secondary power supply in the form of a number of batteries. This means that at least the housekeeping subsystems can be maintained, even if the payload has to be temporarily deactivated. This brings about one of the major trade-offs in satellite design. It is desirable to maintain communications services throughout an eclipse, but there is a limit to the power a single battery can provide and a limit to the number of batteries which can be carried because of their mass. As is often the case, the solution tends to depend upon the type of satellite considered. If it is a low-to-medium-power satellite (the 1 tonne/1 kW class) used for general telecommunications, sufficient battery power for a continuous service is usually made available. The service is referred to as *eclipse-protected*. However, if the satellite is of the high-power category, requiring 2 or 3 kW or more, perhaps to provide high-power direct broadcast transmissions, it is generally considered uneconomic to carry sufficient batteries to maintain the service, and eclipse power is restricted to a kilowatt or so. The ramifications of this for DBS are discussed in chapter B2.

Wherever a quantity is restricted (and what isn't on a spacecraft?), the use of that quantity must be accounted for. As with mass and propellant, discussed in previous chapters, a budget must be compiled for spacecraft power requirements. Table A5.1 shows a typical power budget for a communications satellite with power requirements of just over a kilowatt. A glance at the budget table reveals that the service is partially eclipse-protected, as shown by the reduced payload power in the eclipse column; the battery charge entry in this column reads zero since the batteries are discharging. The difference in battery charge between solstice and equinox is due to the requirement to recharge the batteries after the eclipse, and the increased thermal subsystem consumption is mainly for the replacement of heat lost during the eclipse.

Table A5.1 Typical end of life (EOL) power budget for a 1 tonne/1 kW-class satellite (figures in W).

Subsystem	Transfer orbit (mean)	Solstice	Equinox	Eclipse
Payload	0	761	761	655
TT&C	66	43	43	43
AOCS	5	46	46	46
Power	21	33	33	31
Thermal	26	85	110	10
Losses	1	44	45	72
Battery charge	†	24	85	0
Total	119	1036	1123	857
Solar array capability (EOL)	†	1139	1242	0
Margin	†	103	119	N/A

†Most satellites are designed to survive a greater number of transfer orbits than the batteries alone could provide power for. The outer array panel may provide power to recharge the batteries, or arrays may be partially deployed in the transfer orbit (depending on design). Battery charge depends on the number of transfer orbits, power demands, etc; solar array capability on array type, angle of Sun, etc; margin is always designed to be positive.

The difference in solar array capability between solstice and equinox is due to the seasonal variation in solar flux density. One reason for this is the variation in distance between the Sun and the Earth due to the slight eccentricity of the Earth's orbit; the distance is least at the winter solstice and greatest at the summer solstice. More important though is the declination of the Sun with respect to the solar array surface: examination of figure A5.7 shows this to be 0° at the equinoxes and 23.5° at the solstices. Maximum power can only be derived if the surfaces of the arrays are perpendicular to the solar vector. This angle could be compensated for by incorporating an array-pointing mechanism at the interface with the yoke, but the added complexity is not considered worthwhile. Taking averages of the relative flux densities available at the solstices and equinoxes gives a ratio of 0.917 between the two: the array capability at solstice is therefore 0.917 that at equinox.

The budget shows a margin of about 10% at EOL. Although this seems excessive, since all that is really required at EOL is a 'plus sign', in practice margins have a habit of shrinking. The array capability is usually fixed early in the design programme, but the power requirements of the various subsystems tend to rise towards it as development progresses, eroding the margin.

Figure A5.10 shows the variation of array capacity over the lifetime of a satellite due to the two effects (solar distance and declination) described above. Power is a maximum at the spring equinox (a), reduces sharply to a minimum at the summer solstice (b), then rises to a value less than the previous maximum at the autumn equinox (c), and finally drops again at the winter solstice (d), but not as low as at the summer solstice. This seasonal variation is overlaid on a downward trend which represents the degradation of the arrays discussed earlier.

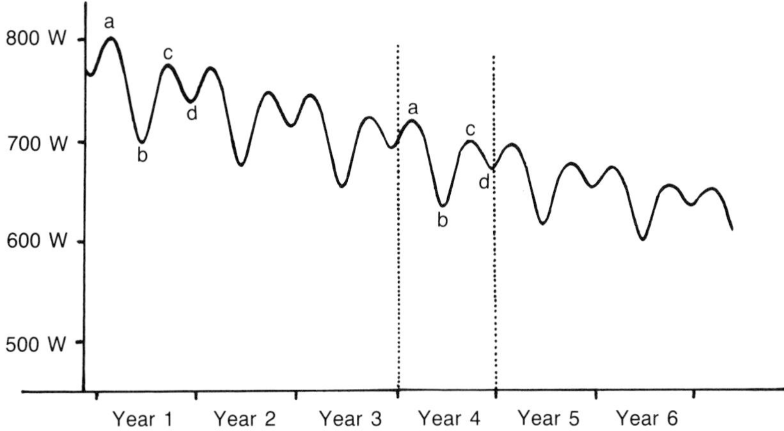

Figure A5.10 Variation of solar array capacity with time in orbit: a, spring equinox; b, summer solstice; c, autumn equinox; d, winter solstice (based on results from the OTS-2 satellite).

5.6 BATTERIES

Batteries are required to store power for use during the transfer orbit, for satellites that have no array capability during this phase, and of course during eclipse. The battery subsystem is often referred to as the secondary power supply for this reason. In the same way that the solar array has a figure of merit which defines its mass efficiency (specific power, W kg^{-1}), the battery is characterised by its *specific energy*, measured in watt hours per kilogram (Wh kg^{-1}). Since this is a measure of the usable energy stored per unit mass, it is sometimes called the *usable energy density*.

The two types of battery commonly used on communications satellites are the nickel–cadmium (NiCd) cell and the more recent nickel–hydrogen cell (NiH_2). Manufacturers currently quote ratings of 30–35 Wh kg^{-1} for the NiCd and 45–50 Wh kg^{-1} for the NiH_2 cell, and respective capacities of 4–50 Ah and 20–100 Ah. The specific energies of NiCd and NiH_2 batteries imply that the NiCd is the heavier and therefore apparently less desirable, but when a choice has to be made there are other factors to be taken into account.

One such factor is the degree to which a battery can be discharged without detrimentally affecting its future performance—this is known as *depth of discharge* or DOD. For example, if battery A has a DOD capability of 50% and battery B 25%, a spacecraft will have to carry twice as many of battery B as battery A to provide the same usable power. If everything else about the batteries is the same, the choice of battery, from the point of view of the mass budget, is obvious. Modern NiCd batteries are capable of about 70% DOD and NiH_2 batteries about 80%. Nickel–hydrogen batteries, therefore, have two main advantages over nickel–cadmium: higher specific energies and higher DOD capabilities. However, where the available volume is particularly restricted, the NiH_2 battery suffers a disadvantage—for equivalent capacities the NiH_2 is larger than the NiCd.

The type of orbit the spacecraft is designed for also has an effect on the choice of batteries. Since eclipse conditions are vastly different in LEO and GEO, the duty cycle of the spacecraft batteries is very different too. A spacecraft in GEO undergoes about 100 charge–discharge cycles per year, with periods varying according to the eclipse duration, whereas one in LEO undergoes around 6000 cycles with a 35 min discharge period and a 55 min charge period. The nominal required DOD in LEO may be typically about 20%, whereas in GEO 60% is common, primarily because the eclipse periods are longer. Although NiCd batteries are likely to be progressively replaced by NiH_2 for applications in GEO, there is some concern that their performance may not be as good in LEO with regard to the large number of charge–discharge cycles [Hudson and Gartrell 1987].

In simple physical terms the NiCd battery has an advantage in that it is easier to assemble and more convenient to package in the spacecraft. Each NiH_2 cell, in contrast, is contained within an individual pressure vessel, a cylindrical metal case containing a stack of electrochemically impregnated nickel and platinum electrodes. As far as battery life is concerned, the wear-out mechanisms of the NiCd, well proven in both

LEO and GEO, are particularly well understood due to its long pedigree. On the other hand, designs are continually being improved and the inherent 'benign chemistry' of the NiH_2 augurs well for its life expectancy.

Battery life is in general dependent upon time, temperature and DOD, but degradation can be retarded by proper temperature control and careful energy management. Part of this management is concerned with making sure that the battery is not overcharged. For an NiCd battery this can be accomplished by monitoring the battery terminal voltage, which shows a rapid increase as the cell approaches full charge. Alternatively, since overcharging generates hydrogen and oxygen, a pressure switch can be used to terminate charging [Stark 1986]. Another aspect of this energy management is *battery reconditioning*, the controlled discharging and subsequent recharging of a battery conducted during the solstice period. This improves battery performance in general and has been known to increase the capacity (Ah) during a mission.

5.7 POWER DISTRIBUTION AND MANAGEMENT

5.7.1 Regulation and voltage

So far, the sources of spacecraft power have been discussed without regard to voltage. Just as battery-operated domestic equipment requires batteries of different voltages, spacecraft equipment must be designed to operate within a certain range of *bus voltages*. Bus, in this case, is not the colloquial term for service module but a short form of bus-bar, a term used for the main electrical power distribution circuit in the spacecraft.

The range of bus voltages the equipment has to cope with is dependent upon whether the power subsystem is *regulated* or *unregulated*. A regulated system provides a voltage constant to within ±1% of the nominal value, whether the satellite is in sunlight or eclipse. An unregulated system feeds the array or battery output, with an inherent wide range of voltages, direct to the power bus. Satellite power subsystems can be fully regulated, totally unregulated, or regulated in sunlight, when power can be derived from the solar array, and unregulated in eclipse, when the output voltage falls to that provided by the battery.

The unregulated supply is simple as far as power subsystem design is concerned, but it makes life more difficult for the payload and other subsystem equipment which must be designed to operate across a range of voltages. The regulated supply is simpler for the power user and is more efficient, but is naturally more complex because of the need for additional power subsystem hardware. The terrestrial power supply and distribution system is, as an analogy, a regulated system, since its aim is to provide a nominal 220 V or 110 V AC power supply for domestic consumers.

The great majority of spacecraft power supplies are of the DC type. An 18 V AC supply was proposed in the early 1970s, but the equipment was heavy and inefficient, and the concept was shelved for a decade or so [Capart and O'Sullivan 1985]. By the 1980s, technology had decreased the mass and the power losses by a factor of 5, but to date its application has been limited to a few scientific satellites and spacecraft payloads. The Hubble Space Telescope, for instance, uses the 20 kHz AC standard for secondary distribution.

DC bus voltages have increased gradually over the years more or less in line with the general rise in available DC power. A 16 V standard was developed in the 1960s and used until the early 1970s when a 28 V system began to appear. From the mid 1970s to the present this has been the predominant system used for satellites manufactured in the USA.

When the European Space Agency began to prepare for satellites with up to 10 kW of onboard power, it was decided to develop a new standard which would cut down on the power losses *and* the consequent thermal dissipation that would be suffered with the 28 V system. Thus it was that the Orbital Test Satellite (OTS) pioneered a new 50 V power bus, as it had many other examples of satellite subsystem technology. The 50 V bus is now the standard for ESA satellites, but there seems to be little standardisation beyond this, even within Europe where 42.5 V is equally common. Some American manufacturers have followed the trend to higher voltages but, just as terrestrial power distribution voltages remain at variance, the majority of American communications satellites retain the previous standard. As far as supply regulation is concerned, a similar dichotomy exists: European satellites tend to use a regulated system whereas American satellite power supplies are generally unregulated [Capart and O'Sullivan 1985]. However, this too should be regarded as a tendency rather than a rule.

Just as three-axis rather than spin stabilisation is the preferred method for future spacecraft with large power requirements [see box A1.1], a regulated high-voltage bus is the most likely contender for the power supply. Regulation allows standardisation and compatibility of components and payloads. High voltage improves efficiency by limiting the current for a particular power ($P = IV$) and thereby limits power loss (I^2R). This is, after all, the reason why terrestrial distribution systems use such high voltages. Higher voltages still are under consideration, but for satellites built and specified during the late 1980s and early 1990s the 50 V supply is unlikely to be exceeded.

5.7.2 The power distribution bus

Early spacecraft accepted power from the total area of the array and then dissipated unused power in a unit called variously a shunt voltage limiter, shunt voltage regulator assembly or shunt dump regulator (because of its power limiting, regulating and dumping attributes). The unused power was converted to heat in the unit's resistors and then 'dumped' or radiated into space. Nowadays it is more common to use an array shunt regulator to switch sections of the array on and off in accordance with demand: it acts as a sort of demand valve between the spacecraft and its arrays. Since power is only drawn into the spacecraft when it is needed, thermal dissipation is kept to a minimum.

As we have seen in earlier sections, demand from the spacecraft is not the only variable: the power available from a given area of array varies with the season because of the Sun–satellite distance and orbital inclination. The accompanying seasonal variation of array temperature, and therefore conversion efficiency, coupled with the array degradation factor, reaffirms the need for a continually active control interface between the power bus and the array.

Figure A5.11 is a block diagram of a typical regulated power subsystem. All spacecraft power equipment operates through a central controller which accepts commands from the Earth and provides telemetry data on the power subsystem for the return transmission. By

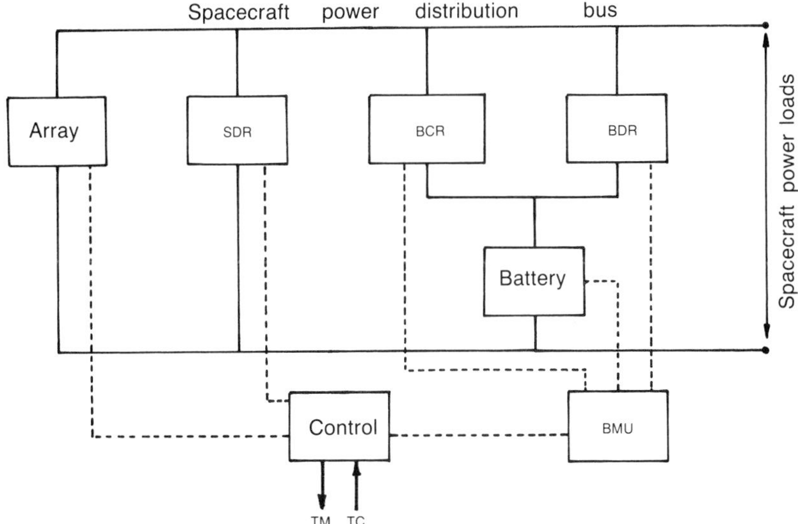

Figure A5.11 Typical regulated power subsystem: SDR, shunt dump regulator; BCR, battery charge regulator; BDR, battery discharge regulator; BMU, battery management unit; TM, telemetry; TC, telecommand.

means of the controller, the shunt dump regulator (SDR) shunts out sections of the array and regulates the raw power to the required bus voltage. Much of the power is of course used immediately by the various subsystem equipment, although some is stored by the batteries for later use. A battery management unit (BMU) controls battery functions and maintains their nominal operating conditions, particularly with regard to temperature. It also oversees the rate of charge and discharge via the battery charge and discharge regulators. The battery charge regulator (BCR) accepts an input of 50 V from the bus and feeds it to the batteries in a controlled fashion. The battery discharge regulator (BDR) converts the variable battery voltage to a stable (regulated) voltage for use during eclipse periods.

5.8 CONCLUSION

This chapter has described the major hardware components of a satellite power subsystem and some of the factors which govern their operation, both in sunlight and eclipse. All active spacecraft equipment requires a power supply. It is, however, an immutable law that no item of equipment uses that power 100% efficiently: power is wasted by conversion to heat. This internally generated heat is one of the problems facing the thermal subsystem designer, as we shall see in chapter A6.

REFERENCES

Capart J J and O'Sullivan D M 1985 *ESA Bull.* no. 42 64–9

Hudson W R and Gartrell C F 1987 *Int. Astronautical Fed. Congr.* IAF-87-07

Pratt T and Bostian C W 1986 *Satellite Communications* (New York: Wiley) p68

Rauschenbach H S 1980 *Solar Cell Array Design Handbook* (New York: Van Nostrand Reinhold) p189

Stark J 1986 *Spacecraft Engineering Course Notes* (University of Southampton) ch.12

A6 Thermal Design

6.1 INTRODUCTION

Like the human body, a spacecraft is sensitive to its environment. Both are complex systems comprising a variety of components which can only operate efficiently within certain temperature limits. Both need the ability to control their internal temperature to relatively tight tolerances, even under the harsh extremes of their external environments. Both require a sophisticated thermal control system. In a healthy body, the control system maintains a nominal internal temperature of about 36.9 °C, by ensuring a balance between metabolic processes and heat loss through evaporation, convection, radiation and conduction. The satellite's environment is far more severe in its extremes, since it inflicts a thermal gradient of several hundred degrees between the sunlit and shadowed sides of the vehicle. However, a similar requirement for balance exists: the satellite experiences both external and internal sources of heat and requires a mechanism for its control. Methods of *heat rejection*, however, are limited to radiation, since the vacuum of space denies the possibility of both convection and conduction to the external environment. In the absence of a conductive or convective medium, the challenge in spacecraft thermal design is to achieve an internal environment which approximates to 'room temperature'.

6.2 ORBITAL ENVIRONMENT

The basic type of thermal design required depends foremost on the operational environment of the spacecraft considered, that is whether it will operate in a low Earth orbit (LEO), geostationary orbit (GEO) or on an interplanetary trajectory.

One of the starting points of spacecraft thermal design is an understanding of the solar spectrum. Since engineers can do nothing to affect the output of the Sun, they must be fully aware of its characteristics.

Figure A6.1 shows how the Sun's output is spread over the frequency spectrum:

approximately 7% is at ultraviolet wavelengths (<0.38 μm);
45.5% is in the visible (0.38–0.76 μm);
47.5% is in the infrared and beyond (>0.76 μm).

[Ishimoto and Herold 1981]

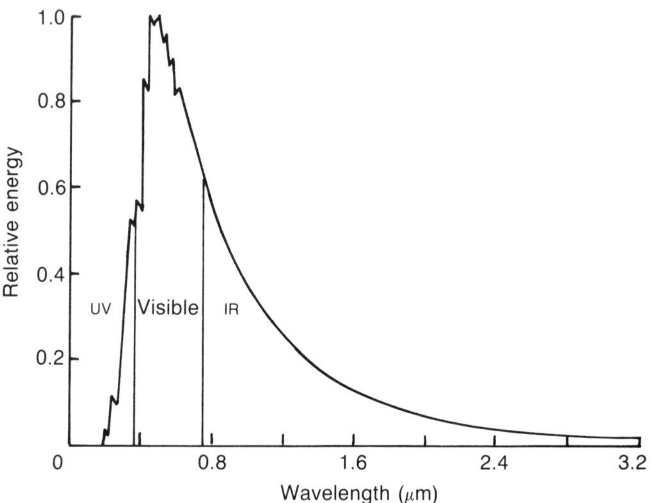

Figure A6.1 Spectral energy distribution of the Sun.

The effect of the solar radiation also depends on the orientation of the surface of the satellite to the source. Only a surface that is normal to the incident radiation receives the maximum heating effect (at the average power density of 1370 W m^{-2} quoted in chapter A5). Thus the north and south faces of a three-axis spacecraft always receive relatively low levels of solar radiation. The relatively low ambient temperature of these faces makes them good places for radiators [see §A6.4.3].

In addition to the direct solar radiation, spacecraft in LEO suffer thermal radiation reflected from and emitted by the Earth. That fraction of incident solar radiation returned to space by reflection is given by the planetary albedo, the average value for Earth being 0.34 [Henderson 1986]. The thermal energy re-radiated by the Earth is known as Earthshine. Although important for spacecraft in LEO, albedo and Earthshine are only significant for geostationary spacecraft carrying devices at cryogenic temperatures (i.e. well below 0 °C).

Interplanetary trajectories may take a spacecraft towards or away from the Sun, both of which significantly alter the vehicle's thermal environment and the design of its thermal subsystem. Since the thermal design of LEO and interplanetary spacecraft is dependent on different environmental parameters, we shall confine ourselves here largely to the communications satellite in geostationary orbit.

6.3 PHYSICAL PRINCIPLES

For simplicity, consider the satellite to be a simple box in geostationary orbit. At a particular time, one side of the box is in direct sunlight while the opposite side is in deep shadow. The external radiation source (the Sun) is only interrupted when the satellite is eclipsed by the Earth or Moon, as explained in chapter A5. The result of this simple situation is a steep thermal gradient: the sunward side of the box gets hotter while the shadowed side gets colder because it radiates into space whatever heat it has. Since there is no convection in space, the only methods of heat transfer between the hot side and the cold side are those of conduction and radiation *within the box*. The relative efficiency of these processes is dependent on the contents and interior characteristics of the box: conduction requires a continuous physical path, whereas radiation prefers an unobstructed void.

The internal heat sources of a real satellite complicate the situation further. The various electronic devices produce heat as a by-product of their normal operation, which must be removed from the immediate vicinity if their temperatures are to remain stable. The equation for the thermal balance of the spacecraft therefore has two principle inputs: an external source and an internal source. The only output is by radiation to space. Thus, if an equilibrium is to be reached, the sum of the absorbed solar energy and the internally generated heat must be equal to the heat radiated [see box A6.1].

The absorption of external radiation and the generation of internal thermal loads lead to an obvious increase in internal temperature. In the attempt to maintain the spacecraft's temperature within certain prescribed limits, there is a continual need to distribute thermal energy to strike a balance between heat sources and heat sinks. At the beginning of the design process, however, only the solar constant and the temperature of the ultimate sink, deep space, are known quantities. The science of arranging this thermal balance can be termed *energy transfer management* or *heat transport*, since the crux of the thermal subsystem is the movement of thermal energy from one place to another. The physical processes of thermal absorption, conduction and radiation embody the philosophy of energy transfer management to form the basis of the thermal design.

Every part of the spacecraft both radiates and absorbs thermal energy to a greater or lesser extent depending on its size, shape and surface

BOX A6.1: SPACECRAFT THERMAL ENERGY BALANCE

For a spacecraft orbiting the Earth, the thermal energy balance may be expressed as:

$$Q_{sol} + Q_{int} = Q_{rad} + Q_{str} \qquad (A6.1)$$

where Q_{sol} is the rate of solar energy absorption, Q_{int} is the rate of internal energy generation, Q_{rad} is the rate of thermal energy radiation and Q_{str} is the rate of energy storage by the spacecraft. If the spacecraft is in thermal equilibrium (energy input = energy output) then $Q_{str} = 0$ and we can simplify the equation to:

$$Q_{sol} + Q_{int} = Q_{rad}.$$

This can be written in terms of spacecraft absorptivity and emissivity as follows:

$$\alpha S A_{sc} + Q_{int} = \varepsilon \sigma T^4 A_{rad} \qquad (A6.2)$$

where α is spacecraft absorptivity, A_{sc} is spacecraft area, ε is spacecraft emissivity, A_{rad} is radiator area, S is the solar constant (1.37 kW m^{-2} at 1 AU) and σ is the Stefan–Boltzmann constant (5.67×10^{-8} W $m^{-2}K^{-4}$).

S : The solar constant is a measure of the solar energy at the top of the atmosphere, the figure given being accurate to ±0.02 kW.

AU : Astronomical unit, the average distance between the Earth and the Sun (1.496×10^{11} m).

σ : Stefan's law of black-body radiation states that the total energy radiated per unit area per second by a black body is proportional to the fourth power of the absolute temperature, which gives the equation

$$\text{energy/unit area/second} = T^4.$$

Equation (A6.2): Of course in practical terms there is not one value of α or ε for the whole spacecraft and the true function is an integral of all the different areas of spacecraft or radiator and their respective values of α or ε. Equally the effective value of S, the solar constant, depends on the angle of incidence of the Sun's rays on the surface.

It is apparent though that Q_{sol} and Q_{int} in equation (A6.1) can change, which will force a change in Q_{rad} to restore equilibrium. Some methods for changing Q_{rad} by altering A_{rad} are discussed in §A6.4.6 (variable heat pipes) and box A6.3 (louvres).

characteristics. Radiation exchanges within the spacecraft may be tailored to distribute heat and reduce temperature gradients by taking advantage of the complex radiative properties of differing shapes. The ratio between the effective absorbing and radiating areas is a geometric property. If the structure is a plane sheet the ratio is 1:2, since one surface absorbs the radiation but both surfaces radiate. A sphere is a better proposition as a radiator, since its cross-sectional area (πr^2) is its effective absorption area while it radiates from its entire surface area ($4\pi r^2$). Its ratio is therefore 1:4.

Not only is radiation the predominant method of heat transfer within the spacecraft, it is the only method of heat exchange between the spacecraft and its environment. Excess heat is either radiated directly from the spacecraft or conducted to another part of the vehicle from where it can be radiated away or used to warm a component which is otherwise in danger of freezing. The balance of heat absorption and rejection is controlled to provide an acceptable mean spacecraft temperature.

Heat loss by evaporation into space is technically possible but would require unacceptably large amounts of coolant for long missions, which would, moreover, tend to pollute the local environment and possibly degrade the satellite's optical surfaces and solar cells [see box A6.2]. However, evaporation in a closed loop, a process used in heat pipes, is a very efficient form of heat transport increasingly used in satellites [see §A6.4.6].

BOX A6.2: COOLING SYSTEMS

Although any device or structure which conducts or radiates heat could be considered part of a cooling system, in spacecraft thermal design the term is usually applied to a device which refrigerates the component it protects. The devices most commonly in need of active cooling are those which are designed to detect heat itself (i.e. infrared sensors) especially when they are used to detect very low levels of heat. The IR detectors used in infrared telescopes, for instance, must be cooled to liquid helium temperatures—a few degrees above absolute zero—so that the received signal is not completely drowned by the thermal noise of the molecules in the detector itself.

The cooling system may be either open or closed cycle. One type of open-cycle system uses an expansion process whereby a high-pressure gas is discharged through a heat exchanger and an expansion valve; another uses a solid cryogen which sublimes when it absorbs heat, the resultant vapour being vented into space [Ishimoto and Herold 1981]. In an open-cycle system the refrigerant is dissipated over the life of the spacecraft but, since this tends to pollute the spacecraft's immediate environment, it is not widely used. A closed-cycle system resembles the domestic refrigerator in that it features a coolant circulated through a closed loop and used as a heat exchange medium. Spacecraft systems, however, are more compact and usually cool to much lower temperatures.

6.4 THERMAL CONTROL TECHNIQUES

Methods of thermal control may be separated into two categories: passive and active. Passive control, embracing reflective and absorptive surfaces, thermal insulation and radiators, is the simplest and usually cheapest method. It also includes the subtleties of geometric design and selection of materials. Early spacecraft required only passive control techniques to maintain their temperature between certain operating limits, but as they became more complex and more powerful, active methods of thermal control evolved to accommodate the increase in complexity. The thermal control hardware in the passive category requires and allows no intervention, whether man or machine derived, in its normal operation. Active control, on the other hand, involves more complex devices which are controlled either by direct ground command or by

BOX A6.3: THERMAL LOUVRES

An early example of the thermal louvre, the *pinwheel louvre,* was used to regulate temperature on the early Atlas–Able and Tiros satellites. It comprised a circular segmented wheel controlled by a bimetallic spring, whose two metals expanded and contracted differentially to produce a temperature-sensitive controller. The pinwheel louvre operates as follows. When the temperature rises, the wheel rotates clockwise to expose a black highly emissive surface, and more of the black surface is exposed until thermal equilibrium is reached. When the temperature drops, the spring forces the wheel to rotate in the opposite direction to cover the black radiator. If the wheel rotates further it exposes a low-emittance highly reflective surface which inhibits radiation into space, thus keeping the spacecraft temperature within its operating range.

Pinwheel louvres, however, were only able to regulate temperatures for spacecraft with relatively low power dissipations. For higher powers a system of *blade louvres,* the type currently used, resembling the domestic venetian blind, was developed. These louvres, first used by TRW on the Orbiting Geophysical Observatory (OGO) spacecraft, also use a bimetallic spring to adjust their angle, although another method developed by General Electric relies on the expansion and contraction of a fluid [Ishimoto and Herold 1981].

Using louvres the emittance can be varied between about 0.2 in the closed position and 0.8 open, meaning that the heat load/rejection can be varied by a factor of 4 while retaining the desired temperature to within ±10 °C or so. Although the louvre is not the most accurate temperature control device, it has proved successful on a wide variety of spacecraft including the Voyager planetary probes which have operated successfully since their launch in 1977.

automatic systems working in the mode of 'feedback loop' or 'governor'. These range from the simple heater to the more advanced forms of variable conductance heat pipe. Another simple example of an active thermal control system is the thermal louvre [see box A6.3]. Although not generally used on communications satellites in geostationary orbit, it is included for completeness since the more advanced versions are useful for spacecraft in LEO.

The various techniques and hardware described below are applicable, in varying degrees, to all spacecraft.

6.4.1 Coatings and surfaces

The choice of materials or coatings constituting the surface of the spacecraft must take into account both the solar spectrum and the characteristics of spacecraft thermal radiation. In this regard, a knowledge of two important properties, solar absorptivity and infrared emissivity, is necessary.

The solar absorptivity, α, is the ratio of the radiant energy absorbed by a body to that incident upon it, in effect the efficiency with which a surface absorbs incoming radiation. The IR emissivity, ε, defines the radiative potential of a surface, being the ratio of the energy emitted by the surface to the energy emitted by a black-body radiator at the same radiation equilibrium temperature. Since a black body absorbs all incident radiation, its absorptivity $\alpha = 1$.

In a definitive sense there are four main types of surface available to the thermal design engineer: the solar reflector, the solar absorber, the 'flat' reflector and the 'flat' absorber [see box A6.4]. The first two

BOX A6.4: SPACECRAFT SURFACE TYPES

In terms of α and ε the characteristics of the four types of surface can be summarised as follows:

solar reflector:	low α, high ε, low α/ε
solar absorber:	high α, low ε, high α/ε (>1)
flat reflector:	low α and ε, $\alpha/\varepsilon \simeq 1$
flat absorber:	high α and ε, $\alpha/\varepsilon \simeq 1$

The solar reflector reflects incident solar radiation and emits IR (e.g. white paint, second-surface mirrors). The solar absorber absorbs solar radiation and emits a small percentage of IR (e.g. most metallic finishes).The flat reflector reflects UV to IR across the spectrum (e.g. a few metallic foils and finishes). The flat absorber absors UV to IR across the spectrum (e.g. black paint, solar cells).

respectively reflect and absorb most of the incident energy in the solar spectrum, while the latter two reflect or absorb practically all incident energy, solar or thermal (i.e. they have a flat response across the band, see figure A6.2).

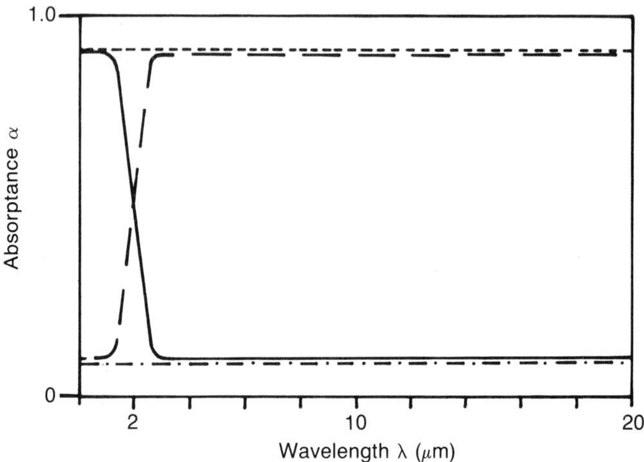

Figure A6.2 Absorptance curves for ideal surfaces: ———, solar absorber; – – –, solar reflector; ----, flat absorber; –.–.–., flat reflector.

One of the thermal subsystem designer's principal tools in the control of spacecraft equilibrium temperature is the ratio between absorptivity and emissivity, α/ε. The solar reflector has a low α/ε ratio since it has a high emittance at satellite temperatures, whereas the solar absorber has a high α/ε ratio since it absorbs far better than it emits. The informed choice of materials with properties such as these is one of the fundamentals of thermal design.

Perhaps the simplest way to coat a surface is to paint it. White paint is useful as a solar reflector, as shown well in figure A6.2 by its high reflectance in the visible and high emittance in the infrared. Black paint fits into the above category of flat absorber since it absorbs almost everything and emits almost everything ($\alpha \sim \varepsilon \sim 1$). These high-emittance paints are equally useful when mixed in varying shades of grey to produce a range of solar absorptance. Black-and-white chessboard patterns have also been used in the past, but simpler paint colour schemes are now preferred. White paint is usually used on the outer surfaces of parabolic communications antennas since an excess of absorbed thermal radiation would distort the antenna by expansion and cause defocusing of the beam. In addition to its external uses, black paint is commonly used on the interior surfaces of the satellite to enhance energy exchange between component parts, which tends to equalise internal temperatures.

Polished metals and foils tend towards the solar absorber category, although there is a great variation depending upon the material. Polished aluminium, for example, has a high α/ε ratio. An aluminium oxide finish, on the other hand, has a low α/ε ratio because of its high IR emittance. Surface textures vary and the number of different materials available is great. The 'toolkit' of passive thermal design includes black anodised metal, silver- or aluminium-backed teflon, mosaics of metals and white paint, and paints pigmented with metal flakes. The variety is virtually endless.

Table A6.1 gives a summary of some average values of α, ε and α/ε for various types of spacecraft surface. One complicating factor is the non-constancy of these factors over the lifetime of the spacecraft: the solar absorptance increases with time because of the bombardment of UV radiation and energetic particles. This degradation of the surface is also due to the deposition of contaminants derived from the satellite itself as a result of the outgassing of materials under the harsh combination of high temperature and vacuum, and combustion products from the apogee boost motor (ABM) and the reaction control thrusters. Contaminants are kept to a minimum by an informed choice of materials [see chapter A2], construction of the spacecraft in a clean-room environment and the inclusion of an ABM plume shield to protect both the radiator surfaces and the solar arrays. Once outgassing has occurred or propulsion products are formed, there is no force available to remove them from the spacecraft's orbital environment, so some deposition is inevitable.

Table A6.1 Spacecraft surface characteristics.

Surface	α	ε	α/ε
Black paint	0.95	0.9	1.0
White paint	0.2	0.85	0.2
Metallic	0.3	0.05	6.0
SSMS	0.1	0.85	0.1
Silicon solar cell	0.8	0.6	1.3
(with silica cover)	0.8	0.8	1.0

According to the results from the Orbital Test Satellite (OTS) degradation of the white paint on the antennas was still occurring in its seventh year in orbit [Howle *et al* 1985]. The magnitude of the effect was shown to level off with time, but the absorptivity, α, had increased from a value of 0.32 at launch to about 0.58 after 7 years. An effect of this magnitude cannot be ignored by the thermal design engineer. White paint was commonly used on early spacecraft for their radiative surfaces, but since it was found to degrade to an unacceptable level within a few years it has now been largely replaced by second-surface mirrors which exhibit a typical end-of-life absorptivity of about 0.25 [Watanabe *et al* 1985].

6.4.2 Second-surface mirrors

Although the mirrored surface of a satellite can evoke to the uninitiated the image of a DIY enthusiast's bathroom, it offers one of the most elegant solutions to passive thermal control, featuring low solar absorptance, high IR emittance and high reflectance. The second-surface mirror, or SSM, comprises a thin sheet of silvered or aluminised glass or quartz which is bonded to the exterior surface of the satellite using high-conductance adhesives. The term *second surface* is derived from its ingenious dual function as a solar reflector and thermal emitter. Glass is transparent over most of the solar spectrum, so most solar radiation reaches the coated rear surface (the 'second surface') and is reflected. Glass is also, however, an excellent emitter over the IR spectrum, which means that thermal energy from the spacecraft interior, conducted via the adhesive to the glass, can be radiated into space from the outer surface (the 'first surface').

A typical SSM has a substrate based on a borosilicate glass doped with cerium dioxide, very similar to the solar cell cover glasses described in chapter A5. A silver coating is vacuum deposited to form the reflective surface to which a protective nichrome overcoating is added. The coatings are protected from micrometeoroids by the glass which is between 0.05 and 0.5 mm thick. Optional exterior coatings, again similar to those used on solar cells, include UV reflectors and a conductive indium tin oxide coating to assist the prevention of static build-up.

SSMs are alternatively known as optical solar reflectors (OSRs), although this term is less definitive and may be used to include other reflectors of the Sun's visible spectrum, such as white paint, silver-coated silica sheets, etc. Naturally, SSMs cost more than the other common external material, paint, but are more widely used, since they are less subject to degradation over the long lifetime of most satellites as discussed above. From beginning to end of life the absorptivity typically increases from 0.08 to 0.25 at rates of 0.01–0.02 per year [Wise 1985]. SSMs are not easy to fit to anything but flat surfaces since they are essentially rigid, but their high resistance to electron and UV irradiation, and their reflective and radiative properties, have led to their widespread use. Continuing studies on thermoresistant polymers may lead to the development of a stable flexible second-surface mirror.

SSMs typically cover areas of the satellite beneath which are the radiators for heat-producing equipment such as travelling wave tube amplifiers, output multiplexers and batteries [see figures A6.3 and A6.4].

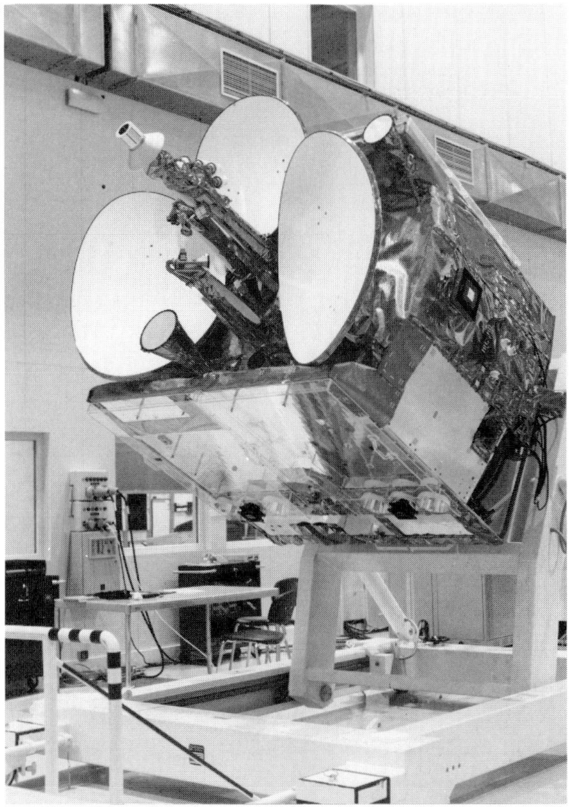

Figure A6.3 Telecom-1, French national telecommunications satellite. Note white-painted antenna reflectors, crinkled foil of the multi-layer insulation (MLI) and radiator panels covered in second-surface mirrors (SSMs). [CNES]

6.4.3 Radiators and heat sinks

It was mentioned in chapter A2 that although the honeycomb panel finds favour with the structural engineer, the appreciation is not shared by the thermal engineer because of the panels' low thermal conductivity. The 'equivalent thickness' of the typical honeycomb panel is 0.3 mm of metal, which is insufficient for more than local conduction and low power densities [Brooks 1985]. As powers increase it becomes necessary to increase the thickness of the panel to improve the conduction path and distribute the excess heat. The additional panel thicknesses are known as thermal doublers. If power densities are higher still, it may be necessary to mount equipment directly onto radiators, or in extreme cases, when the mass of the radiator becomes prohibitive, to use heat pipes.

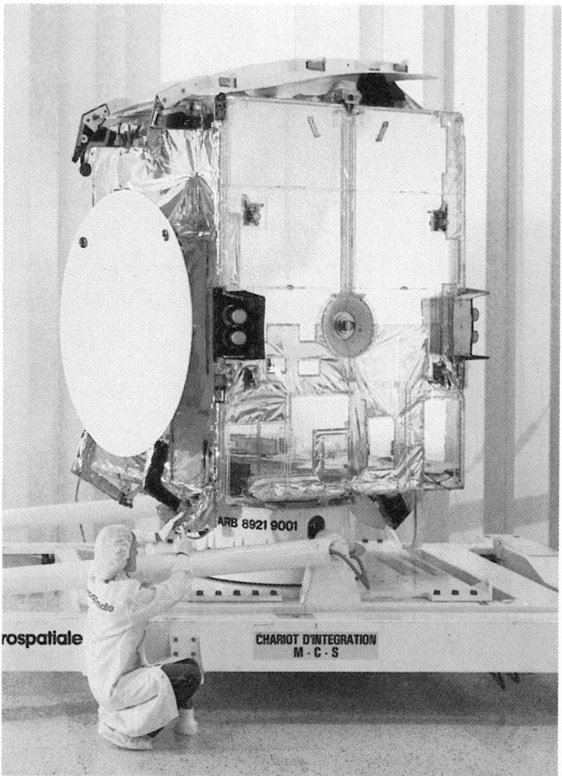

Figure A6.4 Arabsat-2, Arab League telecommunications satellite. Note particularly the mirrored radiator panels, sized and shaped according to the heat rejection requirements of equipment mounted beneath. [Aerospatiale]

In a physical sense every item within a spacecraft can be labelled either a heat source or a heat sink—but they can also be both. The terms are relative and dependent upon the respective temperatures of the items. For example, in the case of a component attached to a radiator which is in turn bonded to an array of SSMs, the radiator is a sink for the component and a source for the SSMs. Although the word radiator can also be used as a general term, it refers here to a specifically designed item of hardware. Similarly, as well as its generic usage, the term heat sink is often used to describe an actual device, otherwise called a *heat spreader plate.*

Radiators are commonly fashioned as panels, which contribute significantly to the structure of the satellite as well as providing mounting points for spacecraft equipment. They may be solid or of honeycomb construction and are often conductively decoupled from the surrounding structure to maintain a known and constant thermal capacity. The heat source is attached or coupled to the interior surface of the panel, which is generally painted black, and the exterior surface is coated with rigid SSMs or a flexible radiative material. Radiator panels are usually located on the north and south faces of a three-axis-stabilised satellite since they receive solar radiation obliquely, as mentioned earlier. The smaller spinners also radiate from the north and south ends of the 'drum'; the larger spinners have, in addition, a radiator panel about the midriff of the drum to increase their capacity for heat rejection.

A spacecraft's heat rejection capability governs its maximum input power from the arrays in the following manner. An average-sized three-axis satellite might have an area of 2 m^2 available for SSMs on each of its north and south faces. Since an SSM radiator typically radiates at a net rate of about 200 W m^{-2}, this gives a heat rejection capability of 800 W. If we assume that the spacecraft equipment is 25% efficient in its overall usage of power, three-quarters of the total input power will have to be radiated to space to maintain the spacecraft's thermal balance. This sets the total input power at about 1.1 kW, a fairly typical figure for a medium-sized communications satellite.

The problems with high-power communications spacecraft are even greater because of the large amounts of heat generated by their high-power travelling wave tubes in their conversion of DC power (from the arrays) to the RF power of the amplified signals. The limited radiator area has placed a practical limit on the communications capacity of contemporary satellite designs. It may, however, be rectified in future designs by the use of deployable fin-shaped radiator panels which would increase the surface area.

Although a heat sink may well resemble a radiator, heat sinks tend to be discrete devices closely matched to a particular source of heat and, moreover, tend to reject that heat by conduction rather than radiation. Indeed they may conduct heat directly to a radiator on the exterior of the spacecraft. In general, metals which are good conductors are also good heat sinks, which is why, coupled with its relatively low density, aluminium is widely used. In fact it is because mass is at such a premium that such items usually serve more than one function: a heat sink may be part of the component or part of the spacecraft structure itself. This philosophy is taken to its limit for the high-powered travelling wave tubes which radiate directly into space.

In order to dissipate heat efficiently, there must be a good path of thermal conduction between items of hardware. Wherever possible dry joints are used for ease of assembly, but the conductance between an 'electronics box' and a honeycomb panel can be as low as 20 W m^{-2} K^{-1}, and even good dry joints can achieve only about 200 W m^{-2} K^{-1}. Where greater conductance is required, interface or interstitial fillers must be used. One such filler is the so-called *wet-joint interface*, which utilises an unprimed thermoelastic compound such as a silicone adhesive; an alternative is a preformed conductive gasket. Although the conductance depends on the pressure on the joint, values between 2000 and 8000 W m^{-2} K^{-1} are typical [Wise 1985].

The primary function of a heat sink on a communications satellite is to spread heat over a wide enough area for it to be radiated away, hence the term heat spreader plate. However, it can also act as a kind of 'thermal capacitor': as the electrical capacitor stores and later releases electrical energy, the heat sink absorbs thermal energy, distributes it throughout its volume and eventually dissipates it into other structures. Heat sinks are, of course, particularly useful for equipment which generates large amounts of heat, such as solid state power amplifiers (SSPAs) and micro-electronic RF components, but their utility may be enhanced by the provision of a dual heat transfer function. Not only do they absorb heat from the component during its powered phase, but may also return some of that heat when it is switched off, which would help to prevent excessive cooling. This is, after all, the principle followed by the domestic storage radiator, which must be the archetypal thermal capacitor.

6.4.4 Multi-layer insulation

There are times, principally during an eclipse, when it is more important to retain the spacecraft's heat rather than radiate it away. Insulation is the answer. Colloquially termed the *thermal blanket*, it is this material which gives a satellite its foil-wrapped 'chocolate box' image [see figure A6.3 and the cover illustration]. The basic design philosophy behind spacecraft insulation is to stop heat getting in on the Sun-facing sides and to stop it getting out on the shadowed sides. Like any insulation, the thermal blanket both limits heat input and output by providing a thermal barrier. The blanket is more correctly known as a multi-layer insulation (MLI) because of its construction. The most straightforward example of MLI consists simply of layers of synthetic polymeric material such as Kapton or Mylar foil (Du Pont tradenames). Each layer is about 0.006 mm thick, aluminised on one or both sides, and acts as a low-emittance shield separated by low-conductance spacers. The low conductivity is produced by crinkling the foil to produce insulating voids. An alternative method uses Dacron netting as a separator.

The peformance of MLI is impressive: a typical 10-layer blanket, with a density of 0.3 kg m^{-3} and a total thickness of 5 mm, would be equivalent to about 0.5 m of conventional insulation. The effectiveness of this type of insulation is shown by the fact that the satellite's internal temperature can be controlled to about ±5 °C when the external temperature ranges between − 150 and + 120 °C [Brooks 1985]. Painting the internal components black and placing them intelligently within the spacecraft, together with a good insulation design, can result in an adequate thermal control design for a simple satellite without resort to active control techniques.

Kapton is the predominant foil used for communications satellites unless the temperature is greater than about 350 °C for prolonged periods. Low-conductivity foam or glass-fibre paper interlayers can improve the insulation properties, but they result in a MLI of an unacceptable mass for most communications satellite applications. Fibreglass is used around high-temperature areas as an interface layer. One example of a high-temperature superinsulating material, particularly useful for rocket nozzles, comprises stainless steel or titanium-foil skins with silica-fibre interlayers.

A typical example of MLI dressing given to a communications satellite is that of the OTS satellite [Howle *et al* 1985]:

- The blanket used to insulate the antenna platform, the reverse faces of the antenna reflectors and parts of the east and west side walls are made from 10 crinkled sheets of 0.006 mm thick Mylar, aluminised on one side, sandwiched between a 0.05 mm and a 0.025 mm Mylar foil with an aluminised rear side.
- The MLI used to protect the spacecraft from the heat of the apogee boost motor and hydrazine thrusters is of a heavier-duty type. It lines the inner surface of the thrust cylinder and the whole ABM end of the spacecraft as well as the outer surface of the motor body. It consists of 18 crinkled sheets of 0.015 mm aluminised Kapton, sandwiched between two 0.075 mm aluminium-backed Kapton foils, and has an operating temperature range of − 200 to + 280 °C.
- Covering all the external blanketed areas except the antenna platform is a VHF shield which protects the more sensitive electronic components from the TT&C radio frequencies. It comprises a 30 μm aluminium foil, 0.006 mm clear Kapton foil and 0.075 mm Kapton foil (rear side aluminised) grounded to the spacecraft structure.

In a broad sense, there are many applications for thermal blankets in and on space hardware. They can be used to maintain the temperature of an isolated component or to insulate launch vehicle cryogenic propellant tanks to minimise *boil-off*. Particularly large components such as astronomical telescopes may be wrapped in MLI to reduce the thermal distortions which could otherwise render a precision instrument useless.

Laboratory experiments have shown that blankets may have an effective emittance as low as 0.005, but in practice 0.05 is more likely due to the imperfections and discontinuities of seams, cut-outs and attachment fasteners [Wise 1985]. Some research suggests that there is a sensible limit to the number of layers in MLI, because extra layers increase the packing density, which increases the contact between layers and therefore the conductivity. The effective conductance is typically in the range ~0.1–0.3 W m^{-2} K^{-1} and improves with the size of the blanket. Certainly the compaction of layers which occurs on bends or corners will increase the conductivity above that of the flat surfaces, but the compaction due to increased weight should not be a significant factor in the 'zero gravity' environment of geostationary orbit. Where the weight of the thermal blanket *is* an important factor and may limit the number of layers is in the overall spacecraft mass budget, the sum which decides whether or not the spacecraft can be launched.

6.4.5 Heaters

The techniques of thermal control discussed so far have been of the passive type; we shall now consider active thermal control. The simplest active controller is the ordinary electrical resistance heater, which is usually small and used to maintain the temperature of an individual piece of equipment. The heater can be used in several different modes according to the degree of control required by the hardware to which it is attached. It may be on continuously, cycled on and off between a maximum and minimum temperature, controlled by a feedback system to maintain a constant temperature, or ground controlled. This flexibility, the power of the heater and its position are all parameters used in the design of the overall thermal subsystem.

Heaters are particularly useful for heating the reaction control thruster propellant, usually hydrazine for communications satellites, which freezes around 0 °C. The extremities of the system, where the volumes are low, are particularly prone to low temperatures, that is in the lines, valves and thrusters, the last of which must be on the spacecraft exterior. To maintain the temperature between about 5 and 50 °C, line heaters are wrapped in spirals onto the fuel lines and fixed in place with aluminium tape; tank and valve heaters tend to be bonded directly onto their respective units, as do battery heaters.

Another application for the heater is the simulation of a device which may not be operating, and therefore producing heat, constantly, for instance the travelling wave tube amplifier (TWTA). If a TWTA is switched off, perhaps during an eclipse, its temperature drops. As a result, the radiator it is attached to, which is designed to remove a large amount of heat, is rapidly cooled. This creates an undesirable temperature gradient which begins to upset adjacent components, either cooling them below their operating limits, or forcing individual component heaters to switch on or heat to be transferred from other areas. To stop this happening, TWTA simulation (or substitution) heaters are bonded onto the radiator or adjacent structure. They go at least some way to replace the heat produced by the amplifier and reduce the thermal gradient.

6.4.6 Heat pipes

So far in this discussion the most effective method of transporting heat from one part of the spacecraft to another has been by simple conduction in heat sinks, radiators and the like. The heat pipe, invented in 1964, provides the thermal designer with a device with a thermal conductance much higher than even the best heat-conducting metal. It is a highly efficient device for transferring large amounts of heat from one place to another, or simply for the removal of hot spots.

The principle of the heat pipe is illustrated in figure A6.5(*a*). The pipe contains a fluid which is vaporised by the applied heat at one end (the evaporator) and condensed at the other end where it relinquishes its heat. The condensed liquid returns to the evaporator end through a porous wick by means of capillary action. The principle is similar to that of the reflux condenser in chemistry, except that capillary action replaces gravity in returning the condensate to its source. An alternative design incorporates a system of axial grooves which operate as the condensate-return mechanism, again using capillary action. The heat pipe presents a prime example of a closed-cycle cooling system as defined in box A6.2.

The operating temperature of the device depends on the working fluid, which for spacecraft applications is usually ammonia or methanol. These fluids fit into the category of 'low temperature' and operate from just above 200 K to about 500 K for methanol; ammonia's range is slightly narrower. Above 500 K the so-called 'high-temperature' heat pipes use fluids such as mercury, sodium and lithium, the last of which may be utilised above 3000 K. Below 100 K cryogenic heat pipes utilise hydrogen, nitrogen and oxygen, the lower operating limit of hydrogen extending heat pipe operation to about 15 K [Wise 1985]. Theoretically then, heat pipes may be used to transport heat from practically any man-made or natural heat source, whatever its temperature. In fact, their utility is exemplified by a terrestrial example where heat pipe technology was used on the Alaska oil pipeline. They transfer heat from the tundra during the warm season to ensure that the ground beneath the pipeline supports remains frozen throughout the year [Ishimoto and Herold 1981].

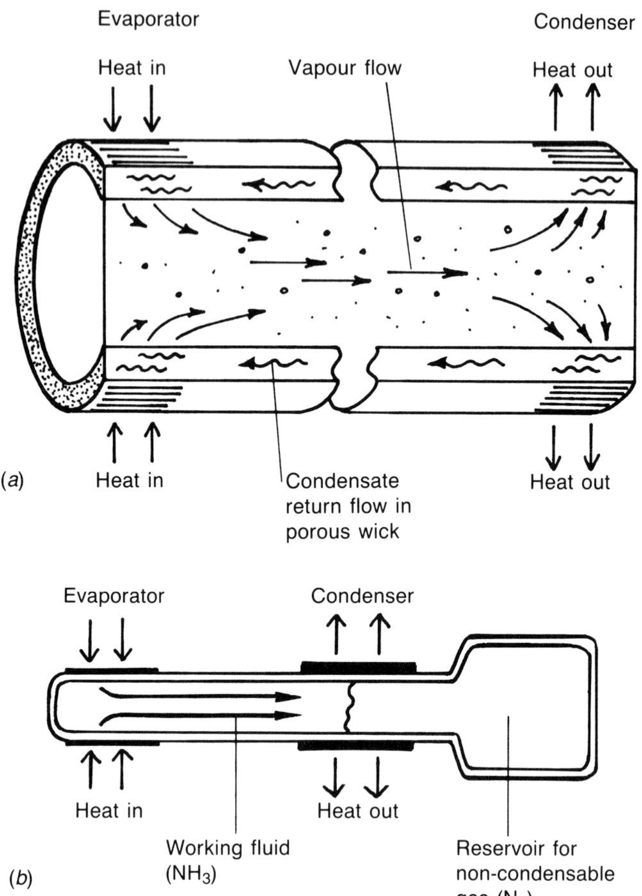

Figure A6.5 (*a*) Principle of heat pipe operation: heat transfer is realised by the evaporation of a fluid at the input (hot) end and its subsequent condensation at the output (cold) end. The condensate returns to the evaporator by capillary action in a porous wick. (*b*) Principle of the variable conductance heat pipe: when the thermal input is high the flow of vapour pushes the nitrogen back into its reservoir, thereby increasing the effective condenser area; when input is low the nitrogen blocks part of the condenser, thereby reducing the heat radiated by the pipe.

In the past, spacecraft designers have opted not to use the heat pipe as widely as they might, because they are complex to manufacture, potentially prone to puncture by micrometeoroids and the generation of non-condensable gas can limit their performance. Fortunately, advances in manufacturing, testing and reliability over the last decade have increased confidence and their use is spreading. For example, TV-SAT, TDF, Olympus, STC-DBS and Satcom-K all feature aluminium axially grooved heat pipes with ammonia as the working fluid.

The Marecs spacecraft design incorporated 11 stainless steel ammonia heat pipes with stainless steel mesh wicks, each of 40 W m capacity, mounted on specially machined radiator panels. A 15 mm aluminium/ammonia pipe can transport about 200 W over 1 m with a temperature difference of 1 °C. With a mass of about 0.4 kg, a device of this type is very mass efficient and has the added advantages that it has no moving parts and uses no electrical power. Today's heat pipes have typical capacities between about 20 and 500 W m, but systems under development have demonstrated capacities up to 20 000 W m [Brooks 1985].

The advantage of the heat pipe over the simple heat spreader plate is one of effectiveness in distributing thermal energy over a radiator panel, as shown by tests on the TWT panels of RCA's STC-DBS satellite: the average base-plate temperature using the heat spreader was 73 °C for a 10 kg panel, whereas using a heat pipe panel of the same mass the temperature was only 45 °C [Wise 1985]. Further tests showed that, to equal the heat pipe's capacity for heat rejection, the spreader would have to weigh 48 kg, nearly five times as much! It is important to note, however, that the use of such devices is not automatic—the advantages of one device are always traded-off against those of another.

So far we have considered only the heat pipe of fixed or constant conductance (FCHP or CCHP respectively); a subsequent development is the variable conductance heat pipe or VCHP.

The advantage of any variable system is that it can adapt to changes in its environment. The FCHP has the tendency to transport heat even when it might be desirable to retain that heat, for example during an eclipse. This tendency leads to an increased requirement for heaters and therefore battery power, which of course means mass! The VCHP has the potential to improve the mass budget by reducing the need for heaters. Its mode of operation is illustrated in figure A6.5(*b*). The variable conductance heat pipe relies upon the deliberate addition of an amount of inert non-condensable gas for its very operation, whereas non-condensable gas in a fixed conductance pipe is considered a nuisance since it limits the heat transport ability of the device. Nitrogen, the gas used in this example, is stored in a reservoir beyond the condenser section of the pipe. If heat is applied to the evaporator the working fluid evaporates and flows towards the condenser. As it does so it sweeps the nitrogen back from the condenser towards its reservoir and the heat pipe operates like an ordinary FCHP, but as the heat load is reduced at the evaporator end, the vapour/inert gas boundary advances over the heat transfer area effectively closing off part of the condenser. The heat lost to the radiator is therefore reduced. Adjusting the variables (reservoir size, condenser area, etc) will adapt the basic design to a specific application so that heater power requirements may be reduced in the 'cold case' and heat rejection capabilities may be retained for the 'hot case'. Although VCHPs have not yet enjoyed widespread use, they have proved successful on the Canadian CTS communications technology satellite and the Japanese ETS III (Engineering Test Satellite) amongst others.

In future it is likely that we will see the development of heat pipes with higher thermal capacities and the increased use of pumped fluid loops, similar in principle to the car radiator. The main disadvantage of pumped loops over heat pipes has to date been the relatively poor reliability of the pumps over the long lifetimes of unmanned space vehicles. Manned spacecraft have, however, used pumped loop technology for many years. The increased use of deployable radiators with heat pipe connections is expected for future missions, along with the development of miniature heat pipes for cooling individual heat-producing components at circuit board level [Wise 1985].

6.5 OVERALL SYSTEM DESIGN

6.5.1 Other mission phases

This chapter has concentrated mainly on the thermal environment of geostationary orbit, but the thermal subsystem must be designed to operate satisfactorily during other mission phases. On the ground, for instance, it is necessary to keep the apogee boost motor warm, since too low a temperature affects its performance. Batteries must be kept cool, otherwise excessive charging on the launch pad could increase their temperature beyond design limits and their high thermal inertia could give problems during the launch and transfer orbit.

In GTO the angle of the satellite with respect to the Sun, known as the solar aspect angle, is important since the Sun can cause excessive heating of the ABM. Solar arrays clamped against the north and south faces of a satellite in GTO block the radiators, so the panels are sometimes equipped with fins, sized to maintain a sufficiently low internal temperature for a number of transfer orbits. The eventual deployment of the solar arrays exposes the radiators, which from then on cool the spacecraft.

An eclipse is one of the most challenging times for the thermal subsystem, since the predominant external heat source is removed for up to 72 min. The internal temperature of the satellite must be prevented from dropping too low, yet the power for heaters is restricted to that available from the batteries. Although the satellite relies to a large extent on the efficiency of its passive thermal design—mainly good insulation—some equipment has a narrow operating temperature range and the use of heaters is inevitable. Since this uses battery power, which is limited by the mass of batteries that can be carried, the spacecraft is heated before it enters the Earth's shadow to ensure an adequate recovery after eclipse.

6.5.2 Operating temperatures

In the introduction to this chapter the spacecraft thermal control system was compared to that of the human body, whose operating temperature we know well. So far no attempt has been made to quantify the temperature regime of the geostationary communications satellite. The true picture is that temperatures vary widely both inside the satellite and about its exterior depending on solar illumination, internal heat generation and the details of the thermal design itself. A reasonable guide, however, is that the interior of the satellite is designed to operate at around 'room temperature' (15–20 °C on average).

The effective average temperature of space itself is around −273 °C, but the actual temperature of a material in space depends upon its thermal characteristics of solar absorptivity and infrared emissivity. The temperatures on the Sun-facing side of an inert Earth-orbiting spacecraft (one which is producing no heat of its own) could reach perhaps 140 °C depending on the materials and design. On the shadow-side of a similarly 'dead' spacecraft temperatures may drop to around −70 °C. Of course the temperature of a body in deep space with absolutely no heat input would drop to a temperature close to absolute zero.

In the case of an operating spacecraft, where there is some electrical dissipation, typical temperatures for a sunlit radiator are 30–35 °C. The cold side may drop to about −25 °C, but this depends on how long the area is kept cold. As any thermal engineer can testify, there are no hard-and-fast answers as far as spacecraft temperatures are concerned.

Before launch a version of a spacecraft, called a thermal model, is subjected to these extremes in a thermal vacuum test chamber which simulates the thermal environment of space. This is done both to verify the thermal design and confirm that the flight model of the spacecraft will operate correctly once in orbit. Figure A6.6 shows the thermal model of the Olympus communications satellite.

Two of the main factors driving the performance of the thermal subsystem are the design constraints placed on the individual components inside the spacecraft and the resultant temperature range they can survive. It is not generally advisable for the interior of the service module to fall below −10 °C even under extreme conditions; indeed 0 °C is preferred as the lower limit. The payload module equipment generally requires a higher ambient temperature, but 35–40 °C is not usually exceeded.

Figure A6.6 Thermal model of the Olympus 1 communications technology demonstration satellite. [British Aerospace]

An examination of the OTS thermal subsystem results, over the spacecraft's 7 years in orbit, provides us with an actual case [below], but it is important to realise that temperature sensors record data only from discrete points throughout the spacecraft and cannot directly predict the existence of other hot or cold spots:

Antenna platform and upper floor:	10–40 °C
Lower floor and thrust cylinder:	2–40 °C
A TWT radiator:	5–40 °C
A control radiator:	– 25 to 15 °C
Solar array panels:	40–60 °C

From these results it can be seen that the interior of the spacecraft was kept above 2 °C, with the payload module (upper floor) showing a slighty higher average temperature than the service module (lower floor). External components, such as radiators and array panels, exhibited more extreme temperature excursions as might be expected.

6.5.3 Special cases

Some items of hardware require special consideration. Batteries are by nature both exothermic and endothermic, depending where they are in their duty cycle. When discharging, a battery gives off heat, but when charging it absorbs heat, except near the end of its cycle when it reverses briefly. The fortuitous part of this cycle is that batteries tend to be used more in eclipse, a time when their production of heat is useful. However, to ensure a long lifetime the operational temperature limits are narrow, typically 5–20 °C. The solutions include isolation from other components, the use of thermostatically controlled heaters and coatings of low solar absorptance and high emittance.

Nickel–hydrogen batteries suffer additional thermal problems, since the hydrogen gas offers no internal heat conduction path. In this case, battery cells have to be mounted on a central flange which serves as a heat sink and radiator. The situation may require the battery system to radiate directly into space in order to accommodate the exothermic part of the cycle [Ishimoto and Herold 1981].

Propulsion systems have further special problems. When they are active they produce large amounts of heat from which the rest of the spacecraft must be protected, mainly by multi-layer insulation and other thermal shields, but when the system is inactive the propellant may be in danger of freezing. Propellant tanks are therefore lagged with MLI and heater wires are woven throughout the system like electrical spaghetti.

The various appendages of the satellite, or *deployables*, deserve equal consideration, since the temperature extremes for items mounted outside the spacecraft body are invariably much greater than those within. Solar arrays, for instance, can reach average temperatures of 50 °C in sunlight, but this can drop to − 170 °C in the shade [Watanabe *et al* 1985]. Their thermal balance is complicated by the operation of the solar cells themselves, which produce heat as they generate electricity from the solar radiation. The three-axis spacecraft has an advantage in that the rear surface of the arrays can radiate continuously to the ultimate heat sink of space. On the other hand, the spinner's cells have a lower average temperature to begin with, since they are not continuously illuminated.

Like the yokes which attach the solar panels to the body of the three-axis spacecraft, antenna subsystems are prone to thermomechanical distortion. The effect is more important for antennas because a change in the relative spacing and alignment of antenna reflectors and feedhorns will cause a loss of focus and alter the beam shape produced by the antenna. Antenna reflectors, whether deployable or not, are subject to frequent and repeated temperature variations between −120 and +110 °C [Dvorko *et al* 1985]. The solutions include painting antenna surfaces white and wrapping practically everything else in MLI. The best results are obtained by using new materials, such as carbon-fibre composites, with very low coefficients of thermal expansion.

The large structures expected to make an appearance in the next decade or so, in the form of space platforms, *antenna farms* and space stations, will suffer more severe structural distortions: one end of a platform could be in sunlight while the other is in shadow, for instance. This could prove particularly important for large antennas and fine-pointing sensors, since any distortion of their structure or that of the platform would severely upset their pointing ability.

6.6 CONCLUSION

Spacecraft engineering has seen a general increase in the size of structures placed into orbit, a trend which is guaranteed to continue. Geostationary orbit offers a finite number of orbital slots which are fast becoming occupied [see chapter B2] and the cost of individual launches and the associated insurance is continually on the increase. The solution to overcrowding which has grown in popularity in the 1980s is to build larger satellites with greater communications capability and extended lifetimes to maximise the return on the initial investment. This trend has an impact on the thermal subsystem in several ways. In order to extend the lifetime, more station-keeping fuel must be carried, which leads to an evolving requirement to minimise the weight of the spacecraft structure. This causes problems for the thermal engineer since light-weight structures are inherently poor thermal conductors. The increased lifetime also leads to the increased degradation of radiator surfaces, which decreases their efficiency and produces an overall temperature rise within the spacecraft. To keep this rise within the operating limits of the spacecraft equipment, thermal systems are deliberately overdesigned in the first place. Of course the problem here is that they weigh more!

Like all other spacecraft systems the design of the thermal subsystem is largely iterative because of its intimate dependence on the payload, the service module components, orbital parameters and expected lifetime, etc. The simplest thermal designs are largely passive in nature. An overall temperature balance is attained by careful design of the satellite's exterior: a thermal blanket reduces the rate of change of temperature by insulating the interior from the external environment, while a selection of spacecraft surfaces control the absorption and reflection of incident radiation and the emission of excess heat. By the same token, an efficient internal distribution of thermal energy is achieved by intelligent placement of equipment and a coating of black paint to enhance the radiative exchange. Although a passive design provides an overall thermal balance and some degree of internal control, local problems remain: propellant systems may require additional heating; electronic components may need individual heat sinks or radiators. If thermal doublers and direct radiation to space are not sufficient, it may be necessary to use heat pipes and other active devices. The integration of all these methods of thermal control constitutes the thermal design of the spacecraft.

Naturally, all design solutions must be compatible with lifetime requirements and mass and power constraints. Surfaces must be specified with a view to their EOL characteristics of absorptivity; the demand for heat sinks must be satisfied within the constraints of the mass budget; and heaters must be similarly limited by the power budget. The basic principles of thermal engineering are well known since they are derived from fundamental physical properties. Whatever the role of advanced methods of active thermal control, the basis of future spacecraft thermal design will rest with the principles of passive control: reflect, radiate, absorb and insulate.

REFERENCES

Brooks P J 1985 *Chartered Mechanical Engineer* Sept. 36–40

Dvorko I M, Marchetti M, Maura G and Rinaldi G 1985 *ESA J.* **9** 351–9

Henderson R 1986 *Spacecraft Engineering Course Notes* (University of Southampton) ch. 14

Howle D, van Holtz L, Leggett P, Strijk S and Ashford E 1985 *OTS: Seventh Year in Orbit* (ESA)

Ishimoto T and Herold L M 1981 *TRW/DSSG/QUEST* Autumn 3–26

Watanabe, M, Nakagawa K, Nakajima K and Tsunoda H 1985 *Int. Astronautical Fed. Congr.* IAF-85-374

Wise P C 1985 *Int. Astronautical Fed. Congr.* IAF-85-373

A7 Communications Payload

7.1 INTRODUCTION

In terms of communications satellite hardware, the subsystems discussed in previous chapters have what could be termed a support role. The lead role is taken by the communications payload, a complex assembly of equipment designed to handle the variety of signals carried by the communications satellite. The role of the payload as part of a satellite communications system is covered in part B. This chapter will describe the components of the payload and some aspects of their operation in the context of the satellite itself.

In the simplest of terms, the communications satellite is an orbiting *repeater*, a device which receives, amplifies and retransmits a signal. The fundamental difference between a satellite repeater and a terrestrial repeater is of course the distance between transmitter and receiver. The simple fact that the satellite in geostationary orbit is some 36 000 km from the earth station presents a challenge from the outset. The signal from the earth station is radiated to the satellite in a beam, which for an antenna beamwidth of 1° is about 630 km wide by the time it reaches the satellite. The spreading of the beam reduces the radiated power density by the ratio of the area of the beam at that height to the area of the earth station antenna, about 10^{10}:1, a phenomenal reduction.

To be of any practical use, the weakened *uplink* signal must be amplified prior to its retransmission on the *downlink* leg. Although amplification is the primary function of the satellite communications payload, it must also filter out any unwanted signals and change the frequency of the wanted signal to a lower value stipulated for downlink transmissions. The precise arrangement of the components which perform these functions is governed by what the payload is required to do, for example transmit a number of broadcast TV channels, provide telephone and data links between and within a number of coverage areas, etc.

Two of the most critical devices in the communications payload, the travelling wave tube (TWT) and the antenna, are given special consideration in successive subsections, but first let us examine the concept of the transponder.

7.2 TRANSPONDERS

A practical satellite repeater comprises several individual chains of equipment called transponders, a term derived from 'transmitter and responder'. Before describing the components of the typical transponder chain, a few definitions are necessary. The *communications payload* can be divided into two parts, transponders and antennas, so a spacecraft transponder chain comprises all the devices of the communications payload with the exception of the receive and transmit antennas. In turn, the transponder chain can be divided in two by function—a *receive chain* and a *transmit chain*. The precise dividing point tends to depend on the particular design and, to an extent, on personal choice, but an example including the channel filter in the receive chain is shown in figure A7.1.

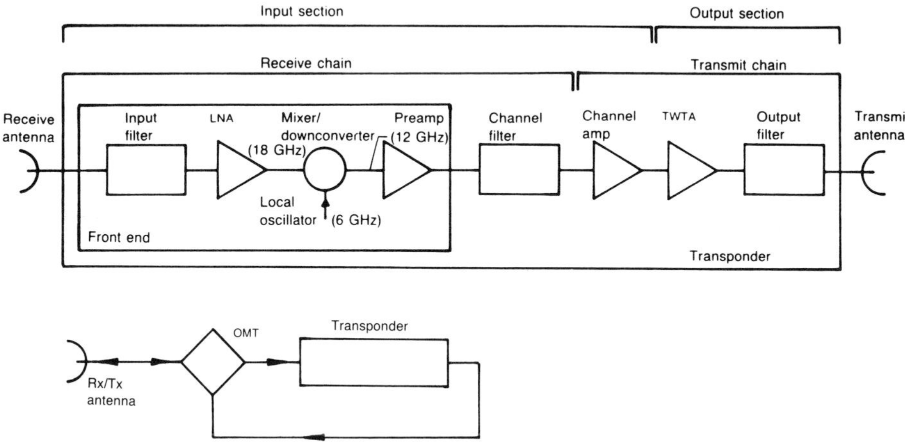

Figure A7.1 Single-conversion payload (Ku-band). LNA, low-noise amplifier; TWTA, travelling wave tube amplifier. The lower diagram shows how a single antenna can be used for both transmitting and receiving; OMT, orthogonal mode transducer.

The transponder may also be divided into an input section and an output section, the former usually extending from the point where the input signal enters from the receive antenna to the point prior to its entry to the TWTA (travelling wave tube amplifier) and the latter from this point to the output to the transmit antenna (note that the antennas are not included in any definition of the transponder). The input devices up to and including the preamplifier are also termed the *front end* or *receiver* and may be combined within a single package on the satellite [see figure A7.2].

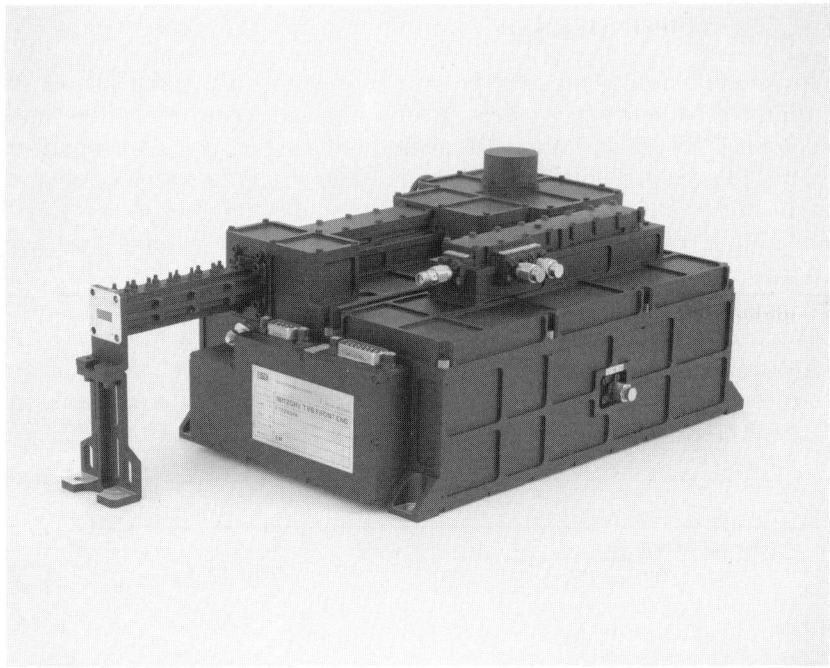

Figure A7.2 Receiver or 'front end' for a TV broadcast satellite payload. [GTE]

It is worth noting that the term transponder is often used by the uninitiated in place of TWTA, which is of course erroneous since there may be several TWTAs in a transponder. Transponders can be divided into two types: single conversion and double conversion.

7.2.1 Single conversion

Figure A7.1 is a simplified block diagram of a *single-conversion payload*, that is one which makes a single frequency conversion from the higher uplink frequency to the lower downlink frequency. These frequencies are known as carrier frequencies since they are used to *carry* the telecommunications *signals*. The distinction between the carrier and the signal is made more fully in box A7.1. Here we use the term 'signal' in a more colloquial fashion simply to indicate that information is being transmitted.

The operation of the simple transponder is as follows. The uplink signal, radiated from an earth station, is collected by the satellite's receive antenna. Some satellites use a single antenna for both receive and transmit: in this arrangement the receive and transmit signals are

BOX A7.1: MODULATION

It may be necessary to clarify what it is that is being handled by a transponder: why do we sometimes refer to 'the signal' and at others to 'the carrier'? In short, the carrier is an electromagnetic wave of fixed amplitude, phase and frequency which, when modulated by a signal, can be used to convey information through a communications system. The modulating signal can be analogue or digital, originating from a telephone, TV camera, computer, etc, and is often referred to as the *baseband* signal.

Modulation is the process whereby a signal is superimposed upon a higher frequency carrier wave, commonly referred to as 'modulating the carrier'. There are two primary modulating techniques in general use: amplitude modulation (AM) and angle modulation (which includes frequency modulation (FM) and phase modulation (PM)). With AM the amplitude of the carrier is varied in accordance with the amplitude of the input signal, the frequency remaining unchanged. With FM the frequency of the carrier is varied, around its nominal or centre frequency, with the amplitude of the lower frequency input signal, and the amplitude of the carrier remains unchanged. With PM the phase of the carrier is varied with the amplitude of the input signal. The PM technique relies on the fact that two or more samples of the same periodic waveform (or carrier) can be arranged to have different time origins (i.e. they are out of phase). In the demodulation process the phase changes can be detected and the original signal reconstructed. For example, carrier waveforms 180° out of phase can be used to represent the digits 0 and 1, a technique commonly referred to as PSK modulation, phase shift keying, or binary phase shift keying (BPSK). Another common form of PM is QPSK (quadrature phase shift keying) which is effectively four-phase modulation and can represent 00, 01, 11 and 10.

Before a carrier is modulated by one of these techniques, an input signal may undergo a modulation of a different kind (a form of coding), for example pulse code modulation (PCM), whereby the amplitude of the original analogue input signal is sampled at discrete time intervals to create a representative digital translation of the signal. PCM was devised specifically to enable (analogue) speech to be transmitted in a digital form, using the principle that if an analogue signal is sampled at a rate at least twice that of the highest frequency present, the original speech signal can be reconstructed with acceptable quality from the discrete sample values. Other methods less commonly used are differential pulse code modulation (DPCM), delta modulation (DM) and adaptive versions of PCM, DPCM and DM (referred to as APCM, ADPCM and ADM, respectively).

typically given opposite polarisations to reduce interference and kept separate by a waveguide device called an orthogonal-mode transducer (orthomode transducer or OMT) [see the lower diagram in figure A7.1]. Since a reduction in the number of antennas constitutes a significant mass advantage, considerable effort is often expended in trying to rationalise a design to fit the single-antenna configuration.

The antenna passes the signal to an input filter, which confines the bandwidth of the signal allowed into the transponder [see box A7.2]. This is the first of several levels of filtering designed to reject unwanted signal frequencies: there is, after all, no point in amplifying what you do not want. These unwanted signals include both interference from other communications systems and harmonics and spurious signals (or 'spurii') generated by the components of the transponder itself. After passing through the input filter, the signal meets the first stage of amplification: the LNA or low-noise amplifier. The LNA is designed to deliver a good signal-to-noise ratio (S/N), which is very important at this early stage since the signal power is extremely low following its passage through 36 000 km of free space [see chapter B4]. Any noise introduced by an amplifier at this stage will be carried through the entire chain, successively amplified and finally retransmitted to Earth. This is a tendency which must be discouraged!

BOX A7.2: FILTERS AND MULTIPLEXERS

Filters are electrical or microwave devices designed to allow a selected range of signal frequencies to pass, while obstructing those outside the range. In addition to confining the bandwidth of the signals entering a communications system, filters reduce the possibility of interference between transmitted signals.

A generic type of filter is the band-pass filter, a filter with both high- and low-frequency cut-offs. Figure A7.3(*a*) shows the characteristics for an ideal filter: a perfectly flat passband and infinite out-of-band rejection (otherwise known as selectivity). Figures A7.3(*b*)–(*d*) show three standard design approximations which highlight the trade-off between selectivity and passband ripple. Butterworth filters have a flat passband but poor selectivity and are rarely used in communications payloads. The Chebychev filter solves the out-of-band rejection problems at the expense of passband ripple, but since a small amount of ripple is negligible compared with the mismatch effects of other equipment this type of filter is commonly used in payloads. The elliptic filter is a compromise, with good selectivity and low ripple, but it has a limited passband-to-stopband rejection ratio and tends not to be used as widely in spacecraft transponders as the Chebychev.

The channel filter in a communications payload is a band-pass filter since it defines the usable bandwidth of a transponder. Other types of filter

Two types of LNA in common use are the bipolar transistor (used up to about 2 GHz) and the field-effect transistor or FET (used between about 2 and 20 GHz). The latter is commonly referred to as a 'gasfet' since it is based on the semiconductor gallium arsenide (GaAs). A more long-standing type of LNA is the parametric amplifier or 'paramp' which, although still used for high frequencies, is gradually being superceded by the FET as devices become available for operation at increasingly higher frequencies.

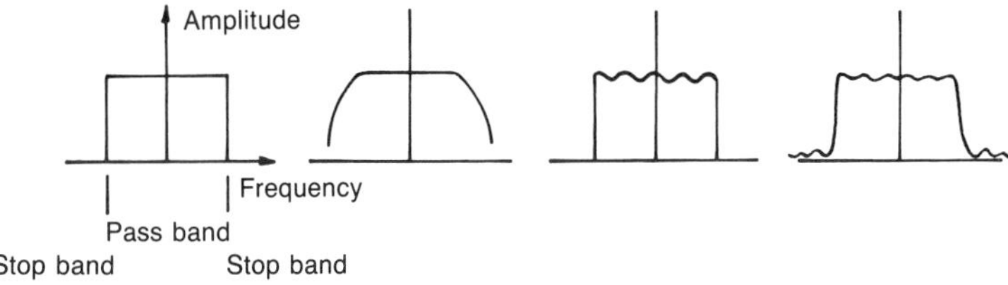

Figure A7.3 Filter characteristics. (*a*) Flat passband, infinite out-of-band rejection; (*b*) flat passband, slow roll-off; (*c*) passband ripple, good out-of-band rejection; (*d*) low ripple, good out-of-band rejection.

include the low-pass filter which has only a high-frequency cut-off (allowing only low-frequency signals to pass) and its opposite, the high-pass filter. The input and output filters in a payload may be either band-pass or low-pass filters.

Filters are often installed in the form of input and output multiplexers or diplexers (a diplexer is a two-channel multiplexer). The multiplexer, a device for combining or separating different signal frequencies is, in effect, a multiple-channel filter. For example, where a number of channels share the bandwidth available in a satellite transponder, an input multiplexer separates the channel frequencies and routes each carrier to its own amplifier chain. Once amplified, the channels are recombined in an output multiplexer for the return transmission. The individual devices are colloquially known as the input 'mux' and output 'mux', or 'Imux' and 'Omux' respectively.

All such devices must be stable over their operating temperature range to maintain the desired rejection characteristics. The thermal coefficient of expansion for aluminium ($22 \times 10^{-6}\ °C^{-1}$) is, however, often unacceptable. Invar, an alloy of iron, nickel and carbon with a coefficient of about $2 \times 10^{-6}\ °C^{-1}$, is a common substitute. Carbon, graphite or boron composites are alternatives, but composites in general do not have the conductivity of metals and need to be plated, which can cause additional problems. Despite this, CFRP filters have been flown successfully on a number of spacecraft and are offered by most spacecraft filter manufacturers.

The next item in the receive chain is the downconverter which, as its name suggests, converts the higher frequency uplink signal to the lower frequency downlink signal. It contains a circuit known as a *mixer*, which mixes the incoming signal with a local oscillator (LO) frequency. The oscillator is a stable frequency source, similar to the quartz crystal in many domestic timepieces, which generates a frequency that is low relative to the carrier frequency (i.e. measured in MHz rather than GHz). It is therefore successively multiplied, typically by a chain of 2×, 3× and 5× multipliers, to produce the desired LO frequency (e.g. 17.3 MHz × 347 = 6.003 GHz). The mixing process, known as the heterodyne process or *heterodyning*, produces frequencies corresponding to the sum and the difference of the two original frequencies; the output of the downconverter is the difference signal. For example, to convert from 18 GHz to 12 GHz, 18 GHz is mixed with 6 GHz: the sum is 24 GHz and the difference 12 GHz. Although the 24 GHz signal is not used in the downconversion process, it is the sum signal which becomes the output of an upconverter [see below].

Having been 'cleaned up' by the input filter, amplified by the LNA and converted to its downlink frequency, the signal can now be amplified

BOX A7.3: TRANSMISSION LINES: WAVEGUIDE AND COAXIAL CABLE

A transmission line is an electrical conductor designed to provide a controlled and protected propagation path for a radio-frequency carrier wave and/or electrical signals (the alternative is propagation through free space). The two commonly used types are waveguide and coaxial cable [see figure A7.5].

A waveguide is a metal or metal-coated tube, of rectangular or circular cross section, which confines and conveys (or 'guides') electromagnetic waves. In simple terms the wave can be thought of as being internally reflected along the tube in a similar way to light in an optical fibre; in reality the wave is propagated in a complex interplay of electric and magnetic fields which induce currents in the surface of the guide. What this means in practice is that the size of the guide is dependent upon the frequency of the radio waves, such that higher frequencies require smaller waveguides and tighter tolerances on other waveguide components. To reduce mass, metal-coated CFRP waveguides have been developed for certain applications.

Coaxial cable, often abbreviated to 'coax', comprises two coaxial conductors known as the 'inner' and the 'outer' separated by an insulator or dielectric material. Two types of coax may be used in communications payloads: *semi-rigid*, which has a solid copper inner and outer and a solid PTFE dielectric; and *flexible*, which has a solid copper inner and an outer formed from a helical copper strip which overlaps itself to form a further. This is the function of the last device in the front end, the preamplifier. Following preamplification, the signal passes through a channel filter—another band-pass filter—which defines the bandwidth of the communications channel handled by the transponder. The channel filter has a narrower pass band than the input filter and it is this filter which defines the 'usable bandwidth' of the transponder, an important quantity as far as the user of the service is concerned.

After filtering, the signal is again amplified, this time by a channel amplifier or driver amplifier. Its purpose is to amplify the signal to a level acceptable to the main high-power amplifier (HPA), which is usually a travelling wave tube amplifier [see §A7.3]. Finally, after high-power amplification the signal is routed to an output filter which rejects any unwanted frequencies that may have been produced during the amplification process. The successively amplified and carefully filtered signal thus passes from the transponder to the transmit antenna which directs it on its 36 000 km journey back to Earth.

continuous conductor even when bent (the dielectric is generally air-spaced PTFE or foamed PTFE). Semi-rigid coax is generally used for connections between transponder equipment and flexible coax for connections to antennas.

The principal advantages of coax are its small size and mass and its ease of use in confined spaces. Its main disadvantage is its high attenuation at high frequencies as a result of increased absorption in the dielectric. Waveguide has a lower attenuation per unit length at a given frequency, but below about 4 GHz is too large and unwieldy to be used in satellite payloads. As a guide to usage, from about 4 to 12 GHz waveguide generally provides the links between the antenna and the LNA in the receive section and between the HPA and the antenna in the transmit section, and coax connects the remaining units. Above about 12 GHz, coax losses begin to lose out over the decreasing size and mass of the waveguide and at this frequency it is possible to produce a payload using waveguide for all inter-unit paths.

Switches designed to isolate one part of a circuit and divert a signal to another are available in both coax and waveguide. Signals carried in coax are always switched electrically in a switching circuit; waveguide switches can be either electrical or mechanical. An example of the electrical type is the ferrite switch now being used more widely in communications payloads. The older mechanical type is still in use. It uses rotating waveguide sections which act like a railway turntable, physically severing one connection and making another.

7.2.2 Double conversion

A double-conversion transponder differs from the single-conversion type in one simple respect: it 'downconverts' the receive frequency to a frequency lower than the transmit frequency for preamplification and filtering, and then 'upconverts' it to the transmit frequency. Figure A7.4 shows a simple double-conversion payload which differs from the single-conversion payload mainly in that it has an upconverter inserted between the driver amplifier (also known as an IF amplifier) and the TWTA.

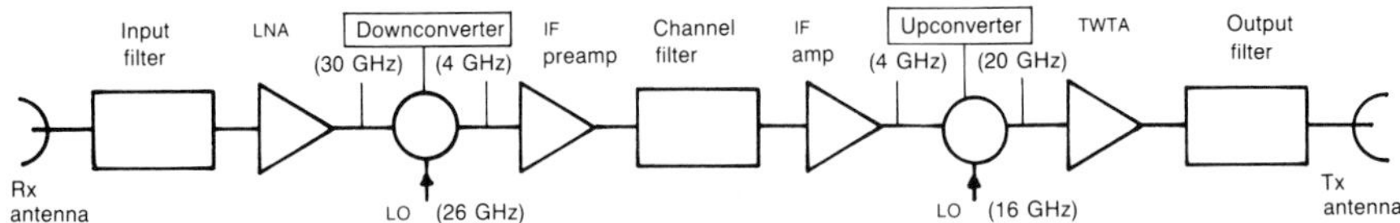

Figure A7.4 Double-conversion payload (Ka-band). LNA, low-noise amplifier; IF, intermediate frequency; TWTA, travelling wave tube amplifier; LO, local oscillator.

Double conversion is the contemporary design solution for payloads operating at Ka-band (20/30 GHz). In such a payload a received (uplink) signal at 30 GHz can be mixed in a downconverter with a 26 GHz local oscillator signal to produce an intermediate frequency (IF) of 4 GHz (the 'difference signal'). The IF is later mixed in an upconverter with an LO of 16 GHz to produce the 20 GHz transmit frequency (the 'sum signal'). This method is usually employed when the transmit frequency is too high to be conveniently handled by the intermediate transponder components, without recourse to significant and expensive redesign. For instance, this may be because the required channel bandwidth is too narrow to be accurately defined by available filters at such a high carrier frequency, or it may be the degree of signal processing required is not feasible at the higher frequency. Not only is 4 GHz equipment more widely available than 20 GHz hardware, the lower frequency allows the use of coaxial cable rather than the heavier waveguide [see box A7.3].

The payload diagrams used so far have been greatly simplified for clarity. It is very unusual for a spacecraft to carry a single transponder (whether single or double conversion) without spare or *redundant* components which can be brought into use in the event of a failure. Every item in the above simplified transponder represents a *single-point failure* (SPF), that is if any single component fails, the whole chain becomes useless. The concept of redundancy is discussed further in chapter A11.

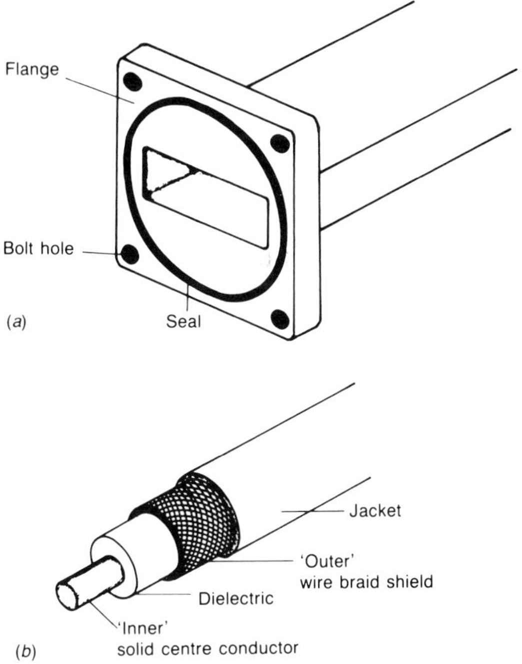

Figure A7.5 Two types of transmission line: (*a*) waveguide and (*b*) coaxial cable.

7.2.3 Example payload: TV broadcast

Figure A7.6 depicts a more practical satellite communications payload—an example of a TV broadcast payload for two TV channels. A single receive antenna is linked to two redundant front ends selected by ground command by means of a waveguide switch. The unused receive chain is terminated by a *load*, a type of attenuator that absorbs any incident RF power. This avoids unwanted reflections inside the equipment which could occur if the redundant receiver was accidentally switched on.

In this example the input filter is a wideband device which passes a frequency band containing both TV channels and a separation bandwidth or *guardband*. DBS TV channels are typically 27 MHz wide, so the input filter will require a bandwidth of 60 or 70 MHz to encompass both channels. Following low-noise amplification the downconverter takes the 18 GHz uplink frequency and converts it to the 12 GHz downlink in a single conversion. Note that the transmission line changes from waveguide to coaxial cable after the downconverter, once the frequency is lower.

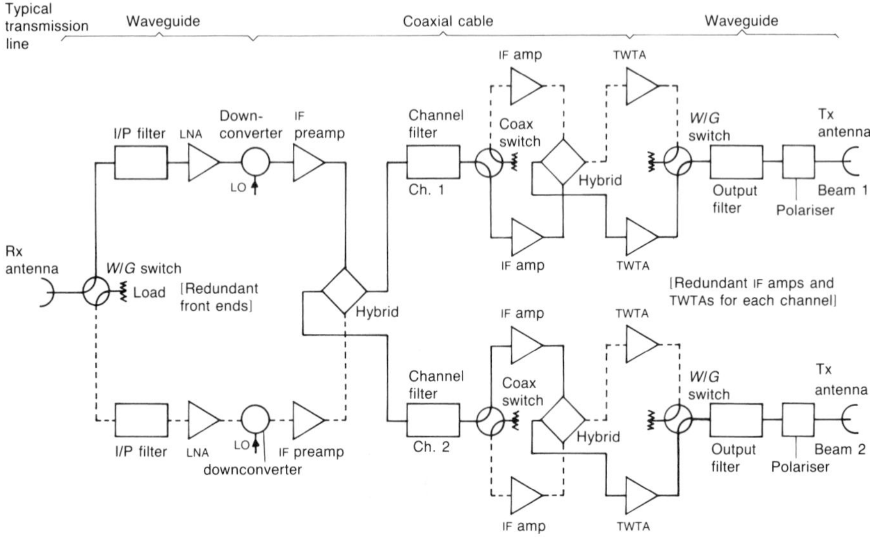

Figure A7.6 Example payload: TV broadcast (two channels); ———, active signal path; – – –, inactive signal path.

The redundant chains are linked together again after the IF preamplifiers by a coaxial *hybrid*. The hybrid is a passive device used to link several input and output paths simultaneously. Thus, whichever input is used, the signal exits from both output ports. At this point the signal still has a bandwidth defined by the input filter. The filters of the two 27 MHz channels, specifically designed for the frequency and bandwidth of a particular channel, take their respective parts of the wider input bandwidth for subsequent amplification. From this point, each chain contains redundant IF amplifiers and TWTAs. The amplifiers are selected by a commandable coaxial switch and their output is routed to either of the TWTAs. The inputs to the TWTAs are in coax and the outputs are in waveguide. A waveguide switch selects the output from the active TWTA and routes it to an output filter. The output signal is then fed to the transmit antenna through a polariser which selects the required band of polarisation (which is circular for DBS; see chapter B2).

Figure A7.7 shows how the transponder hardware might be mounted inside a satellite. Although not an actual design, it does show some of the essential design features. The positioning of items on a spacecraft must take into account the mechanical and thermal properties of the surface, interconnection requirements, electromagnetic compatibility (EMC) and the constraints of integration and testing.

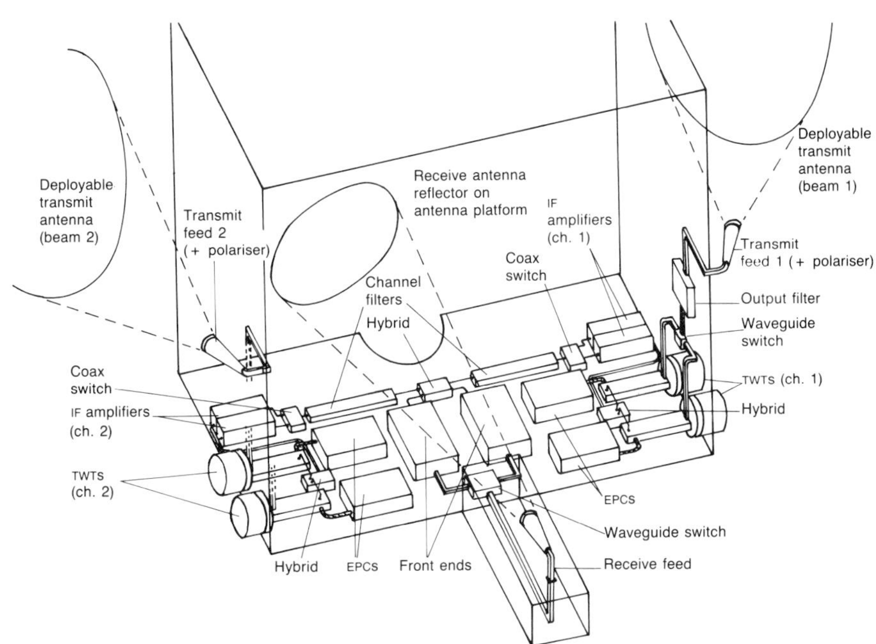

Figure A7.7 Example TV broadcast payload mounted in satellite (for clarity the channel 2 output filter and waveguide switch have been omitted and more space has been allowed between components than would normally be available).

For ease of integration and testing by the manufacturer, the transponder components are attached to a single panel. By comparison with figure A4.6, it can be seen that this is the north-facing panel; the south face would usually have another payload mounted upon it. The high-power TWTs are mounted with their collectors protruding from the east and west faces of the spacecraft to radiate directly to space, and their waveguide outputs are located close to the Earth-facing antenna platform to minimise the waveguide runs to the feedhorns (which reduces RF power losses and thus the initial DC power required). The same philosophy is followed in general for output filters and other waveguide components on the output side.

This example payload features two transmit antennas producing separate beams for two different coverage areas. Feedhorns for the two deployable transmit antennas are mounted directly onto the antenna platform at the foci of their respective reflectors. The horn for the single receive antenna is mounted on a tower above the reflector which is itself attached to the platform.

Although we have covered the main components of a spacecraft communications payload, a single example cannot include all of them. More complex payloads, particularly those for general telecommunications, feature more cross-strapping (the interlinking of redundant chains) and more complex switching networks or *switch matrices* to link transponders with different uplink and downlink beams.

7.2.4 Example payload: telecommunications

Figure A7.8 illustrates the layout of a hypothetical double-conversion transponder for a telecommunications satellite payload designed to provide selected coverage of the eastern United States. It features two elliptical coverage receive antennas, one covering some of the north-eastern states and one covering Florida, and four transmit antennas, one for each of the northern and southern ellipses (NE and SE) and a northern and southern spot-beam coverage (NS and SS) [see figure A7.9]. The spot beams might be designed for a specific application (e.g. television) or simply to provide a greater power flux density (PFD) in the areas where telecommunications traffic is heaviest. The payload is designed to provide complete interconnectivity between beams and a degree of redundancy.

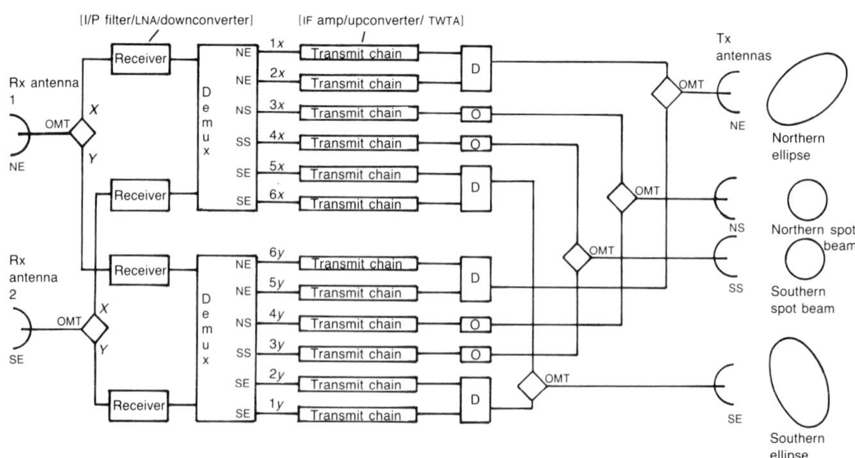

Figure A7.8 Example payload: telecommunications. OMT, orthomode transducer; D, diplexer; O, output filter; X and Y are two opposite polarisations.

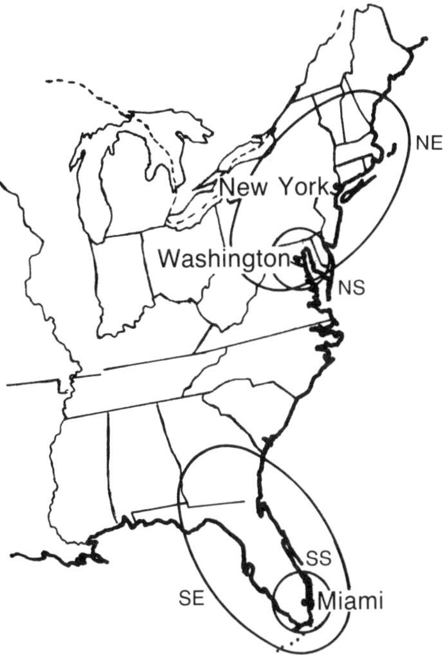

Figure A7.9 Coverage areas for example telecommunications payload. NE, northern ellipse; SE, southern ellipse; NS, northern spot beam; SS, southern spot beam.

As far as the hardware is concerned, the payload is split into two halves for opposite polarisations, *X* and *Y*, the use of which doubles the number of channels in the available bandwidth from 6 to 12, a practice known as *frequency reuse*. An orthomode transducer (OMT) connected to each of the receive antennas separates the polarised signals and routes them to a front end, which in this design contains an input filter, LNA and downconverter. All six channels of a particular polarisation are filtered, amplified and downconverted together in the receiver and passed to a demultiplexer which separates them for high-power amplification in the transmit chain. This so-called 'broadband low-noise amplification and common downconversion' is an alternative to demultiplexing the channels first and then passing them to six individual receivers, which would increase the necessary receiver hardware six-fold. It should be noted, however, that in a real payload the mass saving would be slightly reduced because of the incorporation of redundant components in the receiver, which for simplicity is shown in figure A7.8 as a single box.

Since the demultiplexer is designed to separate the channels, it acts as a channel filter, and no separate channel filter appears on the diagram. The transmit chain consists of an IF amplifier (the output of the receive chain was at the intermediate frequency), an upconverter and a TWTA. Following high-power amplification the signals are routed either to an output filter or a diplexer, depending on the channel. The diplexer filters and recombines two of the channels and passes them to an antenna through an OMT which recombines the separate polarisations.

As far as beam interconnectivity is concerned, any uplink from the northern ellipse (NE) can be routed, through receive antenna 1, to any of the four transmit antennas using either polarisation. Channels 1x, 2x, 5*y* and 6*y* serve the NE coverage area; 5x, 6x, 1*y* and 2*y* serve SE; 3x and 4*y* serve NE; and 4x and 3*y* serve SS. An uplink through receive antenna 2 (from the southern ellipse) has an equivalent downlink access. Any uplink made from within a spot beam area is made using the elliptical beam receive antenna which covers that area.

It is apparent that the channel numbers have been reversed for the *y* channels. This is because channels with the same number generally have the same frequency and, although of opposite polarisation, the discrimination under certain conditions may be insufficient to eliminate interference. For this reason no one downlink contains two channels of the same number.

7.2.5 Regenerative repeaters

The transponders, or repeaters, discussed so far have been alike in one important respect: all operations, such as filtering and frequency conversion, have been performed on the carrier; the type and characteristics of the signal have been of little consequence. This type of repeater is classified as *transparent*. A regenerative repeater, on the other hand, strips the signal (known as the baseband signal) from its carrier, performs various operations upon it, and replaces it on the carrier for the downlink transmission. The stripping and replacing functions are called demodulation and remodulation (or simply modulation) and require the insertion of demodulators and modulators into the transponder, as shown in figure A7.10. This reduction to baseband, which is applicable only to digital transmissions, allows the characteristics of the signal to be 'regenerated' on the satellite, thus correcting errors caused by thermal noise in the preceding equipment, for example, and avoiding the transmission of the errors into the downlink beam. Regeneration therefore isolates the uplink performance from the downlink performance and allows receive and transmit chains to be designed more independently and with improved performance characteristics.

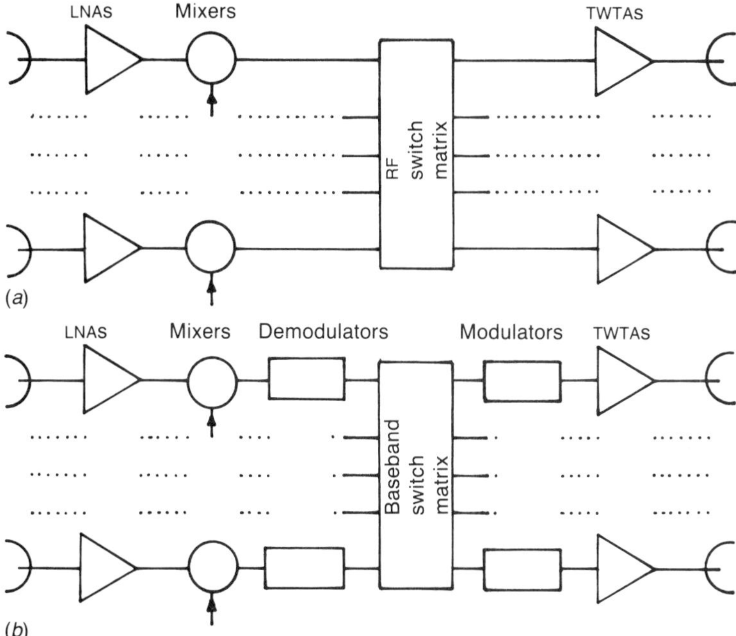

Figure A7.10 Comparison of conventional transparent repeater (*a*) and regenerative repeater (*b*).

Regenerative processing is an advanced feature planned for the satellites of the 1990s. It may take the form of simple channel switching, performed at baseband instead of microwave frequencies, or it may be more complex, including error correction coding, spread spectrum decoding or onboard decryption.

7.3 THE TRAVELLING WAVE TUBE (TWT)

Although every item of equipment so far described has its place in a transponder, the item which has the most fundamental effect on the signal transmitted by the satellite is the travelling wave tube. Its output power is a major factor in the design of a communications link [see chapter B4].

The TWT is part of a family of microwave tubes which includes klystrons, crossed-field amplifiers, magnetrons and gyrotrons, all of which provide the function of high-power amplification or generation in fields as varied as military radar and particle physics. There are two main types of TWT: the helix type and the coupled-cavity variety. The latter is better suited to military airborne applications, so we shall concentrate here on the helix design, which has been used for the majority of existing spacecraft tubes and will be used for many years to come. TWTs are also used in earth stations to amplify the uplink signal [see §A9.4.2].

7.3.1 TWT development

Despite the current importance of TWTs in the satellite communications industry, the TWT is not a new invention: its basic capabilities as a high-power amplifier have been known for nearly 50 years. It was invented during the latter part of World War II (in 1943) by Dr Rudolf Kompfner, an Austrian refugee working on microwave tubes for the British Admiralty, and the first practical device was developed by J R Pierce and L M Field at Bell Telephone Labs (BTL) in 1945. Subsequent development work was done at BTL and Stanford University in the USA, and Standard Telephones and Cables in the UK, with particular emphasis on potential applications in the communications field. Meanwhile, the military services had discovered the potential for TWTs in the newly expanded fields of radar and electronic countermeasures (ECM), for which the high gain and wide signal bandwidth of the TWT were ideally suited.

TWTs have been the mainstay of space communications since the early Intelsat and Syncom satellites and have also been used on scientific planetary probes and manned spacecraft. Much of the early development work on 'space tubes' was done by the Hughes Company (USA), eventually leading to the formation of their Electron Dynamics Division (EDD). Along with Watkins Johnson in the USA, Thomson CSF in France, AEG-Telefunken in West Germany and NEC in Japan, Hughes EDD is now one of the few major space-tube manufacturers.

7.3.2 Component parts

The TWT, often referred to simply as a *tube*, owes its name to its mode of operation: it is designed to cause an RF *wave* to *travel* along its length in a carefully predetermined manner. The energy for amplification is derived from a high-powered electron beam which is made to interact with the RF wave.

Figure A7.11 shows schematically the internal structure of a typical medium-power (20 W) Ku-band TWT taken as an illustration for the subsequent discussion. It indicates the relationship of its component parts: an electron gun, a slow-wave structure and a collector, all housed within a vacuum envelope, part of which is surrounded by a beam-focusing structure. The vacuum allows the electrons to move on collision-free trajectories and ensures long life and efficient operation of the enclosed components. Figure A7.12 shows a cross section through a high-power DBS tube with which figure A7.11 may be compared. Figure A7.13 shows an actual example of a similar device.

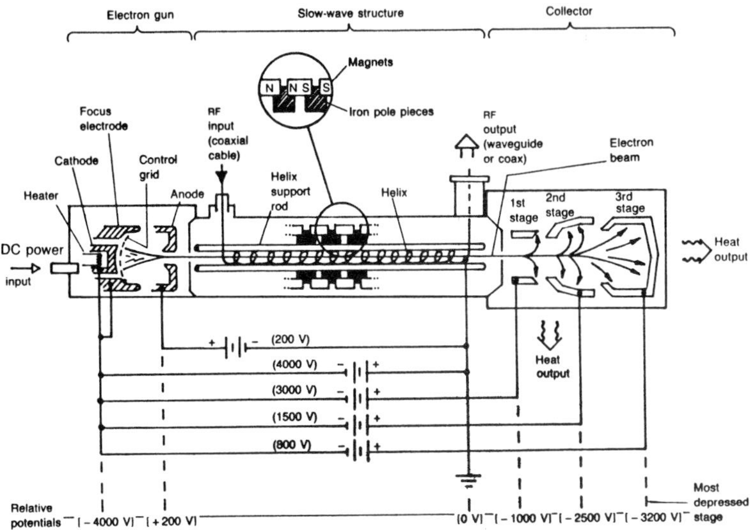

Figure A7.11 Component parts of a travelling wave tube (TWT) in schematic form, with typical voltage supplies for a medium-power (20 W) Ku-band TWT (actual supply input is shown at left).

The electron gun generates the electron beam and injects it into the slow-wave structure. In its basic form the electron gun comprises a heater, cathode, focus electrode and anode. The cathode is the source of electrons and is heated, by the heater, to the temperature required for electron emission. The focus electrode surrounds the cathode and controls the electric field (*E* field) near the cathode surface so that the electrons converge into a well defined beam just under a millimetre in diameter. Since electrons carry a negative charge, they are repelled by the negative cathode and attracted towards the positive anode. The anode potential determines the strength of the *E* field between the cathode and the anode and thus provides the accelerating force in the electron gun. Many TWTs also incorporate a control grid which can be used to turn the electron beam on and off more efficiently than adjusting the voltage on the cathode or focus electrode. A disadvantage of the grid is that it intercepts part of the electron beam and causes some degree of perturbation.

The accelerated electrons pass from the electron gun along the axis of the slow-wave structure, in the form of a helix or a row of coupled cavities. The latter comprises accurately sized and shaped microwave cavity sections, usually of copper, brazed together to form a structure

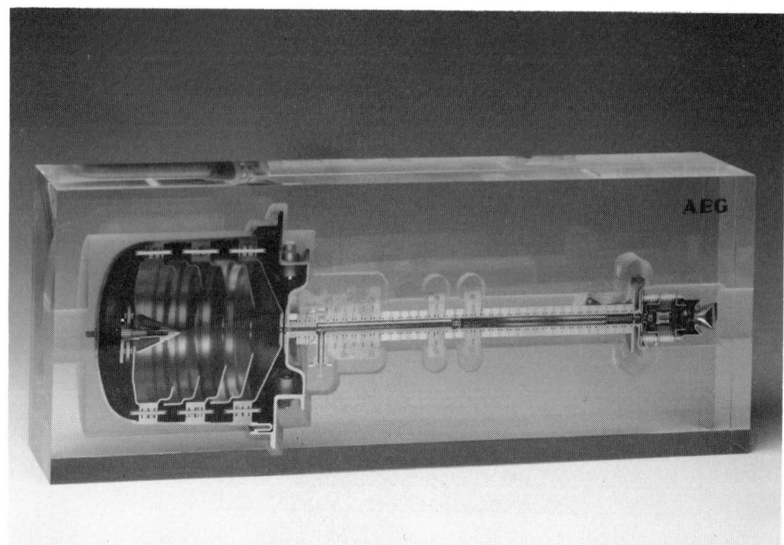

Figure A7.12 Cross section through a high-power travelling wave tube designed for DBS. From left to right, note the five-stage collector, helix slow-wave structure and electron gun. [AEG]

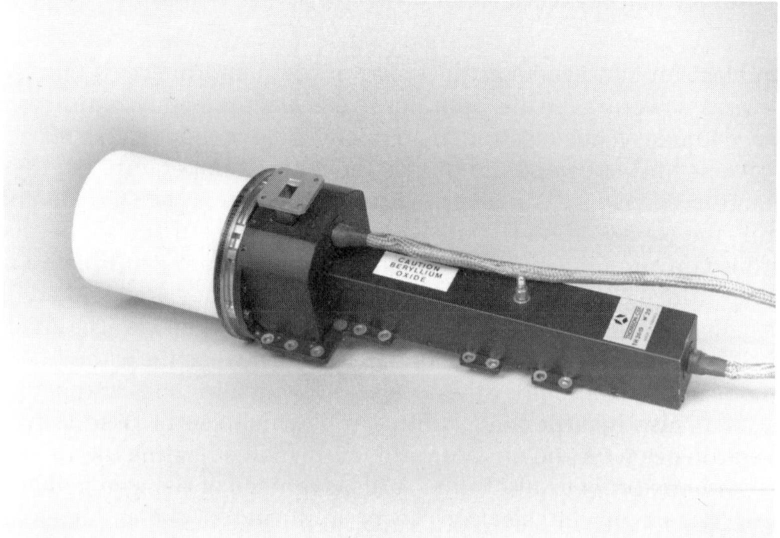

Figure A7.13 High-power DBS travelling wave tube. Collector housing at left is designed to protrude from the spacecraft for direct radiation to space. Note coaxial signal input and waveguide output on top of tube body; braided cables supply DC power to electron gun (right) and collector (left). [Thomson CSF]

of many cavities 'in cascade'. Coupled-cavity tubes tend to be used for very high powers (tens of kilowatts) or for high frequencies at low powers. The slow-wave structure in the helix TWT is commonly made of copper or of tungsten or molybdenum wire wound in a corkscrew shape. The helix used in a Ku-band space tube, for example, is 1 to 1.5 mm in diameter and about 15 cm in length [Smith 1987]. Figure A7.14 shows how, in one design, the helix is supported by three or four ceramic rods to isolate it from the metallic walls of the surrounding vacuum envelope, and figure A7.15 gives an indication of its size.

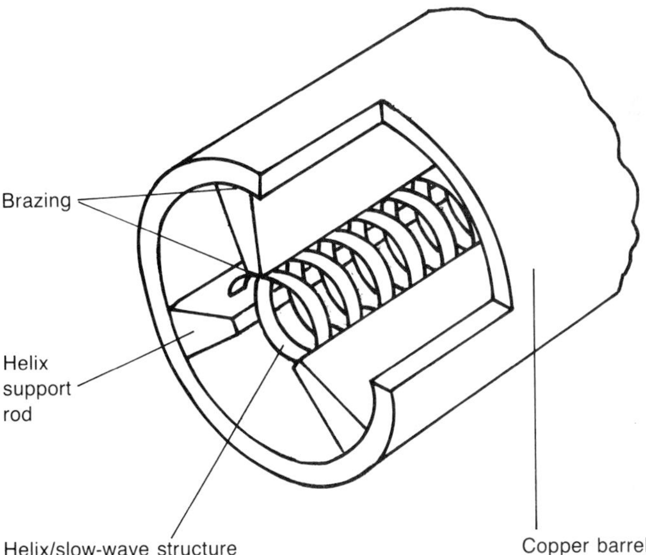

Figure A7.14 TWT helix configuration for 'brazed helix' design; 'pressed helix' design eliminates brazing at the expense of poorer thermal conductivity between helix and barrel.

The helix carries the RF wave (which in turn carries the signal to be amplified) causing it to travel slightly slower than the electron beam. This allows an interaction between the RF wave and the electron beam which, by means of an energy transfer from the beam to the wave [see box A7.4], results in the exponential amplification of the RF wave. Individual TWTs have been built with power gains of more than 10 000 000 (70 dB), although 50–60 dB (100 000–1000 000) is more common.

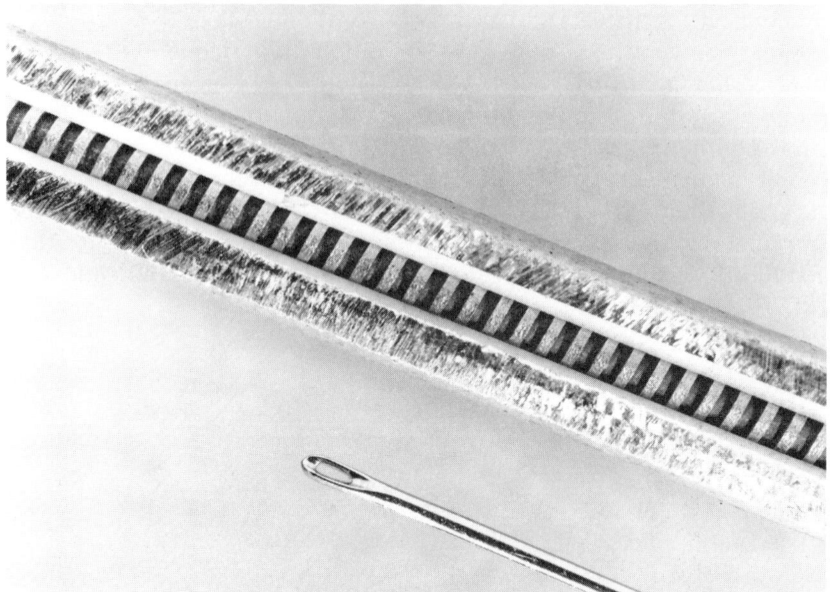

Figure A7.15 Cross section through a TWT helix showing its small size. [Thomson CSF]

BOX A7.4: THE AMPLIFICATION PROCESS OF THE TWT

An RF wave travels at the speed of light, *c*; an electron travels somewhat slower, depending upon the accelerating voltage (16 to 33% of *c* [Kosmahl 1983]). If the two are to interact to allow an energy transfer to take place, the RF wave must be slowed to a similar velocity to that attained by the electron beam. This is the purpose of the helix.

The RF wave travels along the helix at *c*, but because of its helical path the energy of the wave progresses along the axis of the tube at a much lower axial velocity, largely determined by the pitch and diameter of the helix. This axial velocity is engineered to be slightly lower than the velocity of the electron beam, allowing the electric fields of the RF wave to interact with the electrons in the beam. In regions where the *E* field lines are in the direction of electron travel, electrons are decelerated, and where the field is in the opposite direction they are accelerated. The interaction is such that the electrons receive a periodic velocity modulation approximately in phase with the *E* fields of the RF wave.

These alternate regions of acceleration and deceleration produce the phenomenon of *bunching*, whereby electrons tend to concentrate ahead of the accelerating fields and behind the decelerating fields. Figure A7.16 shows that electrons are accelerated into area A from the left and decelerated into area A from the right. Since electrons are also being

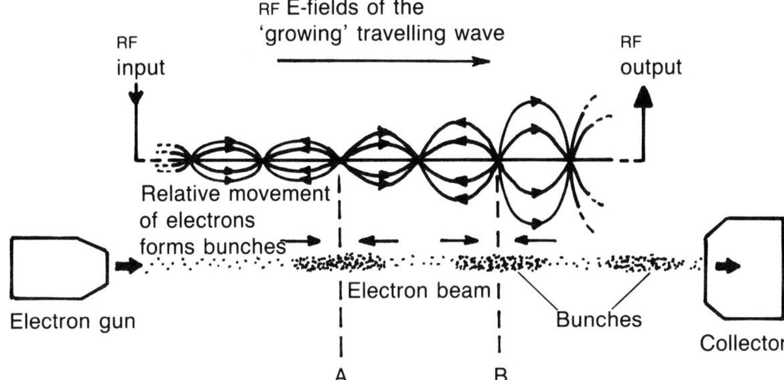

Figure A7.16 The amplification process in a TWT [see text of box A7.4 for details].

The beam-focusing structure comprises a system of magnets surrounding the vacuum envelope along the length of the slow-wave structure. This confines the electrons, which would otherwise tend to repel each other, into a narrow beam for their passage through the slow-wave structure. Once the electrons have passed through the slow-wave structure and completed their interaction with the RF wave, their

accelerated from the left into area B, the deceleration into area A leaves a region containing relatively few electrons. Thus interaction with the RF wave produces bunches of electrons along the axis of the tube. It is not difficult to see that this alternation of more dense and less dense regions of electrons represents a waveform in its own right.

So far, however, it appears that the interaction has been rather one sided with the RF wave dictating the performance of the electron beam. This is where a crucial aspect of the physical structure of the helix comes in: it was stated that the RF wave is made to travel slightly *slower* than the electron beam. This has the effect that, on average, the electron beam is 'dragged back' by the slower RF wave coupled to it.

Initially, as explained above, some electrons are decelerated by the E field into the following bunch. The electrons which are accelerated by the field into the bunch move forward, as a bunch, into the deceleration field and *as a bunch* lose energy by deceleration. Although all the electrons are of course moving in one direction, towards the collector, the critical motion is that of the electron bunches with respect to the RF field on the helix. Most of the electrons are, as a consequence, decelerated. And as they lose velocity they lose kinetic energy to the RF wave, where it manifests itself as an increase in amplitude. The RF signal has been amplified [Williamson 1986].

usefulness is at an end, so they must be decelerated and 'disposed of'. This is the function of the collector, which may consist of several *stages* at various electric potentials. The collector decelerates the electrons, converting their kinetic energy partly into electrical energy which is recovered, and partly into thermal energy which is removed from the TWT either by conduction or direct radiation into space [see §A7.3.4].

The conversion of energy in a TWT is as follows. DC power is used to produce an electron beam which relinquishes some of its energy to the RF signal/wave travelling in the slow-wave structure: the conversion is from DC power to AC power at radio frequencies (i.e. RF power) via the kinetic energy of the electron beam. Part of the energy remaining in the beam is recovered by the collector, the balance being dissipated on the collector electrodes with the consequent production of thermal energy. There are, therefore, just two inputs and two outputs [see figure A7.11]: DC power and the RF signal to be amplified are fed in; the amplified RF signal and the heat from the collector are fed out.

Signal inputs and outputs have already been discussed in the context of the transponder. Power input and thermal output have wider ramifications for the other satellite subsystems and deserve further consideration.

7.3.3 Electronic power conditioner (EPC)

As described in chapter A5, all power for a communications satellite is derived ultimately from the Sun. This is, however, 'raw power' at the standard voltage of the spacecraft power bus. As can be seen from figure A7.11, the TWT requires a multitude of different voltages, the provision of which is the function of the TWT's power supply unit, or electronic power conditioner (EPC). The combination of TWT and EPC is known as a travelling wave tube amplifier (TWTA). The EPC takes the standard spacecraft voltages and transforms them to the numerous closely defined potentials required for the electrodes of the electron gun and the collector. The TWT is, therefore, not a usable device in its own right and can only be used as part of a TWTA.

First of all, a voltage must be supplied to the heater in the electron gun to enable it to heat the cathode. Compared with the other tube voltages this is quite nominal, typically around 5 V. In the self-consistent example illustrated in figure A7.11 the potential differences between the elements of the electron gun, slow-wave structure and collector are much higher. The slow-wave structure is connected to ground. The anode has a potential of 200 V above ground potential and the cathode 4000 V below. This gives a typical accelerating potential (between the cathode and the anode) of 4200 V, which fires the electrons from the gun towards the slow-wave structure. They are, however, decelerated slightly by the 200 V difference between the anode and helix, giving the beam a total kinetic energy of 4 kV (the difference between the cathode and the slow-wave structure potentials).

7.3.4 Collector

In a simplified model of a TWT, the electron beam could be allowed to impinge upon the metallic surface of a single collector at ground potential, but this would cause great problems in the rejection of heat from the device. Since the RF amplification process extracts typically between 10 and 30% of the electron beam's kinetic energy, a great deal of electron kinetic energy would be converted to thermal energy in the collector. A far more efficient method is to decelerate the electron beam prior to collection, which requires a *depressed collector* (i.e. depressed below ground potential).

Ideally, the negative potential of a single (hypothetical) collector stage would be chosen so that the electrons lost all their remaining kinetic energy precisely as they reached the collector surface—but when has physics ever been so kind to technology? For a start, the electrons have a spread of velocities due to the RF interaction process [see box A7.4] and any voltage which slowed the fastest electrons to rest would repel the slower electrons back towards the slow-wave structure. In addition, the inherent charge of the electrons already in the collector, known as the *space charge*, would repel the slower electrons about to enter. Electrons striking the collector surfaces would, moreover, enhance the space charge by secondary electron emission. To obviate the reflection of electrons, a TWT with a single collector stage would therefore be forced to allow electrons to enter the collector with an appreciable kinetic energy which would have to be converted to heat, thus decreasing the overall conversion efficiency of the device.

The solution is to use a multiple-stage depressed collector as in the example of figure A7.11, in which the first stage is 3000 V above the cathode potential, the second stage 1500 V above, and the third stage 800 V above. Since the cathode is at −4000 V, the collectors are also below ground potential. The third stage, the 'most depressed' with a voltage closest to the cathode potential, collects the electrons with the highest kinetic energy. These electrons have sufficient energy to overcome the decelerating potential of the third stage and reach it with very little remaining energy. Those electrons which experienced the strongest interaction with the RF fields on the slow-wave structure enter the collector with very low kinetic energy and are collected by the first stage at a much lower potential.

The collector thereby becomes the antithesis to the electron gun, albeit in a slightly more 'relaxed' sense: electrons liberated by the heater are repelled by the highly negative potential of the cathode and simultaneously attracted by the anode's high relative potential. Subsequently, in the collector they experience a repulsive force from the high negative potentials of the various stages.

Using the multiple-stage depressed collector, overall tube efficiencies of up to 60% have been realised. Advances in tube efficiency produced in the 1980s owe much to the increasing sophistication of computer-aided design (CAD) techniques used to determine the optimum geometry for the helix. For instance, altering the pitch of the helix along its length maintains an optimum synchronisation between the RF wave and the electron beam. By solving the equations of motion of the electrons in the TWT and determining precisely their trajectories and momenta, optimum voltages can be assigned to the helix and collector.

Obviously the collector must be designed to dissipate the high thermal energies resulting from electron collection. For low-power TWTs this dissipation is engineered simply by thermal conduction into the TWT base plate. At higher power levels a heat exchange medium such as water or high-velocity air may be utilised for ground station tubes. The higher-power space tubes, however, provide cooling to the collector by direct radiation into space as discussed earlier. More efficient conduction from the base plate can be produced using heat pipes [see chapter A6].

7.3.5 Tube powers and frequencies

The most commonly quoted parameter of the TWT is its RF output power, measured in watts. Although the most common tube power for contemporary communications satellites is 10 or 20 W, the range is from a few watts to about 230 W. Definitions vary, but they can be loosely divided into three categories based broadly on the different construction techniques used: low power (up to about 25 W), medium power (25–100 W) and high power (over 100 W). Although there are as yet few tubes of around 200 W in orbit, powers of 50–100 W are becoming more common. As the technology becomes more mature, tube powers tend to increase. However, this parameter cannot be taken in isolation: the frequency is also important. In general, the technology to manufacture devices at lower frequencies is more mature, which means that it is easier to produce high-power tubes at relatively low frequencies. In addition, the larger dimensions of hardware designed for lower frequencies make manufacturing tolerances less critical. So, although 30 W is in no way a technical challenge for a tube operating at 4 GHz, it is close to the state of the art for a 20 GHz space tube. One of the driving reasons for the continuing development of tubes at higher frequencies is the ever-increasing overcrowding of the lower frequency bands. Although 20 GHz is the current upper limit for space tubes, the International Telecommunications Union has made frequency allocations at 45 and 60 GHz which will be the target of future-generation tubes.

It is evident that in some cases it would be preferable to use tried-and-tested TWTs rather than incur the expense of a development programme for a new high-power tube. This can be done by linking two or more low-power tubes in parallel, a technique first used on an Intelsat II spacecraft in 1967. For instance, 100 W of RF power can be provided by two 55 W tubes in parallel, and 230 W by two 125 W TWTs. Although there is some loss due to parallelling, it does not usually exceed 10%; if the practice was any less efficient, there would be too much excess thermal energy to get rid of. Parallelled tubes can provide a more forgiving system with regard to failure since it may be possible to continue the service on a lower power or switch in a spare tube in a three-for-two redundancy scheme. Providing redundancy for a system with a single high-power tube would involve carrying an equally massive, equally expensive spare. A fuller discussion of tube reliability is included in chapter A11.

7.4 SOLID STATE POWER AMPLIFIERS (SSPAs)

We have concentrated here on the TWT because the TWTA is the most prevalent satellite high-power amplifier. There is, however, a place for the solid state (semiconductor-based) device, although as yet only at low powers.

Developments in solid state power amplifiers (SSPAs) have shown them to improve the performance and reliability of the transmit section of the transponder, there being no life-limiting hot cathode or complex high-voltage power supply. In addition, the mass of an SSPA is generally lower than that of a comparable TWTA, which makes it very attractive to the payload designer. However, when selecting payload hardware it is essential to ensure a scientific comparison: reliability assessments must be based on sets of data with the same statistical significance, and mass comparisons must be made on equipment with equivalent performance (e.g. in terms of gain and output power, etc). The SSPA tends to be more linear in its output, which increases transponder capacity, but the solid state medium is a much poorer conductor of heat than the copper in TWTs, so as output powers increase larger and larger radiators are required to dissipate the inherent thermal output. TWTs can have as much as a 6.7:1 area ratio advantage over solid state devices (SSDs) [Kosmahl 1983]. At present, above about 20 W of RF output power, the excessive weight of radiator required for SSPAs militates against their use, making TWTAs the favoured option.

It would, however, be unwise to dismiss SSPAs (figure A2.8 shows a satellite with both TWTAs and SSPAs in its payload module). Although in general the SSPA is less efficient than the TWTA, efficiency quotations should be regarded with suspicion since there is no universally agreed method of measurement. A sensible method would compare all power inputs to the tube with the total power output, but some comparative calculations for TWTs ignore the power to the heater, which in some tubes can amount to 10% of the input power. Whether the quoted efficiency is applicable over the whole bandwidth or only at a single frequency is another important difference.

7.5 ANTENNAS

The antenna can be thought of as the input and output interface of a spacecraft; the interface between a free-space electromagnetic wave and a guided wave. There are many different types of antenna and many different variations on the basic types, but their mode of operation is essentially the same: a radio-frequency transmitter 'excites' electric currents in the conductive surface layers of the antenna and it radiates an electromagnetic wave. If the same antenna is used with a receiver, the converse process applies: an incident radio wave excites currents in the antenna which are conducted to the receiver. The ability of an antenna to work both ways is termed the principle of reciprocity. This proves extremely useful when it comes to the analysis of an antenna design and measurement of its performance, since the receive characteristics can be inferred from the transmit performance and vice versa. In practice, for computational analysis the antenna is generally considered in transmit mode, whereas for measurements it is assumed to be receiving [see §A7.5.4].

Four main types of antenna can be identified: wire, horn, reflector and array [see figure A7.17]. Wire antennas range from the simple dipole to more complex forms of helix and conical spiral. A horn at its simplest is a flared section of waveguide with an aperture several wavelengths wide, designed to provide a good match between the waveguide and free space (so-called *impedance matching*). Horns can be circular in cross section, rectangular in either of two orthogonal planes (called E-plane and H-plane horns), or square in section (*pyramidal horns*). They can have smooth or corrugated interior surfaces depending on polarisation requirements.

Both wire and horn antennas can be used alone, usually for wide-angle coverage of the Earth, or as feeds to illuminate reflector antennas which typically provide narrower beams. Early communications satellites carried antennas which were extremely inefficient in their use of the minimal power available from the satellite transponder: they 'sprayed' the signal in all directions at once. The theoretical device which casts an equal illumination in this way is called an omnidirectional antenna or *omni;* in practice there is no such thing as the perfect omni, but the name has stuck. As satellites improved they incorporated an antenna which could face the Earth all the time and focus its radiation accordingly; the so-called Earth coverage horn came into use. The horn antenna is ideal for this application since the Earth subtends an angle of about 17.4° from geostationary orbit. However, it is difficult to obtain a beamwidth less than about 10° and a gain greater than about 23 dB from a horn [Pratt and Bostian 1986]. This brought about the development of the reflector antenna, already used widely for earth stations, which could provide both high gain and narrow spot beams. For example, typical beamwidths for regional and national communications would be 3–4° and about 1° respectively. Reflector antennas are covered in more detail below.

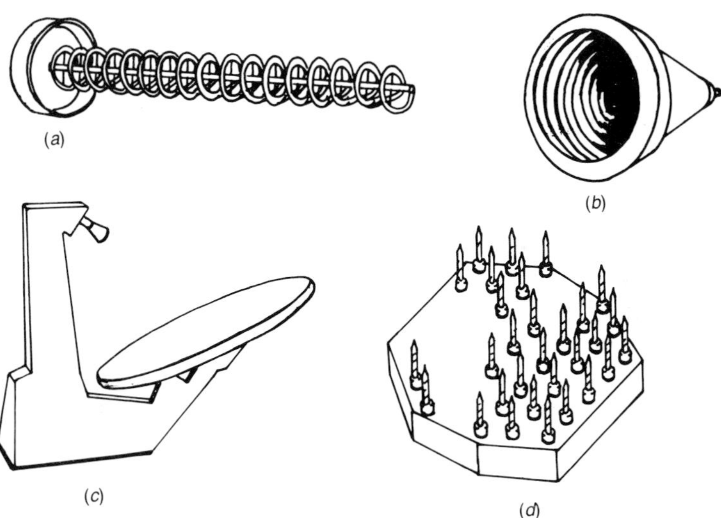

Figure A7.17 Antenna types: (*a*) wire antenna (helix); (*b*) horn antenna (conical corrugated); (*c*) reflector antenna (offset fed); (*d*) array antenna (TDRS phased array helices).

Both horns and reflectors are known as *aperture antennas.* Another type of aperture antenna is the microwave lens, or dielectric lens, which operates in a similar manner to an optical lens in that it can be designed to convert a spherical wave to a plane wave, thereby improving directivity. This type of antenna is, however, rarely used for communications satellites and will not be considered further.

Array antennas, the fourth basic type, consist of a number of radiating elements which can be electrically coupled to form a particular beam, usually a more complex *shaped beam* than would be easily attainable with a simpler antenna design. This type of antenna is often called a phased array. At its simplest the antenna may comprise a number of slots in the wall of a waveguide (a slot-array antenna), but could equally be made from a group of dipoles, helices, horns or reflectors, depending upon the frequency of operation, required beamwidth, etc. An array of feedhorns is usually used with a reflector [see §A7.5.3].

One important fact should be made clear: antenna types and applications do not have a simple one-to-one relationship. Some jobs can be done equally well by several types of antenna; which one is chosen depends on many factors. The antenna subsystem designer knows two main variables at the start of the design process: the required coverage area and the radio frequency. Based on frequency alone, the most common choice for the VHF or UHF bands is the wire antenna, used chiefly for some types of military satellite communications and some TT&C systems. For spacecraft applications at frequencies up to C-band most types of antenna can be used, but above C-band the choice is usually confined to reflector antennas, albeit with a variety of feed systems. The coverage area further constrains the choice since a narrow spot beam requires a large aperture. Intimately related to the frequency and the size and type of the antenna is the gain within the coverage area, a factor which will further constrain the design. The allocation of radio frequencies is discussed further in chapter B2; beamwidth, gain and antenna coverage are discussed in chapters B3 and B4.

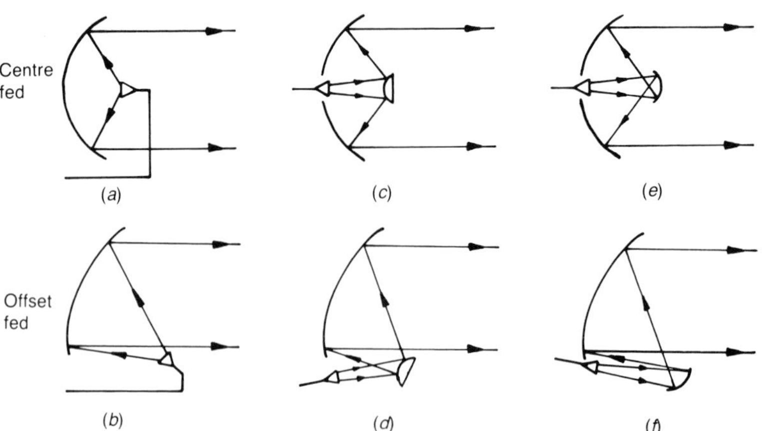

Figure A7.18 Reflector antenna configurations: (*a*) single reflector (centre fed); (*b*) single reflector (offset fed); (*c*) centre-fed Cassegrain; (*d*) offset Cassegrain; (*e*) centre-fed Gregorian; (*f*) offset Gregorian.

7.5.1 Antenna reflectors

Two different types of reflector antenna can be identified depending on the position of the feed. The most straightforward design has the feed mounted at the focus of a symmetric parabolic reflector, so that all parallel rays are reflected from the dish into the feedhorn [see figure A7.18(*a*)]. The chief disadvantage of the *centre-fed* design is *blockage* of part of the useful reflector surface by the feed and its support struts [see figure A7.19]. Obviously the larger and more complex the feed system, the worse the effects of this blockage. This and other problems, including diffraction of the incident wave, led to the development of the offset-fed reflector, a system in which the feedhorn is placed away from the principal axis of the antenna reflector [see figure A7.18(*b*)].

In most designs the horn is placed entirely outside the beam formed by the reflector, thus eliminating the blockage and diffraction effects. To maintain a single focus the main reflector must be a rather more complex section of a paraboloid, since the system is no longer symmetrical, but any difficulties in manufacture are outweighed by improvements in antenna efficiency and sidelobe performance. Both

Figure A7.19 ECS 2 during solar array deployment tests. The antenna platform supports both centre-fed and offset-fed reflector antennas (note feedhorn, secondary and primary reflectors of the three offset antennas). [ESA]

spacecraft and earth station antennas can be either offset fed or centre fed [see §A9.5]. Figure A7.20 shows an offset-fed antenna used for the Telecom 1 satellite, and figure A7.21 shows it and other antennas mounted on the completed satellite.

Figure A7.20 Telecom-1 offset-fed antenna (note multiple feed and waveguide connection). The device at the top of the tower is an 'omnidirectional' TT&C antenna. [Aerospatiale]

In some designs the required ratio of focal length to diameter (*f/D* ratio) would make for an unwieldy antenna subsystem with too great a distance between the reflector and the feed to contemplate either of the above designs. It is, however, possible to interpose a subsidiary reflector, known as a subreflector, between the feedhorn and the main reflector in a *folded-optics* configuration [figure A7.19 shows an example]. Again this is a technique used for both earth station and spacecraft antennas. Figures A7.18(c)–(*f*) show the two most common designs, derived from the Cassegrain and Gregorian optical telescope arrangements, in both centre-fed and offset-fed configurations. The Cassegrain subreflector is a section of a hyperboloid, while that of the Gregorian is a section of a concave ellipsoid. Both offset versions have paraboloidal primary reflectors like the single reflector offset design.

Figure A7.21 Telecom-1 communications satellite in cleanroom. Note the antenna subsystem [shown in figure A7.20] integrated with the antenna platform. [Matra]

For special applications where it is necessary to maintain the pointing of a spacecraft antenna more accurately than the attitude control system will allow (e.g. with the high-power narrow spot beams used for direct broadcasting) the reflector can be mounted on an antenna-pointing mechanism (APM). The APM maintains the pointing of the antenna independent of the spacecraft bus typically by monitoring the power of an RF beacon sited at a known position within the coverage area. A similar pointing device allows an antenna to be steered by ground command to enable it to illuminate an alternative coverage area. Such steerable antennas are useful for applications such as videoconferencing, for which the required coverage area could foreseeably change on a day-to-day basis. The Olympus technology demonstration satellite includes a steerable antenna in its DBS payload to enable broadcasting experiments to be conducted by different European countries.

7.5.2 Deployable and unfurlable antennas

Of course the concern with the overall size of the antenna subsystem is related to one of the most fundamental constraints on a satellite: the size and shape of the launch vehicle shroud. The requirement for narrower spot beams led to the development of larger and larger antennas and there came a point where the antenna required to produce a given spot beam was too large to fit within the shroud of the launch vehicle. The solution was the deployable antenna.

By the time communications satellites had evolved to the point where high-gain narrow beamwidth antennas were the norm, they were being manufactured in the two standard configurations described throughout this book, namely spin stabilised and three-axis stabilised. Antennas were mounted almost exclusively on one particular surface—the antenna platform. The spinner was the most constrained in this respect because of its wrap-around solar array and de-spun antenna platform, and this is still the case. Most spinners carry a single reflector, whose diameter is limited by that of the spacecraft body, which is folded down onto the platform for launch and deployed in orbit. Reflectors can, however, be deployed on booms to increase the *f*/*D* ratio, as in the case of Intelsat VI [see figure A7.22], but this design is rare as yet. Although it is common to think of reflector antennas as the only type to be deployed, examples of deployable wire antennas are those on the Syncom IV/Leasat spacecraft [see figure A7.23].

Figure A7.22 Intelsat VI deployable antennas during tests. Since they are not designed for the 1g environment, the antennas are supported by helium-filled balloons. [Hughes Aircraft Company]

Figure A7.23 Syncom IV (Leasat) spin-stabilised satellite with deployed UHF helix communications antennas and TT&C 'omni'. [Hughes Aircraft Company]

Three-axis-stabilised satellites are more flexible in that antennas can be folded against the sides of the spacecraft as well as over the antenna platform [see figures A6.4 and A10.7 for examples]. The typical deployment mechanism for reflector antennas employs a pyrotechnic bolt cutter to sever the connector which holds the antenna down against the pressure of a spring. When released, the spring forces the antenna into its carefully prearranged position. Although rather crude in concept, this method has proved very reliable in practice.

Despite the flexibility of the three-axis satellite, the solid reflector antenna is still ultimately constrained by the diameter of the launch vehicle shroud, a restriction which has led to the development of the unfurlable antenna. The simplest type of unfurlable antenna is stored and opened like an umbrella. An example 9 m in diameter flew on the ATS-6

Applications Technology Satellite launched in 1974, and the TDRS (Tracking and Data Relay Satellite) carries two 4.8 m unfurlable antennas deployed on booms to either side of the spacecraft [shown furled in figure A10.15]. Using this method, an extremely large-diameter antenna reflector can be stored in a small container and many much larger versions have been proposed, mainly for use with mobile ground receivers whose antennas must of necessity be small and of low gain. The large spacecraft antenna provides a high power density on the ground to compensate for this.

7.5.3 Feed networks

Of equal complexity to the antenna reflectors themselves are the feed systems used to illuminate them. A single feed mounted at the focus of a reflector produces a single beam which illuminates a simple circular or, more likely, elliptical coverage area. A second feed mounted to one side of the first would produce a second beam offset from the first according to the laws of geometric optics (basically, angle of incidence equals angle of reflection). Figure A7.24 shows that this process can be continued until an array of feedhorns produces a corresponding array of beam footprints on the ground. In theory each horn could be connected to a different transponder chain and transmit different signals into different beams using the same antenna reflector. The chief advantage of this type of coverage over a single beam covering the whole area is the enhanced gain of the large antenna producing the narrow spot beams. Another is that an array of spot beams allows the same frequencies to be used in different parts of what would normally be a single coverage area, so-called *spatial frequency reuse*. A practical system of this type requires a complex switch-network on the satellite to direct the telephone calls, for example, into the correct area via the correct feedhorn.

Another example of an antenna designed for frequency reuse (in this case *polarisation frequency reuse*) employs a *dual-gridded* reflector with two surfaces mounted one in front of the other and opposite polarisations at the same frequency. The front grid, which on the uplink reflects one polarity of radiation into its feedhorn, is transparent to the opposite polarity which is reflected from the rear grid into another horn. Angling the grids slightly with respect to each other allows a small separation between the horns while maintaining the same footprint for both polarities for the downlink. Dual-gridded antennas have been widely used on spin-stabilised communications satellites which generally have room for only one antenna on their top platform.

The feedhorns discussed so far have been electrically separate and the individual beam footprints have been isolated from one another. It is, however, also possible to use an array of feedhorns to produce a single beam of complex shape. Figure A7.25 shows how a cluster of five horns can be used to cover an irregularly shaped coverage area. In designing the coverage, altering the relative power levels emitted by the horns can be used to massage the shape of the beam until it provides the best fit, as the inset shows.

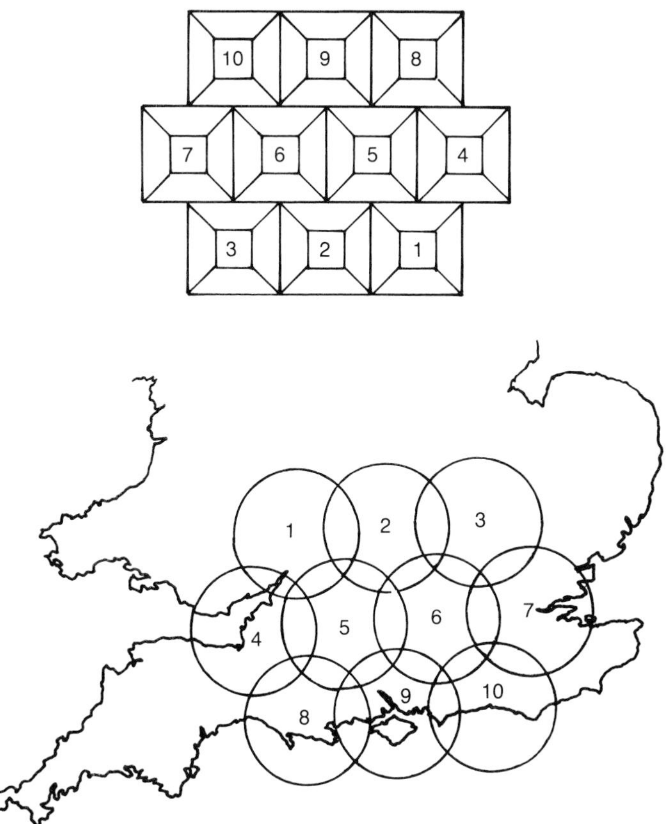

Figure A7.24 An array of feedhorns and the corresponding coverage pattern produced by a large reflector (the reader has the same viewpoint as the reflector).

7.5.4 Analysis and measurement

The incredible variety of antenna designs, which can only be touched on here, lends itself to computer-based analysis techniques. In common with other branches of microwave design, a small change in a single parameter, whether electrical or dimensional, can produce a significant difference in performance. Computer analysis allows a multitude of different designs to be considered before even the tiniest sliver of metal is cut. The basic performance of an antenna subsystem is predicted using standard theories of optics and geometry, but many additional factors germaine to the communications satellite can be easily incorporated (e.g.

pointing errors, thermal gradients, etc). Invariably, once a design has been formulated it must be adapted to conform with the constraints of the other subsystems. Chief amongst these are the limits imposed by the mass budget and the space available both on the surface of the spacecraft and within the launch shroud. Of course the designer must be aware of these factors from the outset, but if a satellite is to carry two or three separate antennas there is likely to be some natural competition for the space available and requirements are likely to change. In addition, it will be necessary to site the antenna components and choose materials for their construction with the thermal environment of the spacecraft clearly in mind. For example, the expansion of a length of waveguide can easily alter the phase length between a feedhorn and reflector and upset a system which relies on the waves from a number of horns to be in phase.

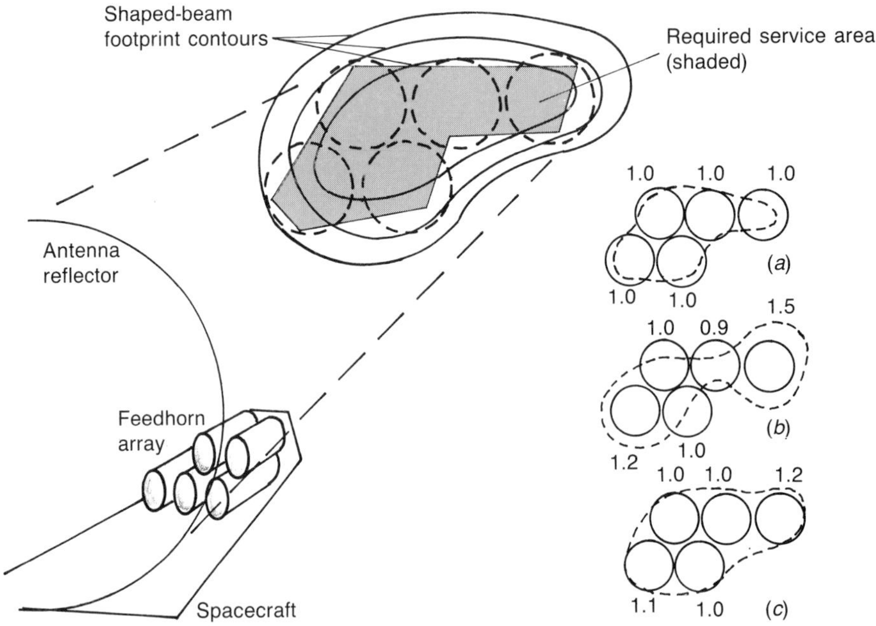

Figure A7.25 Shaped-beam generation: (*a*) equal power levels to all horns; (*b*) higher relative powers can lead to coverage 'lobes'; (*c*) adjusting power to correct levels produces required coverage.

Testing the hardware itself presents problems of its own because of the way a beam is formed by an antenna. Different, but related, beam patterns exist in what are known as the *near-field* and *far-field* zones. The point of transition between the two zones is defined by the Rayleigh distance ($2D^2/\lambda$), where D is the antenna diameter and λ is the wavelength. For example, the Rayleigh distance for a 1 m reflector antenna working at 4 GHz (λ = 0.075 m), is about 27 m; for a 1 m antenna at 20 GHz it is over 130 m. From this it is evident that the size and cost of an RF anechoic chamber for conducting antenna measurements in the far field would be excessive. Open-air ranges are, however, subject to atmospheric degradation, ground reflections, the deformation of structures due to the wind, etc, hardly a good approximation to the space environment. It is here that computer techniques come to the rescue once again.

The solution is to conduct near-field measurements in an anechoic chamber and convert the results to the far field using computer transformations. This technique has been used for long enough and is sufficiently refined to give quite accurate estimates for the far-field performance of contemporary spacecraft antennas. However, once the antennas become so large, as in the case of future unfurlable reflectors, and so flimsy that they only attain their true profile in a microgravity environment, there will be no alternative but to test them in space. Indeed, plans have been made to erect a 20 m diameter reflector mounted in the Space Shuttle payload bay. The Antenna Technology Shuttle Experiment is intended to use a free-flying RF beacon satellite which will be manoeuvred within the far field (5 – 100 km) to measure the antenna's pattern and performance. Antennas up to 55 m in diameter are envisaged, but it is currently uncertain as to when even the 20 m version will fly.

7.6 CONCLUSION

This chapter has described the constituents of the satellite communications payload. All such payloads contain the same basic components (amplifiers, filters, antennas, etc), but the differences arise in the way the transponder is assembled. This chapter has presented both single- and double-conversion transponders with examples of DBS and telecommunications payloads respectively.

The general trend in payload design involves an increasing complexity on board the satellite itself, allowing links with smaller, less complex and less expensive ground stations. Higher power tubes and higher gain antennas contribute to this trend but make ever-increasing demands on the ingenuity of the component and payload designer. And as payload complexity increases, so does the need for positive dialogue with other subsystem designers. More payload means more mass, more panel area, more power and a greater thermal output; enhanced payload capabilities may mean more accurate antenna-pointing requirements and greater TT&C capacity. It is with telemetry, tracking and command, a communications function in its own right, that the following chapter is concerned.

REFERENCES

Kosmahl H G 1983 *Microwave. RF* **22** no. 3 107–9

Pratt T and Bostian C W 1986 *Satellite Communications* (New York: Wiley) p80

Smith B L 1987 *Space* **3** no. 4 22–8

Williamson M 1986 *Phys. Technol.* **17** 218–24

A8 Telemetry, Tracking and Command

8.1 INTRODUCTION

Previous chapters have discussed the various subsystems of the communications satellite and the communications payload which they all support. There is, however, one other major satellite system, without which operation of the communications satellite would be impossible: the telemetry, tracking and command (TT&C) subsystem.

Throughout its lifetime, and especially during the critical phases, a satellite is closely monitored by the staff of a controlling earth station, partly to verify that it is operating to specification but more fundamentally to ensure that contact is never lost. The information received from the satellite is called *telemetry* (literally, measurement at a distance) and the instructions transmitted to the spacecraft are *telecommands,* to give them their full title. Additional use of the equipment provided for telemetry and command allows the position of the satellite to be determined by means of a *tracking* or *ranging* system. Various combinations of these functions give the subsystem alternative titles, such as telemetry, command and ranging (TC&R) and telemetry, tracking, command and ranging (TTC&R). Although the terms are not always used with technical accuracy, TT&C is the most common with TC&R a close second. The TT&C subsystem is, in effect, the satellite's own communications network. The telecommand function allows the Earth-based spacecraft controllers to 'talk to' the satellite subsystems and the telemetry function allows the satellite to reply. Telemetry data are gathered from the subsystems and downlinked to the controlling earth station, where analysis may give rise to additional telecommands which in turn will provide new data. Thus a control loop is established.

Although the TT&C subsystem is often described as a subset of the communications payload, and indeed may use parts of the payload hardware, the two operate independently. In the same way that a man can hold a conversation while his central nervous system manages the conduct of his body, the communications payload can function without reference to the operations of the TT&C subsystem.

8.2 OPERATIONAL REQUIREMENTS

The similarities between TT&C and the communications payload are largely self-evident. The TT&C functions require an uplink and a downlink, so analysis of the link parameters, like gain and carrier-to-noise ratio, follows a similar procedure in both cases [see chapter B4]. In terms of hardware, the requirement for transponders and antennas is similar, although radiated powers and transmission rates are lower and TT&C antennas are generally of wider beamwidth and lower gain. One fundamental difference is that the command receiver must never be allowed to switch off or fail without a back-up, since this would render the satellite incommunicado.

During the launch of a satellite, it is common for the telemetry transponder to be inactive to prevent any interference with the launch vehicle, which has its own telemetry system. As the spacecraft separates from the launch vehicle the telemetry transmitter is activated automatically in preparation for initial acquisition by the controlling earth station. This is the first opportunity to check whether the satellite has survived the launch.

Since, in the transfer orbit, the satellite adopts an attitude different to that of its final orbit, highly directional communications antennas cannot be used. For this reason a satellite's TT&C antenna system is as omnidirectional as it can be within the constraints of antenna design and its placement on the body of the satellite. The intention is that whatever the orientation of the satellite, it will be able to receive the commands to place it in the correct orbit with the right attitude. Figures A6.6 and A7.22 show examples of TT&C antennas on three-axis- and spin-stabilised satellites respectively: the former is mounted on the feedhorn tower and the latter on a deployable boom. Several other examples of both types may be found throughout the book.

For the first 20 years or so of the geostationary satellite era, it was common practice to use *omnis* operating at VHF for the acquisition and transfer orbit stages and then switch to an SHF system (typically at S-band) for the operational phase in geostationary orbit. In-orbit coverage may have been provided by the main antenna(s) of the communications payload, but separate TT&C Earth coverage horns were the more likely alternative. This system was used regardless of the main communications band, which in the early days was usually C-band, but contemporary TT&C subsystems are often designed to make use of a part of the communications band (e.g. Ku-band). There may also be the facility to use part of the communications transponder during the transfer and drift orbit phases.

For these mission phases a worldwide communications network is necessary to maintain a communications link with the spacecraft. Thus networks operating at 173 MHz (VHF) and 2200 MHz (S-band) have been established by regular users such as ESA, NASA and Intelsat. Since only large organisations can afford to run such networks, they are usually hired out from one of these bodies for the launch of an independent satellite. Once on-station, control typically reverts to a single dedicated station, usually procured as part of the satellite system contract [see chapter A9].

The following three sections describe in some detail the functions of the TT&C subsystem: the uplinked telecommand signals; the downlinked telemetry; and the tracking and ranging function.

8.3 TELECOMMAND—THE UPLINK

Throughout the previous chapters the command subsystem has been alluded to many times since most spacecraft subsystems require some degree of ground control. Once the satellite has reached the correct point in its transfer orbit, commands are required for the apogee motor firing, and once on-station for antenna and solar array deployment, attitude and orbital position changes and general payload operations.

As mentioned in chapter A4, the capabilities of largely autonomous on-board control systems are continually being improved. Despite this, there are many occasions when reaction control thrusters, momentum wheels and other actuators need to be commanded from the ground. The same is true for components of the other subsystems. For example, although much of the thermal subsystem is designed to be passive in nature, active elements such as heaters and certain types of cooling system require input from the ground controllers, as do power subsystem operations from battery reconditioning to general power distribution. As far as the communications payload is concerned, typical commands include those to initiate transponder/channel switching, the selection of redundant items and gain control to name but a few. Finally, at the end of the satellite's life, commands are sent to remove it from its position in geostationary orbit and deactivate its communications payload.

Figure A8.1 is a functional block diagram of a satellite TT&C subsystem. On the uplink side is a command receiver and a demodulator which operate in a similar fashion to the receive section of the single-conversion transponder described in the previous chapter [§A7.2.1]. The command signals themselves are modulated onto a radio-frequency carrier in much the same way as any telecommunications signal. They are, however, assembled into a predetermined *command frame structure* which the satellite is able to decode [see figure A8.2]. The frame structure described here pertains to the ESA standard but is typical of NASA systems too [Robinson 1986].

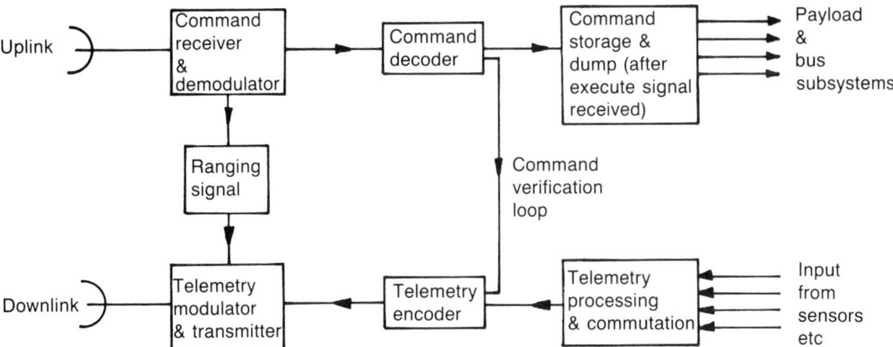

Figure A8.1 Typical communications satellite TT&C subsystem.

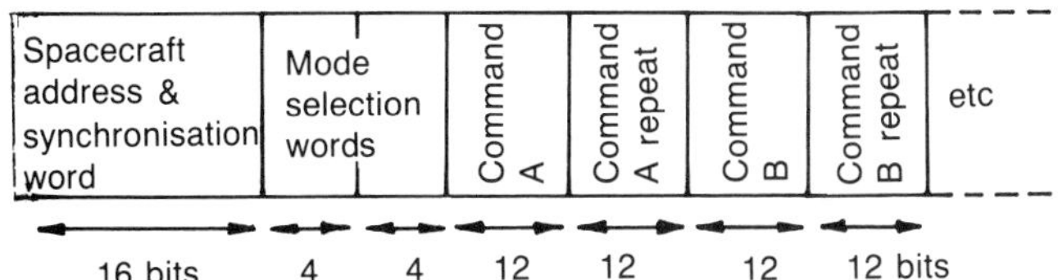

Figure A8.2 Command frame structure: commands can be either on/off commands or memory load commands—see text.

The frame begins with a 16-bit address and synchronisation word which identifies an individual spacecraft and synchronises the telecommand decoder. The system operates such that only the decoder with that particular address responds. All subsequent words are repeated to reduce the possibility of command rejection.

The outputs from the command decoders used on ESA satellites can be divided into three types: low- and high-level on/off commands and memory load commands. The low-level commands are very-low-voltage (logic level) pulses used to set or reset memories in equipment integrated circuits (ICs). The high-level commands are 12 V high-current pulses used to set magnetic latching relays, such as waveguide switches. The memory load commands consist typically of 16 bits of command data transferred serially from the TT&C memory decoder to individual IC memories in the subsystem equipment being commanded. The ESA standard frame structure allows up to 512 on/off commands or 255 16-bit memory load commands, although this can be extended by having more than one decoder. The first word following the address and synchronisation word is a mode selection word which informs the decoder whether the rest of the frame generates on/off pulses or memory load commands. The command data themselves follow this.

The precise operation of command systems vary, but commands are generally coded to prevent the acceptance of spurious or incorrect data, perhaps corrupted by noise on the uplink. This *error correction coding* ensures the suppression of any corrupted words and allows the substitution of their repeats. In addition, commands may be rerouted to the ground station to confirm that they have been properly received and decoded. Figure A8.1 shows a command verification loop through the telemetry encoder, modulator and transmitter which operates as follows. First an enabling signal is sent; then the command is sent and stored; next it is transmitted back to Earth for verification; and finally, if the command is satisfactory, an *execute instruction* is sent and the command is acted upon.

Although such systems make communicating with a spacecraft a much more reliable process, catastrophic errors can still occur. The loss of the Soviet Phobos-1 spacecraft which was en route to Mars in September 1988 provides an example of what can happen: a ground control error caused its antenna to lose *earth-lock* and the mission was irretrievably lost.

8.4 TELEMETRY—THE DOWNLINK

The great majority of engineering creations do not suffer the isolation of the satellite in geostationary orbit, and it is relatively easy to monitor their status and undertake repairs as necessary. The information available regarding the status of a communications satellite's subsystems is restricted to what can be transmitted in the limited bandwidth of its telemetry channel. Of course there are always limitations in terms of the mass and complexity of additional measuring devices, but in the end it comes down to the capacity of the telemetry subsystem. Despite this, some hundreds of 'measurement points' are provided for in the typical communications satellite, measuring such fundamental quantities as voltage, current, temperature, pressure and the position of switches and solenoids. Computer systems in the TT&C earth station monitor continually the more important parameters and sound an alarm to alert the operators if a change in status or a variation beyond set limits is detected. Telemetry transmissions are in general recorded for later analysis.

The lower part of figure A8.1 shows how the processed telemetry measurements are routed through a satellite telemetry encoder to the RF part of the chain—the modulator and transmitter. The encoder modulates the telemetry channels onto a subcarrier and the subcarrier modulates an RF carrier in the transmitter. If required, the carrier can be modulated by a number of subcarriers, each of a different frequency. One common telemetry modulation system uses frequency modulation (FM) at both stages, but there are many alternatives to the FM-FM system, using both phase and amplitude modulation [see box A7.1].

The various types of telemetry data can be categorised as housekeeping data, attitude data or payload data [Robinson 1986]. We shall look at each of these in turn.

8.4.1 Housekeeping data

Housekeeping or 'engineering parameter' data give an indication of the overall status of the spacecraft subsystems and their analysis can preempt the eventual failure of a subsystem component and often allow preventative measures to be taken. Some measurements, such as the determination of power supply voltages and currents, can be considered *direct* in that they produce the desired quantity without the need for analysis. Many others, however, are of a more indirect nature. Temperature sensors, for example, can be one of the most flexible measuring devices on a satellite. For instance, the temperature of a battery can indicate whether it is charging or discharging; the temperature of a radiator may show whether a TWT is operating or not. Since, in a microgravity environment, it is impossible to measure the volume of a liquid directly, the amount of propellant remaining in a tank is inferred from measurements of temperature and pressure.

Once the true thermal environment of a particular satellite has been characterised, by checking the actual temperatures against those predicted as part of the thermal design, any deviation can indicate the nature of a problem. For example, if the temperature of the solar arrays (on a three-axis satellite) is lower than it should be at a particular time of year, the arrays could be mispointed, perhaps due to a fault in their integral sun sensors. In cases where this explanation can be eliminated, perhaps by independent analysis, the low temperature may be attributed to something else, for instance the operating condition of the cells, the shadowing of the array by an antenna reflector, or some other effect—the information will prove useful in one way or another. Of course it may be deduced in the end that the temperature sensor itself is at fault. The analysis of cause and effect is akin to the solution of a 'whodunnit'!

Temperature measurements can also provide direct information in monitoring whether the subsystem components and the structure of the spacecraft itself are being maintained within their operating limits. This is particularly important for reaction control propellants which must be prevented from freezing, especially in the thrusters themselves, since they are usually mounted on the colder extremities of the spacecraft.

Other aspects of housekeeping data include whether an item is on or off, the mode selected, etc (its operating status); whether the prime or the back-up unit is selected (redundancy status); the deployment of mechanisms like antennas and solar arrays; and the confirmation of separation commands. In many ways the list of required telemetry data mirrors that for the telecommands, since it is always desirable to have confirmation that a command has been obeyed.

8.4.2 Attitude data

Attitude data are a particularly important category of telemetry data because of the effect the spacecraft's attitude has on the operation and performance of its communications system. Whereas the temperature of a battery will not directly affect the user of a communications system, a change in the pointing of the spacecraft's antennas most certainly will.

The relative importance of attitude data can be illustrated in terms of the required data rate alone. It is unlikely that the temperature of a radiator, the pressure in a propellant tank or the voltage produced by a battery will alter radically over a period of 10 or 15 s, so their status needs to be updated only infrequently. By comparison, the attitude of a spacecraft can change significantly in a few seconds: if a spurious impulse commands a change in attitude of only 1° per second, the satellite would have lost earth-lock and could be well on the way to an uncontrolled spin before the attitude error was telemetered back to Earth. Of course the communications system would have suffered a detectable pointing loss before this happened, the service would have been disrupted and a reacquisition programme would have been initiated, but this does illustrate the requirement for an enhanced rate for attitude data.

The data themselves may be derived from sun, earth or star sensors, gyroscopes or accelerometers and, depending on the sensor type, may be analogue, digital or a combination of both. If the spacecraft's attitude or velocity are changing quickly during a particular part of a mission (e.g. transfer orbit), a sampling rate of up to four times per second may be necessary. In fact it is during the transfer orbit that extra AOCS data may be transmitted using the payload slots in the telemetry format, since the payload remains inoperative until the satellite is on-station.

8.4.3 Payload data

Data derived from the payload itself constitute the third category of telemetry data. Although a telemetry link is important for the payload of a communications satellite, it is even more crucial for scientific and remote sensing satellites, since the *raison d'être* of their payload is the collection and transmission of data. The data rate is therefore higher for this type of spacecraft, perhaps several Mbits per second compared with a kbit per second or less for a typical communications satellite.

The payload of a communications satellite has its own bandwidth for telecommunications transmissions, but data on the operation of its components have to compete with the other satellite subsystems for use of the telemetry band. It is therefore restricted to data on its operational and redundancy status, and a number of crucial measurements of temperature and voltage, etc. In an ideal world a multitude of parameters could be monitored on a device as complex as the TWTA described in §A7.3, such as all of its operating voltages, the temperatures of its individual components, and so on. As it is, each TWT and its EPC are allowed only a few items of telemetry data. As far as the telemetry subsystem is concerned, the communications payload is just another subsystem.

8.4.4 Telemetry data handling

In addition to classifying telemetry data by their source, they can be categorised according to whether they are analogue or digital.

Digital data are of two types, bi-level and serial. Bi-level data, as the name implies, have simply two levels representing on and off and are the sort of data derived from relay contacts, separation switches, deployment mechanisms and other switch devices. Most digital data are acquired in serial form, whereby an 8 or 16 bit word is stored in a shift register in a sensor or other device and read out serially by the telemetry encoder when required. This serial reading means that a single wire can carry 16 bits of data, as opposed to having a separate wire for each bit, thus simplifying the spacecraft wiring loom or cable harness. For bi-level digital data, the individual bits are grouped into 8 bit words and injected directly into the telemetry data stream.

Analogue data are the type derived from temperature sensors, pressure sensors and devices measuring other continuously variable quantities such as power supply voltages and currents. Before analogue data can be transmitted, they must be converted to a digital standard by scaling the measurements to a common range of voltages, typically 0 to + 5.12 V, so that all determinations can be transmitted in a common pattern. Thus all values of temperature over a range of several tens of degrees and all current values varying by a number of amps are converted to values of voltage. An inverse rescaling process is conducted at the ground station to convert the data back to meaningful measurements, so that temperature, for example, can be displayed on a gauge indicating temperature rather than voltage. When the analogue telemetry has been converted to a single digital standard, it can be amalgamated with the digital data and prepared for transmission.

In designing a telemetry system it is necessary to decide on the data that can be taken from each subsystem or item of payload equipment, bearing in mind the limitations in system capacity. Individual sources are then allocated a telemetry data channel, and a *telemetry list* is compiled. This states the source of each signal, whether it was originally analogue or digital (bi-level or serial), and the required accuracy and sampling rate. The form of the list governs how the data channels are assembled into a telemetry format for transmission [see figure A8.3].

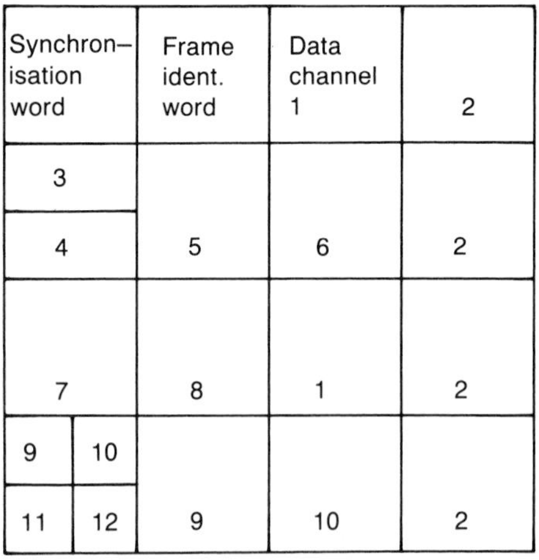

<table>
<tr><td colspan="2">Synchron–isation word</td><td>Frame ident. word</td><td>Data channel 1</td><td>2</td></tr>
<tr><td colspan="2">3</td><td rowspan="2">5</td><td rowspan="2">6</td><td rowspan="2">2</td></tr>
<tr><td colspan="2">4</td></tr>
<tr><td colspan="2">7</td><td>8</td><td>1</td><td>2</td></tr>
<tr><td>9</td><td>10</td><td rowspan="2">9</td><td rowspan="2">10</td><td rowspan="2">2</td></tr>
<tr><td>11</td><td>12</td></tr>
</table>

Figure A8.3 Telemetry format: one frame of 16 8-bit words.

The figure shows a simple example in which a frame of 16 8-bit words is transmitted (from top left to lower right) and then repeated. The frame begins with a synchronisation word which identifies the start of the frame to the ground receiving equipment. The next word identifies the spacecraft and, for more complex formats, the number of the frame in the sequence of several frames. The rest of the frame comprises the actual data, the numbers referring to the different telemetry channels.

The process of sampling different channels in turn is called *commutation* and the inverse process *decommutation*. It is, in effect, a form of time division multiplexing (TDM). A channel which is sampled more than once in the same frame is said to be *supercommutated* and one which is sampled less than once per frame is *subcommutated*. A continuous transmission of frames builds up the required sampling rate for all telemetry sources. For example, figure A8.3 shows that data channel 2 requires twice the data rate of channel 1, and channel 1 twice that of channel 6. Channels 3 and 4, on the other hand, are sampled every other frame, while 11 and 12 are sampled every fourth frame. Channels 9 and 10 are sampled once in three out of four frames, but twice in the fourth frame. The design of formats such as this brings a great flexibility to telemetry sampling.

8.5 TRACKING AND RANGING

The third function of the TT&C subsystem is the determination of the position of the satellite, first introduced in chapter A4. The function described here is one in which the position data gathered from a single controlling earth station are used to *maintain the position* of a satellite. This should not be confused with the other type of earth station tracking, whereby earth stations—perhaps in great numbers—*follow* a satellite's movements to maintain it at the peak of their respective antenna beam. The main types of antenna-pointing system will be described in §A9.7.

One method of position determination is known as *angle tracking*. A satellite in a low orbit generally stores the data collected over a part of its orbit and then transmits them to a ground station on command in what is termed a *data dump*. In this case the downlink (telemetry) carrier, as received by a number of ground stations, can be used to measure the changing elevation and azimuth of the satellite [see figure A8.4]. Successive measurements are used to compute the satellite's orbital parameters, and the more ground stations there are the more accurate the estimate.

The satellite in geostationary orbit is of course nominally stationary with respect to the ground station, so angle tracking alone is not sufficiently accurate. Therefore, in addition to tracking a telemetry carrier or *beacon*, an accurate determination of the satellite's range is made. The knowledge of how far the satellite is from the earth station allows its orbital height to be determined, by three-dimensional trigonometry, and this indicates its position with respect to the known altitude of geostationary orbit. If the satellite is at the correct altitude and coplanar with the equator it will be geostationary. The range of a satellite in geostationary orbit can be measured, using carriers at S-band or higher frequencies, to an accuracy of about 10 m. Impressive though this is, considering the distance involved, experimental laser reflectors placed on the Meteosat P2 weather satellite (launched in 1988) have allowed its range to be measured to within 5 or 10 cm!

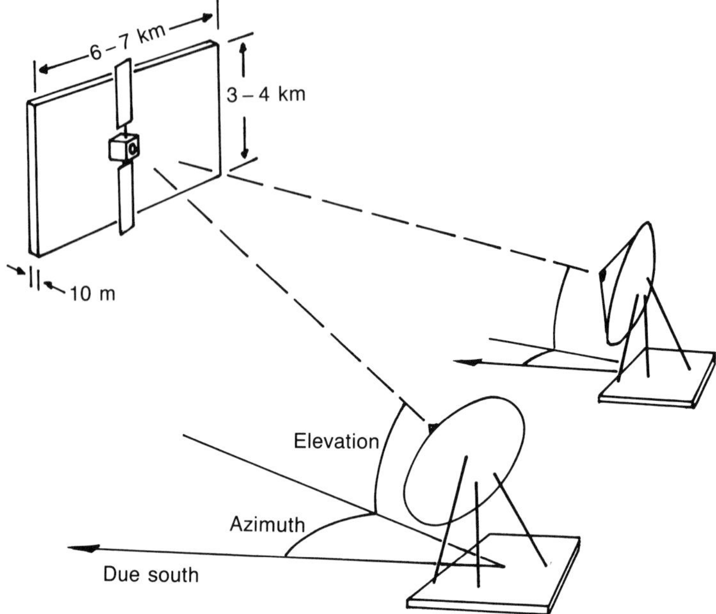

Figure A8.4 Angle tracking; the tracking error box.

A related measurement which can be made is the *range rate*, the rate at which the range changes, which can be determined by measuring the Doppler shift of the signals received from the spacecraft. If the frequency of the signal is shifted to higher frequencies the satellite is moving towards the earth station, and if the frequencies are lower it is receding. This is particularly important for the transfer orbit and for major station-keeping manoeuvres, when the range is changing significantly, but for satellites in geostationary orbit it is normally used only to complement the range measurement, and only then if it can be measured with sufficient accuracy.

Ranging is conducted simply by transmitting a signal to the satellite in the command uplink and arranging for it to be retransmitted on the telemetry downlink. Accurate measurement of the signal travel time determines the range. The process is similar to that of radar, except that the signals are not passively reflected by the subject but actively retransmitted.

In a practical ranging system the uplink carrier is phase modulated with pairs of signals known as *frequency tones* or *side tones*. The tones are detected in the satellite command receiver and passed to the telemetry transmitter which remodulates them onto the telemetry carrier [see figure A8.1]. The earth station compares the relative phases of the transmitted and received tones and calculates the range: the phase lag of the received tone is proportional to the distance travelled. The highest frequency (or major) tone determines the ultimate accuracy of the range measurement and the lower frequency tone resolves any ambiguity in the measurement due to the range being greater than the wavelength of the major tone. Figure A8.5 shows this schematically, using vertical lines to represent the two frequencies. A cursory examination of the high-frequency pulses shows the received signal to be lagging behind the transmitted signal by about half a wavelength. However, the shift in the low-frequency pulses shows that the lag is actually more like one and a half wavelengths. Having established the true order of the delay, the high-frequency pulse separation can be re-examined to give a more accurate range measurement.

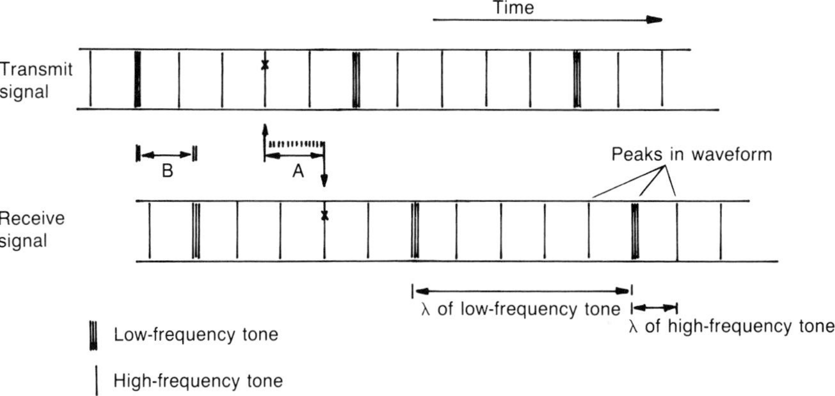

Figure A8.5 Ranging tones; high-frequency tone (A) gives an accurate measure of the frequency shift between the transmitted and received signals; low-frequency tone (B) removes any ambiguity in the number of wavelengths (represented schematically by vertical lines).

To make the best possible use of the limited number of orbital slots in geostationary orbit, it is necessary to confine the satellite to as small a volume of space as practically possible. This limitation of the station-keeping *dead-bands* requires highly accurate determination of the satellite's position. In fact the *tracking error box* of a single earth station [see figure A8.4] is typically 6–7 km long, a few kilometres wide and about 10 m deep [Soop 1980]. (Note: this should not be confused with the station-keeping box described in chapter A4 which depicts the actual movement of the satellite, not the uncertainty in its position.)

Accuracy is not only affected by the manufactured surface quality of the tracking antenna, but also by environmental factors such as thermal distortion by solar heating, distortion due to wind pressure and snow loading, etc. Despite this, an evaluation of the accuracy of a single year's tracking data from ESA's Redu ground station, used with the ECS 1 satellite, showed that most of the pointing errors lay within the manufacturer's specification of 10 millidegrees [Kawase and Soop 1986]. This is equivalent to about 6 km at geostationary height. If more accurate measurement of elevation and azimuth angles than is possible with a single earth station is required, then two earth stations must be used to provide triangulation. This is of course more expensive and therefore less common.

Increasingly important is the worldwide network of laser tracking stations which provide highly accurate range measurements for satellites equipped with optical retro-reflectors. A retro-reflector is an arrangement of three mirrors, set in mutually orthogonal planes, which reflect incident radiation in the direction of its origin. Most practical systems comprise an array of retro-reflectors set in a panel, typically mounted on the Earth-facing platform of the spacecraft. The distance is derived from the propagation time of a laser pulse from the tracking station to the satellite and back, giving a nominal range accuracy of around 10 cm [Holdaway 1986]. Similar arrays were deployed on the lunar surface by the Apollo astronauts and these have allowed the Earth–Moon distance to be measured to within a metre.

In addition to its ground-based tracking systems, NASA is in the process of establishing a tracking and communications network in geostationary orbit, known as the tracking and data relay satellite system (TDRSS). As well as providing communications links between the ground and spacecraft in low Earth orbits, which are poorly covered by the ground networks, the system is designed to provide range and range rate (Doppler) data to at least the same accuracy as those available from ground stations. However, TDRSS is unable to provide these services to the majority of satellites described in this book, those which are themselves in geostationary orbit, and the establishment of a space-based tracking network for communications satellites is not even over the horizon, let alone on it.

8.6 ONBOARD DATA HANDLING

In line with the general move to place more 'intelligence' on the satellite is the development of an onboard data-handling (OBDH) system, an example of which is being standardised for use on ESA spacecraft. The ESA system comprises a data bus of two screened twisted-pair signal lines, called the interrogation bus and the response bus, connecting a bus controller with up to 31 remote terminals [Robinson 1986]. The terminals can be either spacecraft subsystems or experimental payloads, depending upon the type of spacecraft. Instructions sent down the interrogation bus, past all the remote units, are prefaced by the address of a particular terminal and picked off by that terminal. If the instruction is a command, the terminal obeys the command and sends back, along the response bus, an acknowledgment that the command has been received. If the instruction is a request for telemetry, the unit sends the required data to the central controller, which provides the interface between the subsystems and the RF part of the telemetry and command system.

The advantages of the onboard system include the potential to reduce mass and wiring loom complexity and to allow the growth of a TT&C system without the need for substantial redesign. Assuming the capacity of the system would not be exceeded, another terminal could be 'plugged in' at a relatively late stage in the design process or, in the case of retrievable spacecraft, even after launch.

Developments are under way to distribute system intelligence amongst the terminals so that they could send telemetry when it becomes available or necessary, rather than having to wait for a time slot stipulated by the central controller. The extent to which these systems are to be used on the average communications satellite in future remains to be seen.

8.7 CONCLUSION

This chapter has described the communications-based functions of the satellite's service module. The telemetry, tracking and command subsystem allows the ground controllers to monitor the status of the satellite, determine its position and transmit instructions to correct any errors or alter the status of its subsystem equipment. The following chapter discusses the ground-based part of the link.

REFERENCES

Holdaway R 1986 *Spacecraft Engineering Course Notes* (University of Southampton) ch. 5

Kawase S and Soop E M 1986 *ESA J.* **10** 71–83

Robinson J 1986 *Spacecraft Engineering Course Notes* (University of Southampton) ch. 13

Soop M 1980 *ESA J.* **4** 159–69

A9 Earth Stations

9.1 INTRODUCTION

The hardware comprising the space segment of a satellite communications system has been discussed in previous chapters. Equally important, of course, is the earth segment, a major part of which is the earth station itself. On a point of terminology, the terms 'ground segment' and 'ground station' are entirely interchangeable with 'earth segment' and 'earth station' respectively.

9.2 SATELLITE REQUIREMENTS

If we consider an operational communications satellite in geostationary orbit, it is evident that it requires links with an earth station for two separate functions: telemetry, tracking and command (TT&C) and the communications link itself [see figure A9.1].

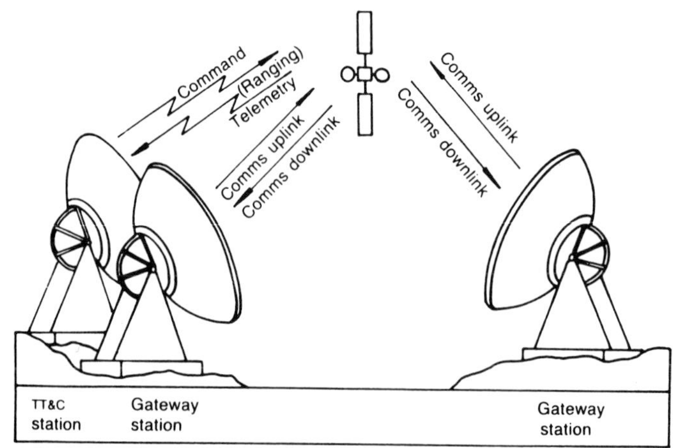

Figure A9.1 Earth station functions: TT&C and communications links.

The TT&C function, described in the previous chapter, is usually provided by a dedicated earth station which can be either co-located with a communications uplink station or completely separate. The size of the TT&C station is governed largely by the uplink (telecommand) performance requirements which are dominated by the need for high reliability: if the system is to operate in all meteorological conditions, a good link margin is a necessity. Thus TT&C earth stations generally have antennas of at least 10 m in diameter, which is also more than sufficient for telemetry reception. In fact, stations for telemetry reception only can be as small as 3 m in diameter.

Secondly, and perhaps most fundamentally, the satellite system needs an uplink station for communications traffic. The station design depends on the type and amount of communications traffic it is required to carry: in the same way that both four-seater light aircraft and wide-body jets can fly from London to Paris, the chosen 'carrier' depends on the traffic density and standard of service required. For communications trunk routes, like the transatlantic link, a large earth station (up to 30 m in diameter) known as a *gateway station* would be the norm, but for routes with lower traffic densities advances in technology have made it possible for relatively small earth stations (with antennas only 3–5 m in diameter) to provide an uplink.

The ground segment hardware required for a satellite downlink is somewhat more varied because of the diversity of communications satellite applications. Of course, to be pedantic, the satellite *requires* no ground segment hardware on the communications downlink: it will transmit whether there is anyone listening or not! In practice, the downlink earth stations vary from the huge gateway stations to the tiny satellite terminals on ships and boats, now increasingly finding application on aircraft and other 'mobiles'. The different types of earth station authorised for use with Intelsat, Eutelsat and Inmarsat spacecraft are listed in box A9.1.

Although most earth stations are required both to transmit and receive, there are applications for stations which only transmit (e.g. for TV broadcast satellite uplinks) and there is a large and expanding market for the so-called receive-only terminal, particularly for the reception of satellite television. Since the transmit-only and receive-only earth stations can be considered a special case, the majority of this chapter will deal with stations capable of both transmission and reception, best illustrated by those with antenna diameters of at least 5 m. Small earth stations are discussed briefly in §A9.8.

9.3 EARTH STATION REQUIREMENTS

One of the main requirements of an earth station is high directivity, that is high gain in the direction of the wanted signal and low gain in the direction of any unwanted signals or *interference*. This is best realised

BOX A9.1: EARTH STATION STANDARDS

A variety of standard earth stations have been authorised by the International Telecommunications Satellite Organisation (Intelsat) to operate within the system. There are three standards for international gateway stations:

Standard A, operating at C-band, originally specified antennas 30 m in diameter or larger, but this has been reduced to 18 m as a result of the general increase in satellite radiated power densities;
Standard B, operating at C-band, for antennas 11 m in diameter or larger, handling lower traffic densities than Standard A;
Standard C, operating at Ku-band, originally for antennas 14–19 m in diameter or larger (now 13–15 m).

There are also a number of standards for smaller earth stations:

Standard D1: 4.5 m diameter at C-band (for use within the 'Vista' low-density SCPC telephony service);
Standard D2: 11 m diameter at C-band (for use within the 'Vista' low-density SCPC telephony service);
Standards E & F: 3.5 – 9.0 m diameter at Ku- and C-band respectively (for use within the international Intelsat Business Service (IBS));
Standard G: wide range of earth terminal sizes down to 0.8 m *microterminals* (for use with the *Intelnet* DATA distribution service);
Standard Z: 3 – 18 m diameter at C- and Ku-band (for use with Intelsat domestic lease services).

EUTELSAT, the European Telecommunications Satellite Organisation, has three standards for use with their Ku-band spacecraft: 15 – 18 m antennas (similar to Intelsat Standard C) for trunk telephony and the Eurovision network; 8 – 12 m antennas for TV transmission; and 2 – 5 m antennas for the Satellite Multiservice System (SMS), TV reception and outside broadcast from transportables.

INMARSAT, the International Maritime Satellite Organisation, has two standards for its L-band ship earth stations (SES): Standard A for the majority of ships (0.85 – 1.3 m antenna above deck, transmitter and receiver below); and Standard C for smaller boats and other *mobiles* (shoebox-sized terminal and 10 cm *omni* antenna). Coast earth stations (CES), operating at C-band, are of Intelsat *gateway* standard. Digital versions of Standards A and C, known as Standards B and M, were announced in late 1989.

by using a large-diameter antenna which, by its nature, has a narrow beamwidth and a high gain [see chapter B3].

The requirement for a high gain relates, of course, to the huge distance between the earth station and the satellite. On the uplink, a large antenna combined with a high-power amplifier (HPA) provides a signal of sufficient strength to be received by the satellite. The satellite transponder, described in chapter A7, reamplifies the signal for the downlink leg, but limitations on the DC power available to satellite equipment means that the RF power density of the downlink signal is necessarily lower than that of the uplink. The weakness of the downlink signal favours the larger earth station antennas, since they intercept a greater percentage of the highly dispersed satellite beam. In addition, a larger antenna is more effective in rejecting interference from other satellites because of its narrow beamwidth. This is not to say that a large earth station is mandatory for satellite communications—we have already said that it depends on the traffic. It does, however, provide the high carrier-to-noise ratio required for trunk telecommunications and the gain margin required to overcome high rain attenuation, thus providing the continuity of service expected of such links. In fact, gain increases with the antenna diameter and beamwidth decreases proportionally. Naturally, one of the disadvantages of the larger antenna is its higher cost, which increases in proportion to a factor between the square and the cube of the diameter [Pratt and Bostian 1986].

Hand in hand with the high-gain requirement is the need for the communications equipment to introduce as little noise as possible to the system. The low noise-temperature requirement of the earth station amplifiers and the technology of the antenna system relevant to noise reduction are discussed in §§A9.4.2 and A9.5.1 respectively.

Another important requirement for an earth station is that it should be accurately pointed towards the satellite, and the narrower the beamwidth of the antenna the more critical this requirement becomes. The apparent motion of a satellite in geostationary orbit, as viewed from the ground station, has been described in chapter A4. Large earth stations need to track the satellite continually to ensure not only that the spacecraft does not wander out of the uplink beam, but also that it remains at or near the peak of the beam in order to achieve the maximum gain from the earth station antenna. The requirement here can involve holding the gain steady to within a few tenths of a decibel which, in the case of a 30 m antenna operating at 4 GHz, would dictate a pointing accuracy of 0.01° [Claydon 1987]. The pointing accuracy of an earth station antenna depends partly on the stiffness of its mount and its ability in resisting wind deflections, but mainly on the accuracy of the tracking system [see §§A9.6 and A9.7 respectively].

9.4 COMMUNICATIONS EQUIPMENT

9.4.1 Transmit/receive chains

Since the earth station transmitter and receiver carry the same signal as the satellite, it comes as no surprise that their configuration is similar to that of the spacecraft transponder. Figure A9.2 is a block diagram of a typical earth station system configured for both transmit and receive functions. On the far left is the interface with the terrestrial telecommunications network which accepts inputs by landline or microwave link and converts them to a suitable form for transmission to the satellite. This typically involves multiplexing the various channels onto a subcarrier which modulates an intermediate frequency (IF) carrier generated by the earth station equipment (typically at 70 MHz or 140 MHz). Following modulation [see box A7.1], the IF is amplified and then mixed with a local oscillator frequency in the upconverter to produce the uplink carrier frequency. The carrier is amplified by the HPA, filtered, and routed through the waveguide feed to the earth station antenna. This process is analogous to that which occurs in the transmit chain of the typical double-conversion satellite transponder, as depicted in figure A7.4.

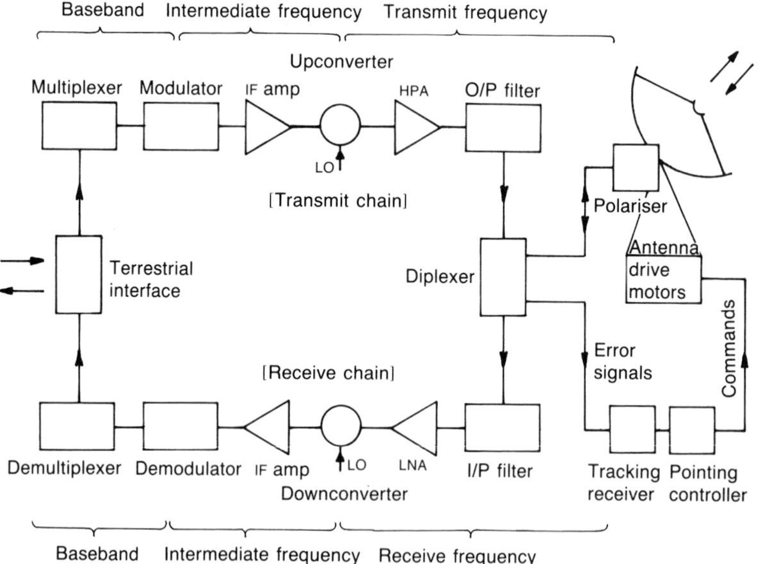

Figure A9.2 Typical earth station block diagram (including tracking).

On the downlink, the antenna routes the signal through an input filter, a low-noise amplifier and a downconverter (which converts the RF carrier to the low frequency IF) to an IF amplifier. The signals pass from there to a demodulator which extracts the baseband signals, then to a demultiplexer which extracts the required channels, and finally to the terrestrial network. Between the antenna and the diplexer, which separates the transmit and receive frequencies, is a dual-band polariser designed to provide a high discrimination between orthogonally polarised signals. Some systems have a separate polariser for each of the receive and transmit chains located between the diplexer and the respective filter. This arrangement offers active polarisation control to counteract the effects of changing propagation conditions through the atmosphere. For simplicity, figure A9.2 shows transmit and receive chains for one polarisation only: if two polarisations are to be used, the diplexer is replaced by an orthomode junction (OMJ) to separate the frequency bands, and an orthomode transducer (OMT) on its input and output sides to separate the orthogonally polarised signals for the double transmit and receive chains.

The figure shows a typical arrangement for a telecommunications application in the fixed-satellite service (FSS). In the broadcasting-satellite service (BSS), the uplink earth station transmits only a limited number of TV channels and receives nothing for routing to the terrestrial network, since the downlink is to individual receive-only terminals located at the users' premises.

9.4.2 Amplifiers

The two main amplifiers in an earth station are the low-noise amplifier (LNA) and the high-power amplifier (HPA). The LNAs in the early large earth stations were generally parametric amplifiers, or 'paramps', which had to be cryogenically cooled to produce the required low noise temperature (typically 15 K at 4 GHz [Maral and Bousquet 1986]). They required frequent maintenance and, as a consequence, operating costs were high. Subsequent improvements in antenna efficiency and other aspects of the design made the noise temperature requirements less stringent and thermoelectric cooling (with noise temperatures of around 35 K) became the norm for the large stations. For the smaller domestic stations uncooled paramps at about 50 K are now considered sufficient.

More recent improvements have brought the GaAs FET amplifier into the fray. It is more compact, more reliable, cheaper and requires virtually no maintenance, which makes it ideal for the smaller domestic stations. Typical noise temperatures at 4 GHz are 50 K (cooled) and 75 K (ambient). For smaller earth terminals still, the LNA has been combined with a downconverter in a compact unit known as an LNC (low-noise converter) which is usually attached to the antenna itself.

HPAs for earth stations, sometimes referred to simply as power amplifiers, are similar in function to those used in spacecraft. The main differences are the higher power requirement and the relative ease of maintenance. The choice for large earth stations, transmitting for example multi-carrier TV and telephony, is between the klystron and the travelling wave tube (TWT), both of which are capable of producing output powers from several hundred to several thousand watts. Figures A9.3 and A9.4 illustrate typical examples of the two types of earth station HPA. The respective devices can be differentiated by the bandwidth they are capable of handling: perhaps as low as 40 MHz for the klystron and as much as 500 MHz for the TWT for the lower frequency bands. The bandwidth of the service handled by the earth station has a bearing on the design of the transmitter as follows. Using klystrons, individual channels may be amplified separately by individual devices and then combined before connection to the antenna feed; with TWTs, the carriers can be combined at low power and amplified by a single wide-band TWT.

Figure A9.3 An earth station klystron. [Thomson CSF]

For a small terminal transmitting a single carrier (e.g. up to 256 kbits per second), a solid state (FET) amplifier with an output power of only a few watts may be sufficient. The development of solid state devices for ground use is similar to that for space use in that devices are as yet available only for relatively low powers (several tens of watts, dependent upon the frequency).

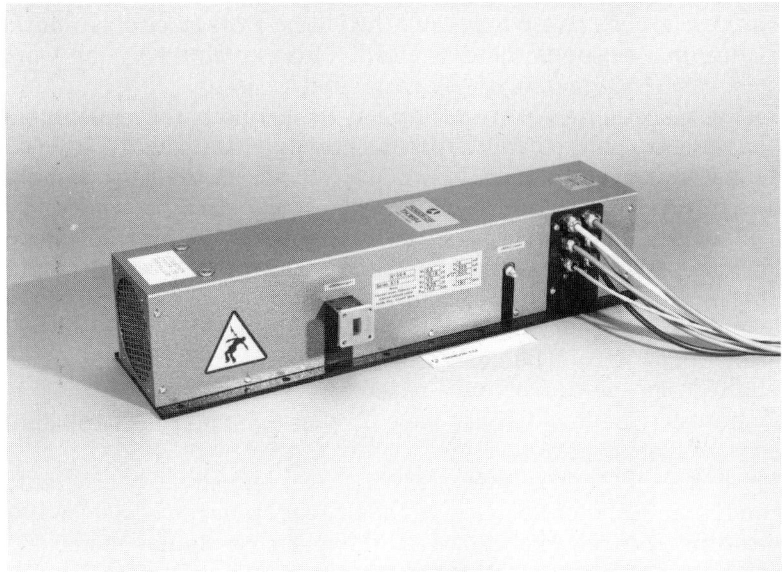

Figure A9.4 An earth station travelling wave tube. [Thomson CSF]

Apart from the advantage of accessibility for ground-based HPAs compared with space-based devices, cooling is much easier since both air- and water-cooled systems are available. Power supplies for earth stations are primarily derived from the national grid, but automatically activated diesel generators and/or batteries are usually provided in case of a mains power failure.

9.5 ANTENNA TYPES

Almost without exception, antennas for satellite ground stations have reflectors with a paraboloidal cross section, largely because horn antennas have limited gain and phased arrays are, as yet, too costly. Thus the ubiquitous upturned dish, commonly known as a parabolic reflector, has become an everyday sight throughout the developed world. There are, however, several variations on the theme. Like the spacecraft reflector antennas discussed in §A7.5, earth station antennas can be either centre fed (axisymmetric) or offset fed (asymmetric).

9.5.1 Axisymmetric antennas

The centre-fed design is the best known and the most widely used because of its straightforward geometry and mechanical simplicity. The circular symmetry of the reflector makes its construction easier and therefore cheaper than asymmetric designs. This tendency is reflected in the relative simplicity of the antenna mount, which supports and steers the antenna as a whole, and the lattice-work of struts or *backing structure* that supports the dish itself. This explains why the world's largest earth stations are of the centre-fed variety.

The simplest form of this design has its feedhorn at the prime focus of the reflector (the so-called front-fed configuration), but the larger the antenna the longer the run of waveguide necessary to link the feedhorn to the receiver and/or transmitter. It is now generally considered that the RF losses incurred in this arrangement are too high for antenna diameters greater than about 3 m. This was not always the case, as the first large earth station to be installed at Goonhilly Downs in Cornwall, England, in 1962, shows: its reflector is 26 m in diameter and fed by a prime focus horn. The design of Goonhilly Aerial 1 was developed from the large aperture antennas used for radio astronomy and it is the only front-fed antenna to have been used as a Standard A earth station [Pratt and Bostian 1986].

Most large antennas now use the dual-reflector system, with a secondary reflector at the focus of the primary reflector and a feedhorn, protruding through the centre of the primary, at the focus of the secondary. As with spacecraft antennas, the dual-reflector designs are of two main types, Cassegrain and Gregorian, based on the optics of the

Figure A9.5 ESA earth station at Villafranca, Spain: centre-fed Cassegrain dual-reflector design. [ESA]

astronomical telescopes of the same name. The Cassegrain has a hyperboloidal subreflector located within the focus of the paraboloidal main reflector, and the Gregorian has an ellipsoidal subreflector outside the focus of the main reflector [see figure A7.18]. Both systems provide similar RF performance, but the Cassegrain design is by far the most common [see figures A9.5 and A9.6].

Figure A9.6 Earth station antenna with Cassegrain optics and beam waveguide. Note also terrestrial microwave link antennas.

Although housing the RF equipment close to the back of the primary reflector and reducing the waveguide run has the advantage of reduced RF losses, this arrangement is not always convenient, especially where a large amount of equipment would have to be rotated with the earth station antenna. The solution is the beam waveguide, an enclosed set of reflectors which channel the transmitted wave from a feedhorn at ground level, using free-space rather than guided-wave techniques, to the subreflector (the earth station in figure A9.6 has a beam waveguide). Both the two- and the more common four-reflector systems are possible as shown in figure A9.7. Both arrangements allow free rotation of the antenna in azimuth and elevation by aligning the optics along these axes as shown.

The performance of a centre-fed dual-reflector design is, however, compromised because the subreflector and its support structure obscure part of the main reflector—so-called *aperture blockage*. Although the dual-reflector design is preferred for large- and medium-sized antennas, its advantages disappear when the diameter of the main reflector is less than about 60 wavelengths (λ). Since the subreflector must be at least 8λ in diameter to prevent diffraction around it, there comes a point where it obscures an appreciable proportion of the main reflector, thus reducing its efficiency. Therefore, for small earth stations, sometimes defined as those with apertures less than 60λ across, the front-fed configuration is the most common design solution since a feedhorn creates less blockage than a subreflector.

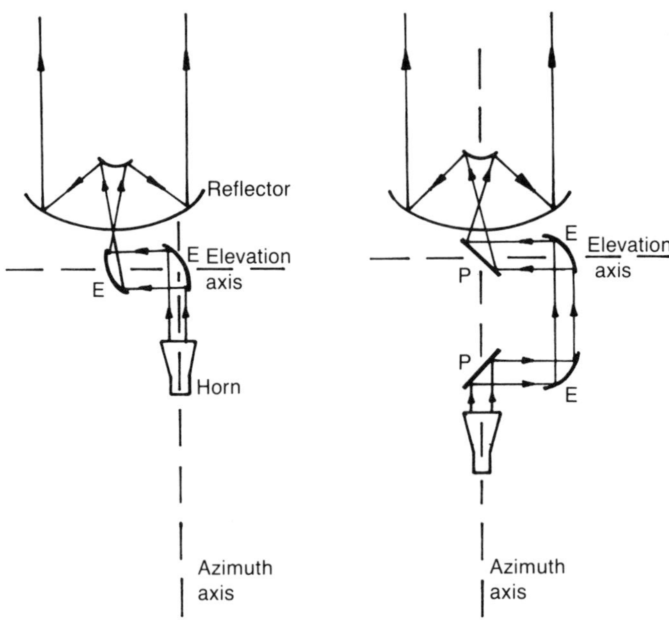

Figure A9.7 Beam waveguide configurations: P, plane reflector; E, elliptical reflector (or offset paraboloid).

The diffraction of the incident radiation around the obstructions also degrades the antenna radiation pattern. An antenna's radiation pattern illustrates its directional sensitivity when receiving and the relative strength of the radiated signal when transmitting. This equivalence of antenna properties on receive and transmit is known as the principle of reciprocity. Figure A9.8(*a*) shows a simplified radiation pattern, or amplitude distribution, for a small antenna and figure A9.8(*b*) shows the same pattern in polar coordinates to indicate better the directional sensitivity of the antenna. Of course, in reality the pattern exists in three dimensions.

The patterns comprise a main lobe and a number of sidelobes. Figure A9.8(*a*) is plotted logarithmically to emphasise the details of the sidelobes, which on a linear scale would be very small. The main lobe defines the antenna's beamwidth, as discussed in chapter B3. Most of the antenna's radiated energy is in the main lobe, but some is diverted to off-boresight directions. If the earth station is transmitting, this means that some of its power is going to waste and, worse, may cause interference with another spacecraft. Since there is only a finite amount of transmitter power available, any power lost to the sidelobes causes a reduction in the peak gain, typically between 0.1 and 0.5 dB [Claydon 1987]. On receive, the sidelobes may collect interfering signals from other satellites.

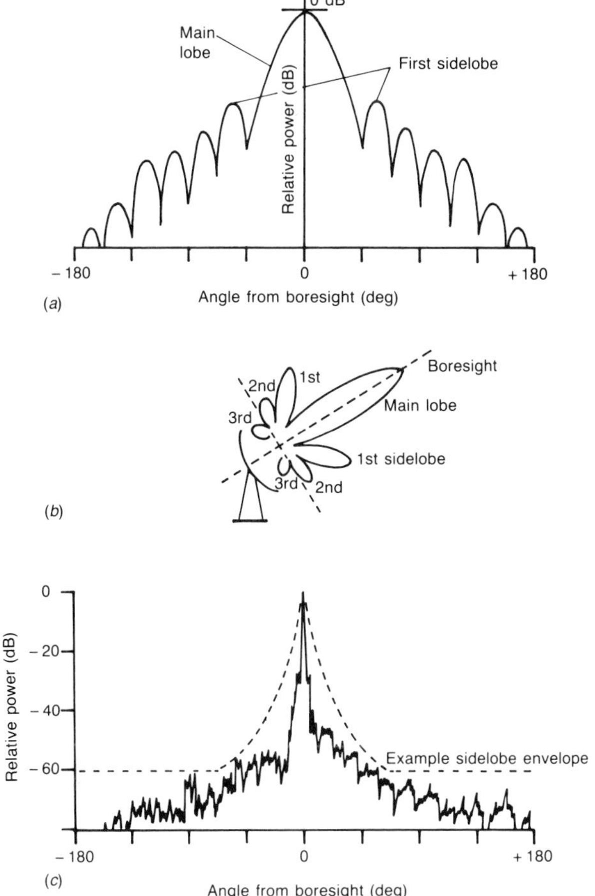

Figure A9.8 Antenna radiation patterns: (*a*) conventional linear plot (simplified); (*b*) pattern in polar coordinates; (*c*) example pattern for a 3 m earth station.

Figure A9.8(*c*) shows a more realistic pattern (for a 3 m Ku-band antenna) with a highly directional main lobe and a far more complex sidelobe structure. Sidelobes are due to diffraction around the circular edge of the antenna (which is analogous to the diffraction of light passing through a small circular aperture). All antennas have sidelobes to some degree and their reduction is one of the prime considerations in antenna design. For earth stations installed after 1987, the sidelobes are constrained according to the rule that the gain, G, of 90% of the sidelobe peaks should not exceed a value given by the expression $G = 29 - 25\log_{10}\theta$ (dBi) where θ is the angle from boresight. This rule, for what is known as sidelobe performance, applies to any off-axis direction between 1° and 20° [Maral and Bousquet 1986]. It is meant to be applied to earth station antennas greater than 100λ in diameter as part of a policy to reduce the nominal spacing of satellites in geostationary orbit to 2°, but a modified specification, which includes a term dependent upon the diameter, is used for smaller antennas. The rule is also primarily concerned with transmitting earth stations because of their ability to interfere actively with other services; for receive-only stations it is only a recommendation.

An additional problem is *spillover* which can occur with any antenna arrangement. The feed at the prime focus of an antenna simply wastes transmitted power if its radiation pattern causes power to spill over the edge of the reflector [see figure A9.9(*a*)]. When receiving, having part of the ground in the feedhorn's field of view causes an increase in noise temperature in the receiver, which reduces the overall carrier-to-noise ratio of the signal [see chapter B4].

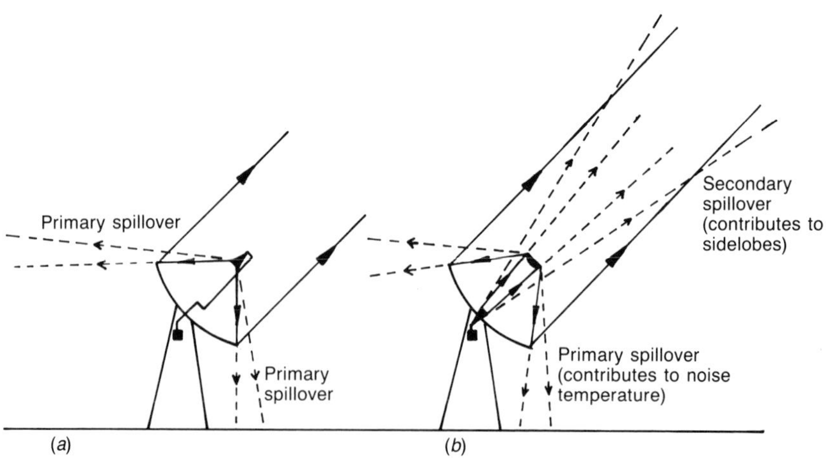

Figure A9.9 Antenna reflector spillover (antennas depicted in transmit mode). (*a*) Prime focus feed; (*b*) dual-reflector antenna.

For a dual-reflector antenna any radiation not intercepted by the subreflector causes an increase in the amplitude of the sidelobes (a degradation in sidelobe performance and a consequent loss of efficiency), since the beamwidth of the feedhorn alone is greater than the antenna as a whole [see figure A9.9(*b*)]. Spillover of the primary reflector may also occur as with the front-fed design. However, when receiving, the feedhorn views the sky rather than the ground, so any spillover round the secondary has less effect on noise temperature.

Spillover can be reduced, as with most other degradations, by careful design, first by ensuring that the feedhorn itself produces low sidelobes and secondly by increasing the percentage of energy intercepted by the reflector or subreflector. The reduction of the sidelobes, typically achieved using a corrugated or dual-mode feedhorn, is quite straightforward, but matching the feedhorn's radiation pattern to the area of reflector surface is more involved. The amplitude distribution across the reflector can be tapered towards the edge, so that there is less power to spill over the edge, but this also produces a reduction in antenna efficiency and with it a decrease in gain. It can, however, be improved by complex shaping of the reflector surfaces, or *profiles*, to maintain a uniform energy distribution across the main reflector.

The importance of sidelobe reduction tends to be related to the size of the earth station. For a large station, where the emphasis is on maximum gain, the sidelobes resulting from a near-uniform amplitude distribution are tolerated because the beamwidth is small. However, for small earth stations the sidelobes are further from the boresight as a consequence of the wider beamwidth, making it more desirable to taper the amplitude distribution. This is one aspect in which the design of antenna systems has become more critical since small earth stations were introduced for domestic telecommunications, cable TV, etc.

The surface accuracy of the reflector is another important consideration, since imperfections of the same order of magnitude as the wavelength can cause reflected radiation to be out of phase at the feedhorn. As an example, the surface accuracy of a typical 30 m earth station operating at 6 GHz (λ = 5 cm) is usually better than 1 mm RMS. This is especially important for antennas working at higher frequencies, since the wavelength is even shorter (e.g. 1 cm at 30 GHz).

9.5.2 Asymmetric antennas

Despite the ingenuity of the design solutions for improving efficiency whilst reducing sidelobes, the performance of the axisymmetric antenna is still limited by aperture blockage. Although the geometry of an asymmetric, or offset, antenna is more complex, the removal of the feed or subreflector from the antenna aperture produces significant improvements, not least that of eliminating diffraction effects due to the blockage itself.

Again as figure A7.18 shows, the offset antenna can be fed from a prime focus feed or using the dual-reflector solution in Cassegrain or Gregorian configuration. Shaped reflector profiles and beam waveguides can be used with asymmetric antennas as they are with the axisymmetric designs. The efficiency of an offset antenna may be as much as 70% compared with a typical 55% for centre-fed versions [Claydon 1987]. Using shaped reflectors, the efficiency of the offset antenna has been shown to exceed 80%, and the centre-fed antenna can achieve 65% if high spillover is acceptable, but it usually is not.

Figure A9.10 shows an example of a dual-reflector offset Gregorian antenna which has been produced with a 5.5 m main reflector. A variation on the design has shown it to offer a particularly compact solution for transportable earth stations.

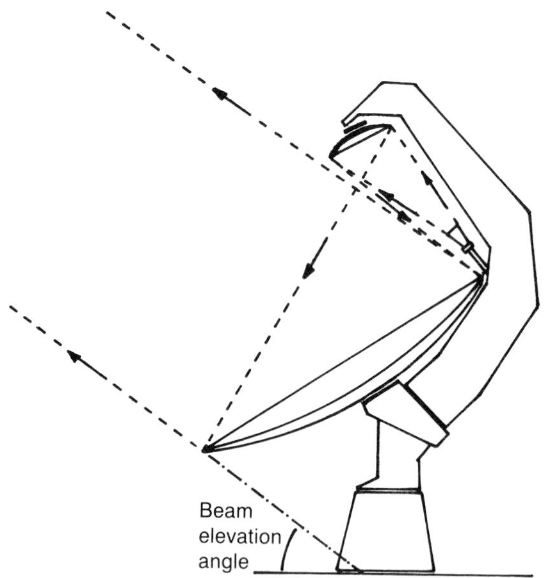

Figure A9.10 Dual-reflector offset Gregorian antenna configuration.

Although the asymmetric configuration is well suited to antennas for small earth stations, the mechanical problems and requirement for steerability make it less attractive for very large earth stations. However, with increasing demands for tighter sidelobe specifications this may change in future.

9.5.3 Toroidal antennas

One antenna configuration not yet mentioned is the toroidal antenna depicted in figure A9.11. The reflector is a section of a parabolic torus which allows the formation of a number of separate beams with different boresights. Each feedhorn in an array uses the reflector to form a separate beam to intercept a number of different satellites on the geostationary arc. One axis of the antenna is aligned with the arc, as shown in the figure, which makes it a non-steerable polar mount [see next section]. The toroidal antenna is a useful alternative to a number of separate conventional dishes and has been used particularly in the United States where cable TV receive stations require a view of several satellites at the same time.

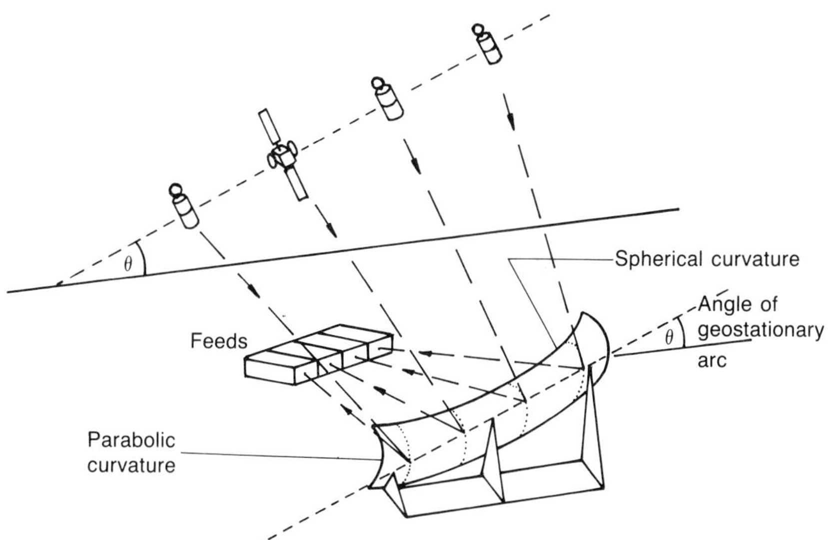

Figure A9.11 Toroidal antenna configuration.

9.6 ANTENNA MOUNTS

Just as there are different types of antenna reflector there are several generic types of antenna mount, once again largely derived from those used for astronomical instruments. Figure A9.12 shows the three most common dual-axis mounts, usually known as *elevation-over-azimuth*, *X–Y* and *polar*. All three have their beam axis or pointing direction perpendicular to the upper axis and each system is theoretically capable of pointing in any direction. Single-axis mounts, derived from these dual-axis mounts but with one of the axes set at installation, are also available. Although their simplicity makes them cheaper, they tend only to be used where a flexibility in pointing is not important, that is for relatively small earth terminals.

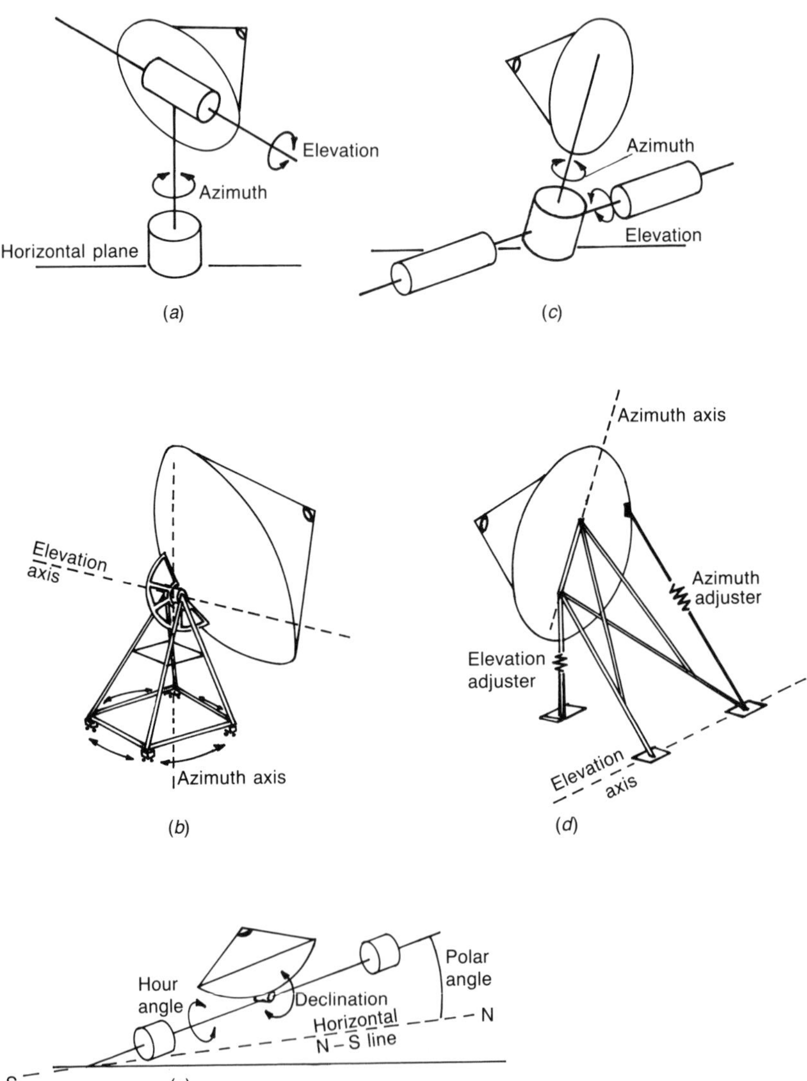

Figure A9.12 Earth station antenna mounts: (*a*), (*b*) elevation-over-azimuth; (*c*), (*d*) *X–Y* (azimuth-over-elevation); (*e*) polar (HA-DEC) in northern hemisphere.

The elevation-over-azimuth mount [figures A9.12(*a*) and (*b*)] is the one used for the majority of large earth stations since it lends itself to the support of large movable structures [see e.g. figure A9.5]. As its name suggests, the elevation axis is mounted above the azimuth axis, the two being orthogonal to each other. When used for an astronomical telescope this type of mount is called an altitude-azimuth or *alt-az* mount. Altitude or elevation is defined as the angle between a line joining the earth station and the satellite, and the horizontal plane at the earth station (i.e. the angle above the horizon); azimuth is the angle of rotation about an axis perpendicular to the local horizontal [see figure A8.4]. The zero reference for azimuth depends on whether one is following navigation terminology, in which it is clockwise from due north, or astronomical practice, in which it is clockwise from due south. The latter is suggested for satellite applications but not universally agreed upon.

The *X–Y* mount [figures A9.12(*c*) and (*d*)] is used for smaller earth stations and is conceptually more difficult to picture in operation. Figure A9.12(*d*) shows a practical example in which the antenna is supported on a tripod whose two back legs are aligned along the elevation axis. Adjustment in elevation is produced by altering the length of the single front leg. Azimuth adjustments are made by rotating the reflector about an axis perpendicular to the elevation axis, but whose angle relative to the ground changes with elevation: if this axis is anything but vertical, any swing in azimuth will also cause a change in elevation. Although the effect is negligible for small azimuth adjustments, it can prove important for initial alignment of the earth station. This type of mount is alternatively known as an azimuth-over-elevation mount, but this is really a misnomer since true azimuth is measured about an axis *perpendicular* to the local horizontal.

The polar mount [figure A9.12(*e*)] is based on the astronomical *equatorial mount*, a system devised to simplify pointing by having one axis parallel to the Earth's axis of rotation. This is called the hour-angle axis due to its origins in astronomy: it is usually expressed in units of time where one complete 360° rotation, measured towards the west, is equivalent to 24 h. The axis is inclined to the horizontal in the north–south plane by an angle equivalent to the latitude of the site, making it parallel to the ground at the equator and perpendicular at the poles. A telescope driven about this axis will automatically follow the path of a star across the sky. The other axis, declination, indicates the angle north or south of the celestial equator (north is positive, south is negative). Another name for the polar mount is therefore hour-angle–declination or HA–DEC.

9.7 ANTENNA POINTING AND TRACKING

Once an earth station has been installed, it is necessary to determine the elevation and azimuth angles to which the antenna must be set to point towards a particular satellite. Figure A9.13 shows a family of curves depicting the geostationary arc from earth stations anywhere in the world below a latitude of 75°. Above that latitude, the orbit is at such a low elevation that the ground is within the beamwidth of the antenna's main lobe, or at least the sidelobes, and severely degrades the performance.

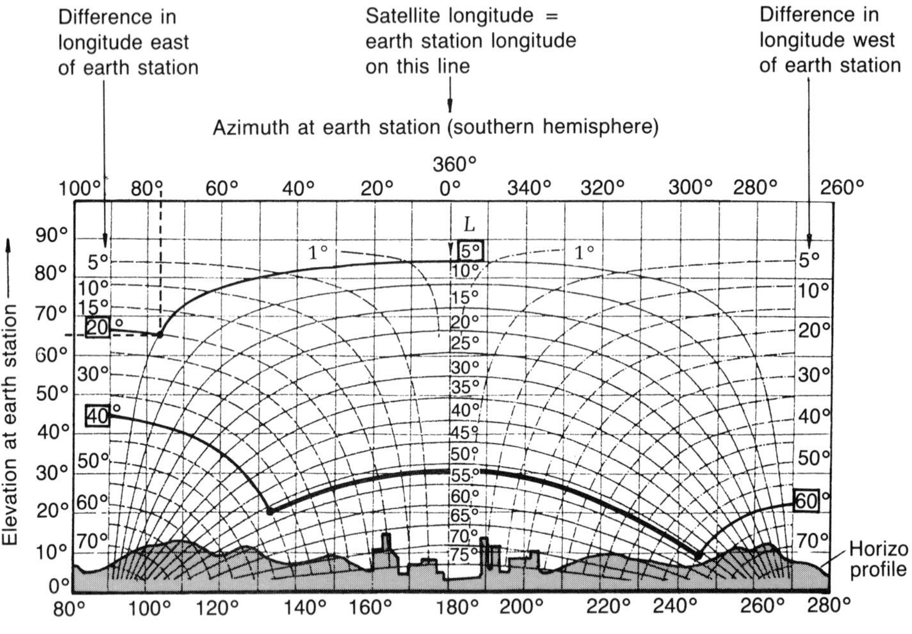

Figure A9.13 Elevation and azimuth pointing angles for geostationary arc—worldwide [annotations explained in §A9.7; shaded horizon profile discussed in §A9.8]: ———, arc of geostationary orbit visible from earth station at terrestrial latitude *L*; – – – –, – · – · – · – ·, difference in longitude between earth station and sub-satellite point.

In using the figure, it is first of all necessary to know the latitude of the station so that the arc of interest can be identified. Take, for example, the top curve (latitude 5°). Next it is necessary to know the orbital position of the satellite, which is quoted as the line of longitude of the sub-satellite point. The difference between this and the known longitude of the station can then be calculated and located on the diagram. Say the satellite is 20°E of the earth station. The 5° latitude curve is then followed down to the left until it reaches the broken line for 20°E. The required elevation (about 65°) and azimuth (about 76°) can then be read off the outer scales of the chart. As another example, the heavy line shows the location of satellites from 40°E to 60°W as visible from London, England (longitude 0°, latitude 51.5°).

Once the satellite has been located, the earth station antenna must be moved to keep the satellite at or near the peak of its beam. This activity is known as *tracking*, but it should not be confused with the type of tracking described in chapter A8 as part of the satellite's *telemetry, tracking and command* subsystem. The latter is concerned with controlling the position of the satellite in orbit, whereas the tracking systems described here are simply a means of causing the earth station to follow the satellite's movements and offer no benefit to the satellite itself.

The earth station tracking circuit is shown schematically in figure A9.2. Although there are variations between systems, tracking is achieved in general by detecting the instantaneous level of the satellite's signal and noting how this relates to the signal level at boresight when the antenna is correctly pointed. This information is used to decide how the antenna should be moved to improve the signal level and instructions are sent to the antenna's drive motors. Depending on the complexity of the system, pointing may be achieved in the manner of a simple feedback loop as described above, or using a computer system programmed with data on the expected movement of the satellite based on previous analysis.

Apart from the programmed system there are three main methods of tracking: monopulse, step-track and conical scan. The most complex system is the monopulse method, which employs either *static-split* or *multimode* techniques. The former, in which additional feedhorns are clustered around the main earth station feedhorn, is conceptually the simplest. Since the horns are mounted off axis, they have their peak sensitivities away from the boresight of the main antenna. So if the signal level received in one or more of the horns is greater than the others, the antenna must be mispointed. A comparison of the signal levels produces an error signal which is used to drive the antenna servomechanisms until the signals in all horns are the same. The name static split comes from the diverging radiation patterns of the horns which, considered together, appear to produce a split beam.

The multimode method involves the excitation of higher order asymmetric waveguide modes in the main feed aperture when the antenna points away from an RF beacon radiated from the satellite. In this case, error signals are derived by extracting the extra modes from the ordinary communications signal. This RF *sensing* arrangement offers a high tracking sensitivity without unduly compromising the performance of the communications system. The monopulse RF sensor is also used on board satellites which require a higher degree of pointing accuracy than can sensibly be derived from their attitude control system.

A less expensive method of tracking is the step-track or *hill-climbing* technique used extensively for all types of earth station antenna. A station equipped with this device tracks the satellite in a series of steps, according to the received signal level. By comparing the amplitude of the signal with the previously detected value the antenna moves either in the same direction if the signal is stronger, or in the opposite direction if it is weaker. This is like the step-by-step process of a climber continuously trying to attain the summit of a hill, passing over it only to find he/she is descending, and turning round to ascend again . . *ad infinitum*. With this method, the earth station's pointing accuracy depends mainly on the step size, which varies from system to system.

The third method, conical scan, is less common. It involves a tracking beam offset from the main axis of the antenna such that if the feedhorn is rotated the offset beam describes a cone about the axis. As long as the antenna is correctly pointed, the signal level derived from the tracking beam will remain constant, but if it becomes mispointed the signal will vary, an error signal will be generated, and the antenna drives will be activated.

9.8 SITING AN EARTH STATION

In these days, when transmit/receive earth stations can be towed behind vehicles and erected in car parks or on buildings, it may be difficult to imagine the constraints on siting an earth station. It is, however, a highly regulated business. Amongst the constraints on the early large earth stations were a clear line of sight to the satellite, good foundations to carry the weight of the hardware, reasonable environmental conditions and a situation which minimised the risk of terrestrial radio interference to the system. Today the wavebands are crowded, especially in the lower frequency ranges, and sharing between terrestrial and space users (i.e. microwave links and satellites) makes the siting of an earth station a much more involved affair. The physical constraints are still important for the large stations: factors such as rain and snowfall, temperature, humidity and prevailing wind must be taken into account; particularly corrosive atmospheres and earthquake zones are best avoided. But more important still is the need for frequency coordination.

Frequency coordination is a process which ensures the simultaneous operation of both satellite and terrestrial communications systems without mutual interference. The process, which begins when a new site is proposed, involves a submission of technical details to the International Frequency Registration Board (IFRB) through the relevant national radio regulatory administration. The details include channel frequencies, the orbital position of the target satellite, the geographical location of the intended station, a polar diagram of the antenna radiation pattern and other parameters such as transmitter power and modulation characteristics of the proposed signal. Using these and other data, an interference analysis is performed to decide whether the earth station will interfere with existing communications systems and vice versa.

Both uplink and downlink are subject to coordination. Of the two cases the transmitting earth station has the greater effect on other users, but even a receive-only earth station can have more of an impact than first envisaged. The coordination process is, in effect, an appeal for protection from interference from other existing services or those which may be installed in the future. It involves the stipulation of a required *coordination area* within which the service will be protected. If this has not been applied for, there is nothing to stop another installation transmitting and interfering at a later date. Equally, if a smaller coordination area is stipulated at first, and subsequent enhancements to the station require a larger area, the station must be re-coordinated. This would certainly be the case if transmitting facilities were later fitted to a receive-only station.

One of the many requirements of the frequency coordination process is the horizon profile, a graphical representation of the horizon (both natural and man-made) as seen from the site. Figure A9.13 includes an example profile, showing how buildings and terrain can obscure parts of the orbit, especially for earth stations at high latitudes. On receive, the small station has an advantage since the antenna's centre-line is closer to the ground, providing better shielding due to natural terrain and reducing the necessary coordination area.

Most large earth stations have been based on the same general design: the axisymmetric Cassegrain geometry, often with a beam waveguide feed. Small earth stations are under continual development as more and more satellite services become available to small users. The majority of designs have, until fairly recently, tended to be scaled-down versions of the larger earth stations, but it is becoming increasingly common to see offset antennas where once the centre-fed antenna was the norm. Figure A9.14 shows a selection of small antennas.

Nowhere is this more evident than in the growing preponderance of antennas less than a metre in diameter. One application of these is direct broadcasting by satellite (DBS) which promises dishes at least as small as 60 cm and even 30 cm in some cases. Another is the very small aperture terminal or VSAT, used for data transmission and distribution between a central 'hub station' and a potentially large number of outlying stations, for example for corporate communications between the head and branch offices of financial institutions, supermarkets, etc. Although the hub station would be equipped with an antenna from 4 to 10 m in diameter, the VSATs would use 0.8–1.2 m dishes and be capable of transmitting at perhaps a sixth the data rate of the central station.

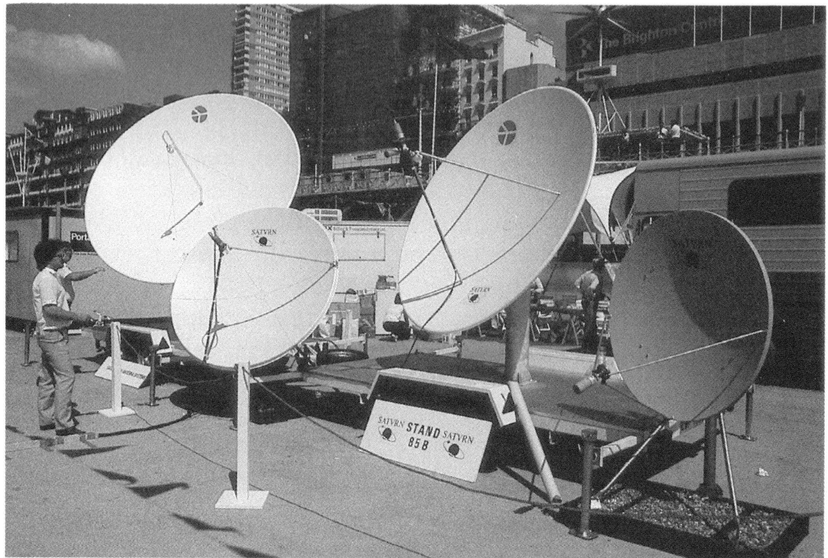

Figure A9.14 A number of small earth terminal antennas. [Mark Williamson]

9.9 CONCLUSION

This chapter has described the basic building blocks of an earth station: the transmitting and receiving equipment, antenna reflector, mount and tracking system, and the variations in available systems. As the most visible, and arguably the most important, part of an earth station, the antenna has received particular attention. Factors such as gain, directivity, sidelobes, spillover, efficiency and pointing have been discussed. Continual improvements in earth station design concentrate on RF performance while incorporating the moves in satellite communications to higher operating frequencies and a general increase in available bandwidth. But as technology improves, the regulatory requirements evolve to keep pace. The need for frequency coordination and related procedures shows that the installation of even a relatively small receive-only earth station is more complicated than it first appears—far more than simply choosing a plot, laying foundations and pointing the antenna!

REFERENCES

Claydon B 1987 *Satellite Communication Systems* ed. B G Evans (London: Peter Peregrinus) ch. 12

Maral G and Bousquet M 1986 *Satellite Communications Systems* (Chichester: Wiley) ch. 8

Pratt T and Bostian C W 1986 *Satellite Communications* (New York: Wiley) pp 127, 369

A10 Launch Vehicles

10.1 INTRODUCTION

As far as the satellite designer is concerned, the launch vehicle is a necessary evil: since there is no other way to get a satellite into orbit, it must be designed with the constraints of the available launch vehicles in mind. It is therefore necessary for a spacecraft engineer to become acquainted with launch vehicle operations. Although it is a vast subject, usually accepted as the province of the propulsion engineer, a good working knowledge of the field is well within the grasp of the non-specialist, as this chapter will attempt to show. All manner of spacecraft require a launch vehicle, and a variety of orbits and trajectories are attainable, but since this book concerns the communications satellite in geostationary orbit, it is this we shall concentrate on here.

10.2 AVAILABLE LAUNCH VEHICLES

The great majority of launch vehicles comprise a number of 'throw-away' or *expendable* stages which are used for a short period of flight and then discarded; these are the *expendable launch vehicles* or ELVs. Launch vehicles capable of lifting relatively large and heavy payloads are termed heavy lift vehicles (HLVs). The term is imprecise, but a vehicle that delivers a payload of 4 tonnes or more to geostationary transfer orbit could be considered an HLV. The fully reusable launch vehicle has not yet been developed, but the American and Soviet space shuttles *are* partly reusable and are therefore not classed as ELVs.

Most contemporary launch vehicles for geostationary satellites have three stages. The reasoning behind multiple stages is that the entire mass of the launch vehicle, including empty propellant tanks and associated structure, does not have to be carried all the way to orbit, with the result that a larger payload mass can be carried. Many conceptual designs for a single-stage-to-orbit (SSTO) launch vehicle have been proposed, but it is unlikely that a working example will be produced before the next century.

The American ELVs, Atlas, Delta and Titan [see figures A10.1–A10.3], have all been derived from military missiles, as indeed have the major Soviet vehicles. The Atlas was developed from an intercontinental ballistic missile (ICBM) which became operational in 1959; the Titan was developed as a back-up to the Atlas; and the Delta was developed from the Thor, another 1950s missile launcher. Although the three vehicles have progressed through a number of variants in the last 30 years, augmented by cryogenic stages and/or strap-on boosters, their origins remain in the 1960s. This long service record has given these vehicles a certain pedigree and the Delta, in particular, is considered a very reliable satellite launcher.

Figure A10.1 Atlas–Centaur launch vehicle. [General Dynamics]

Figure A10.2 Delta launch vehicle. Note strap-on solid boosters. [McDonnell Douglas]

Although versions of the same Atlas, Delta and Titan launch vehicles began commercial operations in the late 1980s, a change can now be observed in launch vehicle development: they are being designed specifically to launch satellites into geostationary orbit. The Ariane launch vehicle [see figure A10.4] was the first to be designed in this way, progressing through a number of versions with increasing capability and culminating in the Ariane 4, which has six variations depending on the number of solid or liquid strap-on boosters used. Ariane was, from the outset, intended to be a commercial launch vehicle.

However, the definition of what constitutes a truly commercial venture is continually under dispute since most launch vehicles have historically been subsidised to some extent by a national government. A similar practice is, incidentally, followed for certain aircraft manufacturers because of the high cost of such endeavours for a single company.

Figure A10.3 Titan III launch vehicle (this photograph of a USAF Titan III has been retouched to show the larger 4 m diameter payload shroud used on commercial Titan III flights) [Martin Marietta]

Figure A10.4 Ariane 3 launch vehicle. Note strap-on solid boosters. [CEF, Bernard Paris]

A term often heard in the years following the *Challenger* disaster was 'launch vehicle crisis', since for a while all the western world's major launch vehicles (Space Shuttle, Titan, Delta, Atlas and Ariane) were grounded pending investigations of launch failures. By 1989, there was no longer a crisis, although the backlog of launches remained, and a programme of American launch vehicle commercialisation was well underway. Progress was, moreover, matched in the eastern bloc with the USSR offering the Proton and China offering the Long March 3 launch vehicle to the West. Here again the question of what constitutes a commercial launcher arises, and this time with political overtones, as the respective governments offer cut-price launches. Not surprisingly, Japan has been keen to develop a launch vehicle industry, based on assistance from the United States, which allowed the development of the N-series booster from the Delta. There followed the H-I, whose use was restricted by agreement, and the H-II, which is fully Japanese in origin. The latter offers competition to Ariane and the American launch vehicles for placing satellites in geostationary transfer orbit.

10.3 LAUNCH VEHICLE SELECTION

Of all the available launch vehicles, how does one decide which one to use for a particular satellite launch? There are of course many factors, including price, availability and, where the eastern bloc vehicles are concerned, political will. The USA, in particular, is concerned about the possibility of 'technology transfer'. However, in engineering terms the decision is based on the constraints imposed by the launch vehicle on the satellite, which were first introduced in chapter A1. The most fundamental is the lifting capability of the launch vehicle, which places an upper limit on the mass of its satellite payload. The second constraint is that on the size and shape of the satellite which must be designed to fit within the launcher's payload shroud.

Other constraints are largely a result of the environment produced by the launch vehicle in operation, for instance vibration and acoustic effects, acceleration forces, separation shocks, etc. Mechanical effects such as these can be accounted for in the structural design of the spacecraft, as discussed in chapter A2. The mass and volume constraints have a bearing on all satellite subsystems and are worth considering further.

10.4 MASS CONSTRAINT: LAUNCH VEHICLE PAYLOAD CAPABILITY

The main performance criterion of a launch vehicle is the mass it can lift into orbit, and table A10.1 compares a number of contemporary launch vehicles in this regard. It is essential to ensure a like-for-like comparison since, naturally, a given vehicle can lift a much greater mass to low Earth orbit (LEO) than to geostationary orbit (GEO). However, just as likely to be quoted in a manufacturer's brochure is the payload capability to geostationary transfer orbit (GTO), which includes the perigee burn but not the apogee motor firing [see chapter A3].

The inclusion or exclusion of these motor firings shows the designer of the satellite propulsion subsystem what must be provided, in terms of thrust, to place the satellite in its final orbit. The simplest possible brief would present a launch vehicle which delivered the satellite to GEO. The designer would then have only to provide a propulsion system to position the satellite at its allocated orbital slot and maintain that position—in other words an attitude and orbital control system (AOCS) [see chapter A4]. This is, however, far from standard practice. A version of the Titan 3 booster *is* available to do this, but it is an expensive solution and tends only to be used for US Government military satellites. By far the most common practice is that followed by the Ariane, Atlas–Centaur and Delta vehicles, which inject their payloads into GTO and leave them to their own devices. This is why most communications satellites carry an integral apogee kick motor (AKM) [see figure A1.5]. The alternative is to fit a combined bipropellant propulsion system, for both apogee motor firings and AOCS, as explained in §A3.9.

Table A10.1 Launch vehicle payload capability (kg).

Nation	Vehicle	Payload to GTO	Payload to GEO
Europe	Ariane 1	1850	1110
	Ariane 2	2175	1305
	Ariane 3	2700	1620
	Ariane 4	4200	2520
	[Ariane 5]	[6800]	[4080]
USA	Atlas G/Centaur (C)	2350	1220
	Delta 3920/PAM (C)	1270	730
	Titan III + PAM-D (C)	1270	
	Titan III + Transtage		1814
	Titan IV + IUS		2360
	Titan IV + Centaur	9080	4540
	Shuttle + PAM-A	1990	1000
	Shuttle + PAM-D	1250	550
	Shuttle + IUS		2270
	(Shuttle + Centaur G)		(4540)†
	(Shuttle + Centaur G′)		(5910)†
Japan	N-II‡	700	350
	H-I		550
	[H-II]		[2000]
China	Long March 3 (C)	1400	
USSR	Proton (C: three stage)		1900

[ ] to be confirmed by flight results.
(C) version offered commercially.
† currently excluded from the Shuttle.
‡ equivalent to Delta 2910.

Notes: Figures are taken from contractor's literature where possible but should be used with care (see discussion in §A10.4). Inclination of orbits attained depends on the launch site latitude (see box A10.1).

In cases where the American Space Shuttle [figure A10.5] has been used to launch a geostationary satellite, an extra rocket stage—the perigee stage or perigee kick motor (PKM)—is employed to inject the satellite into its transfer orbit since the Shuttle delivers its payloads to low Earth orbit or *parking orbit*. Shuttle deployment methods are discussed in §A10.7.

Figure A10.5 Space Shuttle launch. [NASA]

A further problem in a like-for-like comparison of vehicle performance is that the true performance depends on the particular mission, that is on the payloads carried, the initial orbital parameters, the latitude of the launch site [box A10.1], the launch profile [§A10.6] and the launch window. The launch window itself is constrained by the desired orbit, the point of injection into that orbit, and the possibility of a collision with other orbiting spacecraft or debris. In addition, if a selection of payload shroud sizes are available, what appear at first sight to be identical vehicles may have different masses due both to the shroud and the extra payload it houses. Apart from this, the launch vehicle may have slightly larger propellant tanks, more efficient engines, added strap-on boosters, etc. And just as the performance of the average family saloon car improves year by year, so too does the lift capability of the average launch vehicle, as engine performance is improved and structural masses are pared by the introduction of lighter alloys or composite materials.

A case in point is the American Space Shuttle, whose external propellant tanks for the first two missions were painted white. When this cosmetic coating was eliminated from subsequent flights, the vehicle's total launch mass was reduced by about 270 kg. However, the destruction of the orbiter *Challenger* in 1986 led to the introduction of additional safety measures and a general increase in margins, which caused an overall reduction in performance.

Looking at a launch vehicle's performance from the point of view of a satellite designer, it is useful to talk in terms of *payload fractions*, that is the proportion of the total weight of the launcher/payload combination that can be given over to the payload itself. A related engineering yardstick is the launch vehicle *mass ratio*: the ratio of the mass of the fully fuelled vehicle prior to launch to the mass of the vehicle structure and payload (*dry mass*); or initial mass to final mass.

In the early days of rocketry the typical payload fraction was very small—in fact engineers were happy to get anything at all into orbit. One early endeavour used a stripped-down Atlas booster to deliver a tape recorder, carrying a message from the President of the United States, to low Earth orbit! Payload fractions in the early 1960s were around 1% to LEO and 0.2% to GEO. In the 1980s they were closer to 2–3% to LEO and 1% to GEO, but this is low compared with typical payload fractions for aircraft, which are around 12–15% [Pardoe 1986]. This is largely because of the way rockets are launched (i.e. vertically) instead of making use of lift generated by a wing (one reason in favour of the reusable horizontal take-off launch vehicles planned for the early twenty-first century). One well known launch vehicle, the Saturn V, had a particularly high payload fraction of 5%. If NASA had continued its development, instead of scrapping it at the end of the Apollo and Skylab programmes, they would have had an ideal vehicle for launching the heavy military satellites and space station modules needed in the late 1980s and 1990s.

By way of a practical example, table A10.2 shows a mass breakdown for a typical contemporary launch vehicle, the Ariane 3, which delivers about 2700 kg to GTO. Compared with the total launch mass of the vehicle (about 240 tonnes) this amounts to a payload mass fraction of just over 1.1%. Although this is a good mass fraction for a launch vehicle, it is like saying an 8 tonne truck could carry only its driver!

It is interesting to note the relative masses of structure and propellant in all of the stages: the propellant mass is about nine times that of the structure which contains it. Taking the vehicle as a whole, the propellant and pressurant account for some 210 tonnes, leaving about 30 tonnes for the structure of the launcher and its payload. A knowledge of the vehicle's mass with and without propellant allows a determination of the vehicle's mass ratio, as the table shows.

Table A10.2 Ariane 3 mass breakdown (tonnes).

First stage:	structure	15.6
	propellants	145.0
Two strap-on solid boosters:		
	structure	4.8
	propellants	14.6
Second stage:	structure	3.6
	propellants	34.0
Third stage:	structure	1.2
	propellants	10.7
Vehicle equipment bay (VEB)		~0.3
SYLDA		~0.2
Payload fairing		~0.85
Other (e.g. pressurant)		~6.45
Payload		2.7 (to GTO)
	Total	~240.0

Payload fraction = 2.7/240 ≃ 0.11 (1.1%)

(Mass of propellant/pressurant: 210.75 tonnes)
(Mass of structure and payload: 29.25 tonnes)

$$\text{Mass ratio: } \frac{\text{Mass before launch}}{\text{Vehicle dry mass}} = \frac{240}{29.25} = 8.2$$

Based on figures for the Ariane V19 mission.

Payload mass is, not surprisingly, directly related to launch costs, although the relationship can depend on anything from government sponsorship to market forces. The Space Shuttle was intended to reduce the cost-to-orbit but failed to meet expectations, partly because the eventual design was for a vehicle only partially reusable, but also because of the decrease in expected launch rate. Original intentions had dictated a launch rate of once a week or greater; the closest approach to this came in 1985, the year before the *Challenger* disaster, when 10 Shuttle missions were launched. In 1985, projected flight rates for 1986 and 1987 were 14 and 19 respectively. Following the resumption of shuttle flights in 1988, they were placed at a more conservative average of 9 per year throughout the early 1990s. Since the reinstatement of the system the launcher market has undergone substantial changes, not least in the exclusion of commercial satellites from the Space Shuttle manifest following the *Challenger* accident. Under current thinking, Shuttle flights are reserved for large government and military payloads, some science payloads, and component modules of the US/international space station.

10.5 VOLUME CONSTRAINT: LAUNCH VEHICLE PAYLOAD ENVELOPE

As mentioned earlier, mass is not the only constraint on the satellite: its size is constrained by the payload accommodation provided by the launch vehicle.

The structure at the top of a conventional launch vehicle, which carries the payload, is known by a variety of names, payload fairing and payload shroud being the most common. In engineering design terms, the volume available within the shroud is called the *payload envelope,* and it is this which defines precisely the maximum available volume and constrains the size and shape of the satellite and its various appendages, such as antennas and solar arrays. Figure A10.6 depicts the payload envelopes for the Ariane 3 and 4 launch vehicles compared with the volumes available in the shrouds of the competing Delta vehicle. Figure A10.7 shows how a three-axis-stabilised satellite (Intelsat V) can be packaged within the confines of a shroud: the solar array panels and deployable antennas can be seen folded against the sides of the spacecraft.

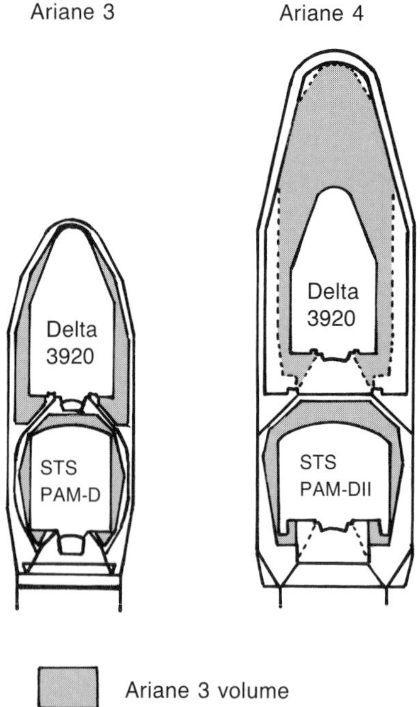

Figure A10.6 Ariane 3 and 4 payload envelopes compared with those for the Delta 3920 and Space Shuttle/PAM. [Arianespace]

Figure A10.7 Intelsat V-F7 satellite and Ariane shroud. Solar arrays and deployable antennas are folded to fit the payload envelope. [Centre Optique du CSG/CNES]

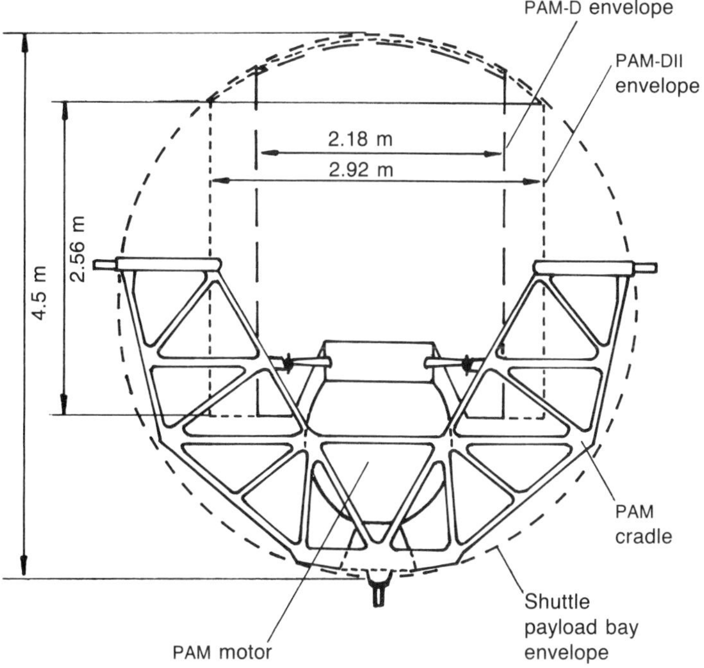

Figure A10.8 Space Shuttle/PAM-D and PAM-DII payload envelopes.

Apart from its provision of payload accommodation, the shroud is designed to protect the payload from the atmosphere and the weather before launch and from aerodynamic forces during the early phases of the launch itself. It also goes some way towards minimising the various mechanical and thermal loads to which the spacecraft is exposed.

There is an historical as well as a technical aspect to the development of the launch vehicle envelope. In the 1970s the predominant launcher of commercial communications satellites was the Delta, whose main volume constraint was the 2.18 m inner diameter of its fairing. This predominance brought about the term 'Delta-class payload', which is still in use despite increasing launch vehicle competition. Launching a Delta-class payload from the Space Shuttle introduced an additional constraint by way of its payload bay, which has a usable diameter of about 4.5 m. Figure A10.8 shows the vertical mounting position of a satellite attached to its PKM. The requirement for a PKM constrained the height of the satellite to about 2.56 m, only just over half the total available, and was partly responsible for the telescoping solar array design used on many spin-stabilised satellites [see figure A1.4].

The satellite/PKM combination could have been mounted horizontally in the bay, but due to the mechanisms of Shuttle pricing this would have

Figure A10.9 Space Shuttle payload bay containing two HS376-series satellites and a Syncom IV/Leasat spacecraft. [NASA]

been more expensive (charges are based on the proportion of bay volume or payload mass capability used, whichever is the greater). Moreover, since most customers required the option of both Shuttle and Delta launchers, the satellite designer was obliged to respect both the Delta diameter and the Shuttle height constraints, forcing satellites to be particularly compact [Renner 1983]. Figure A10.9 shows three spin-stabilised satellites in the payload bay prior to launch.

Throughout the 1980s a number of alternative launch systems were developed and, although volume constraints must still prevail, the restrictions of the Delta-class payload no longer apply. In a synergy between increasing satellite size and mass and improved launch vehicle performance, payload envelopes have expanded to house the new generations of spacecraft. This process has continued to such an extent that a significant proportion of launches now carry two satellites at once. The dual-launch system designed for the Ariane is a prime example. The system used originally with the Ariane 1 and 3 versions (the Ariane

Figure A10.10 Ariane shroud and SYLDA (Systeme de lancement double Ariane—Ariane dual-launch system). [Aerospatiale]

2 launched only single spacecraft) is known by its French acronym, SYLDA (systeme de lancement double Ariane, or Ariane dual-launch system) [see figure A10.10]. It was mounted above the third stage, fully inside the shroud, and contained one satellite while the other was mounted on top.

A similar development for Ariane 4 is the SPELDA (structure porteuse externe pour lancements doubles Ariane, or external support structure for Ariane dual launches). The SPELDA, mounted above the third stage and below the shroud, differs from the SYLDA in that it is part of the external structure or *fairing* of the vehicle, and not simply a container mounted inside the shroud. A similar structure used with the Titan III launch vehicle is called an 'extension module'. To accommodate triple payloads a SYLDA can be mounted on top of a SPELDA. A structure known as a SPELTRA (structure porteuse externe pour lancements triples Ariane) is under development for Ariane 5 triple launches.

10.6 TYPICAL EXPENDABLE LAUNCH VEHICLE MISSION

To describe a typical launch vehicle mission, the Ariane 3 and 4 vehicles are again taken as representative examples of contemporary ELV technology. Table A10.3 shows a typical launch event sequence for the launch of two satellites into GTO by Ariane 3. It can be read in conjunction with figure A10.11(*a*), which is a typical launch profile showing the height of the vehicle in relation to its range from the launch pad, and figure A10.11(*b*) which shows the associated ground track and the coverage of the tracking stations mentioned in the launch sequence table.

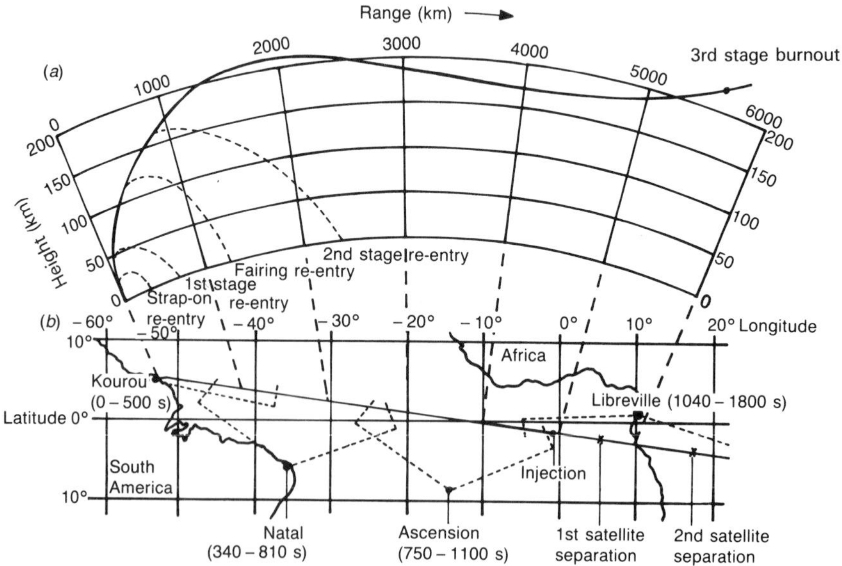

Figure A10.11 Typical Ariane launch profile and ground track (Ariane 4), including tracking station coverage times.

Table A10.3 Ariane 3 launch sequence (V19 dual launch with SYLDA).

Time (min:s)	Event
0:00	First-stage ignition
0:03.4	Lift-off
0:07.2	Solid strap-on booster ignition
0:10	End of vertical ascent; begin pitch-over
0:40.2	Solid booster jettison
2:16	First-stage burn-out
2:20	First-stage separation
2:24	Second-stage ignition
3:31	Fairing jettison
4:26	Second-stage burn-out
4:31	Second-stage separation
4:36	Third-stage ignition
5:40	Launcher acquired by Natal tracking station
11:30	Launcher acquired by Ascension Island station
13:50	Launcher acquired by Libreville station
16:39	Third-stage shutdown sequence commanded
16:40	Injection into GTO
18:27	First satellite separation†
21:59	SYLDA jettison
22:04	Second satellite separation‡
22:07	Third-stage avoidance manoeuvre
24:38	End of Ariane mission

† First satellite apogee motor firing about 26 h after launch at the third apogee; solar array deployment about 12 h later.

‡ Second satellite apogee motor firing about 37 h after launch at the fourth apogee; solar array deployment about 8 h later.

Note: Based on the Ariane V19 mission. In general, timings vary by 1 or 2 min depending on payload.

The table indicates that lift-off occurs several seconds after the first-stage engines are ignited. This is standard practice for liquid-fuelled rocket engines which need time to build their thrust to that necessary for lift-off. Ariane's solid propellant strap-on boosters are ignited at *T*+7.2 s to minimise damage to the tower and launch pad. The Space Shuttle, which uses both liquid and solid propellants, has a slightly different timing sequence: the liquid-fuelled main engines are ignited at *T*–6.6 s and it leaves the pad at '*T*–zero' as the solid rocket boosters are ignited. Most launch vehicles begin their mission with a period of vertical flight and then undertake a pitch manoeuvre which gradually adjusts their trajectory to match that of the elliptical transfer orbit or circular LEO, for whichever they are destined.

Towards the end of the Ariane vehicle's subsonic flight, the arms holding the solid strap-ons are severed by pyrotechnic cutters and the boosters are pushed aside by powerful springs. The liquid propellant strap-ons used with the Ariane 4 version are released pyrotechnically at about Mach 5 and separated by a cluster of tiny rockets. Stages are separated by a pyrotechnic cutting cord which conducts the blast from a charge around the circumference of the stage structure and severs the used stage from the rest of the vehicle, typically in less than a millisecond. Stage separation occurs during a period of near-zero acceleration, between lower stage burn-out and upper stage ignition. Small rockets on the upper stage, called ullage rockets, fire to settle the effectively weightless propellant near the tank outlets and, 2 seconds later, at the moment of separation, retro-rockets on the lower stage fire to decrease its velocity and prevent it from colliding with the upper stage before its engine ignites.

The payload fairing is jettisoned as soon as the atmosphere is sufficiently rarefied. Most fairings are constructed in two equal halves and attached to the top of the third stage with a clamp band. With the Ariane, the fairing is ejected by releasing the clamp band and pyrotechnically separating the two halves. Following the release of similar clamp bands, the satellite payloads are separated from the launch vehicle by the force from a number of compressed springs. In the case of a dual launch the uppermost satellite is released from the top of the SYLDA (or similar device), then the top of the SYLDA is released, and finally the second satellite is ejected.

It has been known for launch vehicle third stages to catch up with a payload after its release, due to an unplanned residual thrust from its propulsion system. It is now standard practice to initiate an avoidance manoeuvre, whereby the trajectory of the third stage is altered to prevent a collision.

The modern launch vehicle carries a computer to coordinate its subsystems, giving the vehicle total in-flight autonomy, except for the telecommand destruct signal. When necessary, launch vehicles are destroyed by the ignition of a pyrotechnic cord running the length of a propellant tank which splits the tank open, disperses the propellant and naturally triggers an explosion. Such arrangements are often called propellant dispersal systems for this reason.

The flight-control computer is housed in or above the vehicle's uppermost stage so that it remains with the vehicle throughout its mission. For example, Ariane's vehicle equipment bay, or VEB, is located on top of its third stage. Ariane's first stage flies according to a pre-programmed *attitude law* which cannot be amended once the vehicle is in flight. Its second and third stages, however, are guided by an inertial platform which monitors axial accelerations and senses the attitude of the vehicle.

The onboard computer determines the vehicle's instantaneous position and computes the adjustments necessary to optimise its trajectory. The vehicle is steered by gimballing the engine nozzles. The computer also controls the launch sequence and generates commands for ignition of the stage's main engines, propellant flow control, ullage and retro-rocket firings, and the separation of stages, fairings and payloads. Concurrently, the launch vehicle's telemetry system transmits some 600 measurements to the ground stations: about 60 for real-time diagnostics to indicate how the flight is progressing, and the rest for later analysis as a flight post-mortem. Launch vehicle telemetry is usually transmitted at S-band (about 2.1 GHz).

Ground tracking of a typical launch vehicle is performed using radar techniques by locking onto a transponder in the equipment bay. The general safety requirement which dictates that launch vehicle trajectories should not pass over centres of population leads, for most western launch sites, to a trajectory that is predominantly over the sea. This can make the siting of tracking stations difficult, but as the example of figure A10.11(*b*) shows, a station can often be sited on an island to fill in the gap.

The third stage of a typical three-stage launch vehicle injects its payload into GTO as shown in figure A3.1. This is an elliptical orbit which, for Ariane, has a perigee of about 200 km, an apogee of about 36 000 km (geostationary height) and nominal inclination of 7°. The inclination of the transfer orbit depends to an extent on the latitude of the launch site, but in Ariane's case is greater than the expected 5.23° as a result of an obligatory 'dog-leg' manoeuvre designed to avoid populated areas on the north coasts of Brazil and French Guiana. It can, however, be reduced by the launch vehicle to limit the propellant consumption of the satellite in reaching its final 0° inclination orbit,

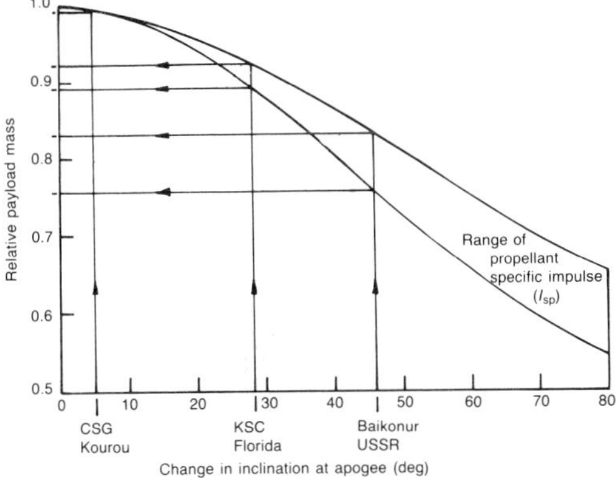

Figure A10.12 Variation of launch vehicle payload capability with latitude of launch site.

thereby increasing its orbital lifetime. Box A10.1 discusses the dependence of the velocity increment (ΔV) provided by the launch vehicle on the latitude of the launch site and the bearing this has on its payload capability.

BOX A10.1: EFFECTS OF LAUNCH SITE LATITUDE ON LAUNCH VEHICLE PERFORMANCE

If a launch vehicle is launched to the east, it can benefit from the Earth's rotational velocity to an extent governed by the latitude of the launch site [see appendix C1]. At the equator this velocity is 465 m s^{-1}; at the latitude of Kennedy Space Center, for example, it is only 409 m s^{-1}, which means that the launch vehicle will have to provide more of the orbital velocity itself.

For a spacecraft launched into a geostationary transfer orbit from the equator (0° inclination) the velocity at a perigee of 200 km will be 10 234 m s^{-1} and that at apogee (35 786 km) will be 1597 m s^{-1} [Maral and Bousquet 1986]. The launch vehicle supplies the initial velocity increment (ΔV) to attain the transfer orbit; the spacecraft apogee motor provides the ΔV for injection into geostationary orbit. The velocity of a spacecraft in GEO is 3075 m s^{-1} [see appendix C2], so the ΔV required at apogee is 1478 m s^{-1} (3075–1597) for a launch in the equatorial plane.

For actual launch sites not on the equator the required ΔV is greater because of the need to change the inclination of the satellite's orbit as well as to circularise it. The following table shows a few examples. It does, of course, assume that the transfer orbit will have an inclination equal to the latitude of the launch site (which is not always the case).

Launch site	Latitude	ΔV(m s^{-1})
Equatorial	0°	1478
CSG Kourou	5.23°N	1492
KSC Florida	28.5°N	1837
Tyuratam/Baikonur	46°N	2277

CSG = Guiana Space Centre; KSC = Kennedy Space Center.

This means that for the same launch vehicle, about 12% more payload can be launched from CSG than from KSC and nearly 30% more from CSG than from Baikonur [Maral and Bousquet 1986].

Figure A10.12 shows the percentage of mass in the transfer orbit that can be injected into GEO as a function of the required change of inclination at apogee. The 'expanding curve' is in fact a family of curves relating to the specific impulse (I_{sp}) of the propellant being used and covers a range of I_{sp} from about 300 to 450 s [see chapter A3]. The trend is, however, clear.

The accuracy of the orbital injection is of fundamental importance. Precision is an aim second only to success in the satellite launching business, since an improvement in deployment accuracy (as the jargon has it) leads directly to an increase in the satellite's lifetime by reducing the amount of thruster propellant required for orbital adjustments. With a revenue-earning capability of $1.5–2 million a month, a well placed satellite can earn its sponsors a sizable bonus.

10.7 THE SPACE SHUTTLE

So far we have concentrated on ELVs which deliver their payloads to GTO. As mentioned previously, the Shuttle assumes a low-altitude parking orbit and releases its payloads from there. Although the Shuttle has been designed to operate between altitudes of 185 and 1110 km, it typically operates from an orbital height of around 300 km; so for geostationary satellites a perigee motor or *upper stage* is required to inject the spacecraft into transfer orbit.

10.7.1 Upper stages

There have been two main types of upper stage used with the Shuttle: the payload assist module (PAM) and the inertial upper stage (IUS) . The PAM is a solid rocket motor designed for use as a PKM with both the Shuttle and the Delta ELV. It is also known as a spinning solid upper stage (SSUS) because it spin stabilises its payload during the transfer orbit. It is available in two main versions: the PAM-A for payloads in the Atlas–Centaur class and the PAM-D for those in the smaller Delta class [see table A10.1]. A variant, the PAM-D II, has a similar payload capability to the PAM-A.

The IUS is used with the Shuttle and the Titan ELV. It comprises two solid propellant stages to provide the functions of both PKM and AKM and is three-axis stabilised and inertially navigated in the transfer orbit. Another upper stage designed originally to launch larger payloads from the Shuttle was derived from the Centaur stage used with the Atlas launch vehicle. However, after the *Challenger* disaster the cryogenic propellants, liquid oxygen and liquid hydrogen, used by the proposed Centaur G and Centaur G-prime (G′), were considered too dangerous to be carried by the Shuttle and use of the Centaur was confined to the expendable Titan IV launch vehicle.

Throughout the Space Shuttle era, a significant number of other upper stages have been proposed to supercede the present stages. However, their future is uncertain, due to the removal of commercial satellites from the Shuttle manifest and the fact that there is no need for this type of upper stage on an ELV which injects its payload directly into GTO. A more promising course of development is that of the orbital transfer vehicle (OTV), a reusable vehicle based permanently in space and designed to transfer spacecraft between orbits and then be refuelled, either by the Space Shuttle or at a space station. A similar concept, the orbital manoeuvring vehicle (OMV) would perform the function of a 'space tug', repositioning payloads in LEO.

10.7.2 Deployment methods

For safety reasons, a satellite launched by the Shuttle is released from the payload bay about 45 min before the PKM is required to fire. This allows a sufficient separation to be obtained before the motor is ignited. A similar safety limitation prohibits the firing of a satellite's attitude control thrusters for 3 min after separation [Renner 1983]. This restriction can constitute a stability problem for satellites with bipropellant combined propulsion systems, which are generally not spin stabilised in the transfer orbit. Since they are released from the Shuttle without a spin their propellant is prone to sloshing, and as the propellant can account for up to 50% of the satellite's mass, this may seriously affect the satellite's stability. Although it is tempting to avoid the problem by designing solely for ELV launches, the limited availability of launch vehicles has tended to force manufacturers to offer compliance with both expendable and reusable types (except of course for payloads which are prohibited from using the Shuttle).

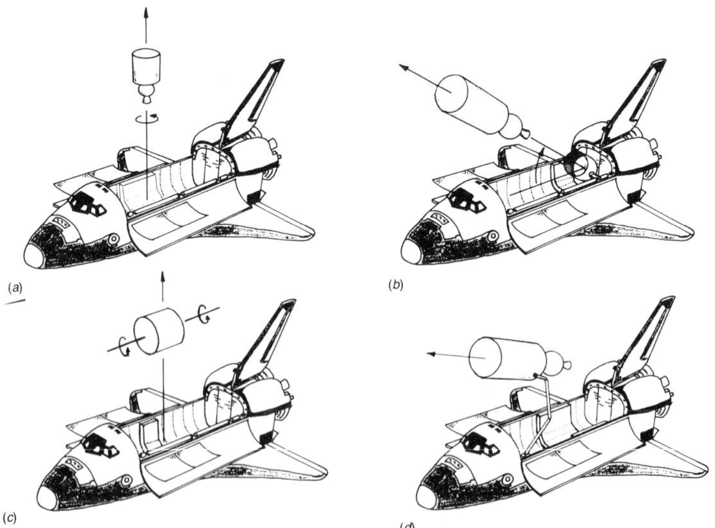

Figure A10.13 Space Shuttle satellite deployment methods: (*a*) vertically mounted vertical spin deployment; (*b*) horizontally mounted inclined deployment (no spin); (*c*) frisbee deployment; (*d*) RMS deployment.

There are four main methods of releasing satellites from the Shuttle orbiter's payload bay [see figure A10.13]. The most widely used to date is the vertical release deployment of satellites with attached PAM stages [see figure A10.14]. This method entails rotating the combination on a spin table at 50 rpm, firing a number of pyrotechnic bolts and allowing the payload to be pushed from the bay by a set of compressed springs which impart a velocity of about 1 m s^{-1}.

(a)

(b)

Figure A10.14 Vertical release deployment of (a) a spin-stabilised HS376 satellite and (b) an Insat three-axis-stabilised satellite, both with attached perigee kick motors. [NASA]

The second method is for payloads mounted horizontally in the bay, which is the typical configuration for large spacecraft mated to an IUS stage. In this case the combination is first tilted up so that its axis points over the orbiter's flight deck, then, using a similar release mechanism to that used with the PAM, it is ejected from the bay [see figure A10.15].

Figure A10.15 Release of a Tracking and Data Relay Satellite (TDRS) from the Shuttle orbiter's payload bay. Note unfurlable antennas at top and solar array folded around spacecraft. [NASA]

The third launching method is one used for satellites designed to utilise the diameter of the payload bay in the most efficient manner. The Hughes spin-stabilised Syncom IV (Leasat) was designed to occupy a minimum bay length but the entire available diameter [see figure A10.9], a philosophy followed also for the Intelsat VI design. Both satellites are deployed using the *frisbee* method, named after the plastic discs since it imitates their motion [see figure A10.16]. The spacecraft is mounted in a U-shaped cradle with its spin axis parallel to the Shuttle's longitudinal axis. An attachment to one side of the payload bay is released and a spring forces the satellite to rotate about a pivot point on the opposite side. As the satellite rises out of the bay the pivot attachment releases it and allows it to 'roll' out of the Shuttle. The simultaneous translation and rotation about the spacecraft's centre of mass have typical values of 0.36 m s^{-1} and 2 rpm.

Figure A10.16 Syncom IV/Leasat 'frisbee' deployment from Space Shuttle. Note deployable UHF communications antennas folded against top platform [compare with figure A7.23]. [NASA]

The fourth method uses the remote manipulator system to remove the spacecraft from the payload bay.

10.8 CONCLUSION

This chapter has been able to provide only an introduction to the technology of the launch vehicle and its operation, but it has presented the aspects most important to the satellite engineer, namely the mass and volume constraints imposed by the launch vehicle on the satellite. As satellites become larger and heavier, launch vehicle capabilities will grow to accommodate them and the market for the heavy lift launch vehicle will expand. Whatever the size of the payload, the launch vehicle industry seems certain to develop worldwide in the 1990s, as access to orbit becomes ever more desirable to developed and developing countries alike.

REFERENCES

Maral G and Bousquet M 1986 *Satellite Communications Systems* (Chichester: Wiley) ch.7

Pardoe G K C 1986 *Space* **2** no. 3 52–3

Renner U 1983 *Space Commun. Broadcast.* **1** 145–54

A11 Satellite Reliability

11.1 INTRODUCTION

Reliability is one of those words we are all familiar with and have some practical understanding of, even if it only extends as far as whether the car will start and, if it starts, whether or not it will keep going! To the engineer, however, reliability is a formal discipline, a subject worthy of study in its own right.

Although the foundations of reliability, such as life tests, failure statistics and materials fatigue, have a longer history, the discipline of reliability engineering is not much older than the space industry itself. It is traditional to set its birthdate in 1952 when Robert Lusser, then an engineer at the Redstone Arsenal in the USA, presented a formal definition of reliability at a San Diego symposium: 'The reliability of an object is the probability that it will perform correctly for an assigned period of time and under specific conditions' [Perrotta and Somma 1983]. Of course, this begs the question of what is 'correct performance' and what are the 'specific conditions', but these are factors which can be defined before a reliability assessment is made. In brief, the required performance of a spacecraft is written into the design specifications, and the operating conditions include the space environment in general and such variations as eclipse conditions in particular. The key to reliability is the word *probability,* which places the subject firmly in the realm of mathematics. The relatively small statistical sample for most space-qualified components and subsystems, compared with aviation or marine technology for example, is a disadvantage in reliability analysis, but the increase in the number of payloads placed in orbit brings a natural maturity to the subject.

In common with most fields of human endeavour, the prevention of potential problems is sadly overlooked. The need for a scientific approach to failures in the space industry was prompted mainly by the dismal record of American launch vehicles in the 1950s (15 of the first

25 Atlas launchers were failures, for example). The situation improved, but in January 1967 the deaths of three astronauts in the Apollo 1 fire prompted a reassessment, with a particular regard to safety. Such was the reliability of NASA hardware during the following decade and a half that NASA became renowned for its technical esteem and attention to detail. Unfortunately, there seems to be a necessity for a periodic re-evaluation of reliability, as the 1986 Shuttle disaster showed only too well.

This chapter will present a mainly qualitative description of satellite reliability by describing the main factors which limit the life of an operational satellite and the methods used to improve its overall reliability.

11.2 RELIABILITY OF THE COMMUNICATIONS SATELLITE

Once again it is the isolation of a communications satellite which sets it apart from most other technologies. The satellite is required not only to withstand the *g* forces, vibrations and acoustic environment of a launch and the propulsive forces associated with its orbital transfer, it must also survive for a period of 10 years or more, totally unattended (physically if not electronically), in the vacuum and temperature extremes of the space environment. Over this period all electronic and mechanical components are expected to operate perfectly without adjustment, servicing, parts replacement, lubrication or replenishment of consumables. This high reliability requirement leads to the most exacting standards of manufacture, selection of materials and components, and electronic and mechanical design. To ensure that the standards are maintained, exhaustive testing of all components and circuits is undertaken before integration of the spacecraft, at intermediate stages and after final integration.

As implied by the definition above, the reliability of a satellite is the probability that it will perform satisfactorily throughout its specified lifetime. Another term which sports a formal definition is *availability*, that is the probability that a satellite is operating satisfactorily at any particular point in time [Maral and Bousquet 1986]. The availability of a satellite *system* depends on a variety of factors such as the success of the launch, the number of operating satellites and spares (in orbit or on the ground), and the replacement time, which depends primarily on the availability of a launch vehicle. The availability of an earth station, or any other ground-based equipment for that matter, depends upon its reliability *and* its maintainability (how soon it can be mended if it does go wrong). Whereas satellites in low Earth orbits may be tended by a space shuttle, the availability of an individual geostationary satellite is solely dependent upon its reliability, since maintenance is impossible. This makes reliability of paramount importance to the communications satellite.

11.2.1 Failure rates

The probability of failure is generally illustrated by a plot of failure rate against time, for obvious reasons known as the bathtub curve [see figure A11.1]. The same plot may also be found with the ordinate labelled probability of failure, $P(t)$, as the figure shows. Two general curves can be identified: one for electronic components, the other for mechanical devices. Looking first at figure A11.1(*a*), it is evident that the failure rate is initially high but decreases to a constant level. The early phase is known as the burn-in phase, when manufactured devices can be expected to fail in larger quantities.

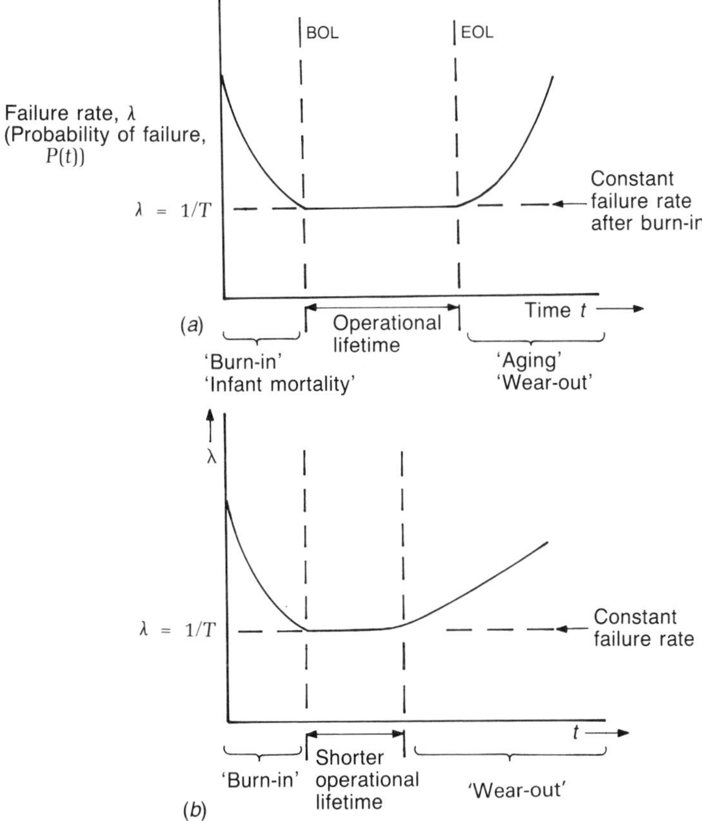

Figure A11.1 Component failure rate against time: (*a*) electronic components; (*b*) mechanical components.

As we come to expect in everyday life, if a new TV or hi-fi unit is going to fail electronically, it will do so in the first few weeks or months of ownership. This is why it could be worth buying the TV that has been working in the shop window for the past few weeks—it has been 'burnt-in'. This is, effectively, the philosophy applied to the communications satellite. By the time it is launched, it has been tested sufficiently to rule out most infant mortalities, as they are known. The burn-in process itself can enhance reliability by a factor of 100 at very little extra cost [Wilson 1983]. At the other end of the age scale, electronic devices suffer, not surprisingly, from aging or *wear-out* and the failure rate increases again. The failure rate of an item can be improved by *de-rating* the component, that is by reducing the maximum power consumption, voltage, current, operating temperature, or a combination of these, to retard the aging process.

It is evidently advisable to place the service or operational life of the component between the two sides of the bath tub, so to speak. Thus the period between a satellite's beginning of life (BOL) and end of life (EOL) exhibits a constant failure rate, due simply to random unpredictable failures [see figure A11.1(*a*)].

Figure A11.1(*b*) indicates how the curve is adapted for mechanical devices, which have a similar burn-in phase but begin to wear out much sooner, thus decreasing the optimum operational life. This again is evident in everyday life: the largely mechanical motor car has its teething troubles during the early months of its life and then, if the owner is lucky, enjoys a few years of reliable service, but after that it rapidly begins to fall from grace as breakdowns occur at an ever-increasing rate. In the case of a satellite, mechanical degradation is less of a problem, since most of the mechanical functions (e.g. antenna and solar array deployments) are carried out early in the mission. There is, however, no opportunity for periodic preventative maintenance: the concept of the 10 000 mile service is, as yet, unheard of in the field of satellite technology!

The curves in figure A11.1 are universal. They can be used as general indicators of the lifetime of a single component or of a whole system. Actual failure rates are derived from data gathered during testing and throughout the operational life of a system. However, since in statistical terminology these can only ever represent a sample of the total population, it is necessary to attach a *confidence level* to the quoted failure rate. The confidence level, quoted as a percentage, indicates how close the sample data are to the total population.

Table A11.1 shows some typical failure rate requirements for various satellite subsystems. Reliability engineers use an unusual unit for failure rates called the 'fit' or 'failure in 10^9 hours'. This incredibly long period of time (about 114 000 years) allows a very low failure rate to be expressed as a conveniently small number. For example, a length of waveguide is so inherently reliable that it is assigned a failure rate of 1 fit (one failure in 1000 000 000 hours!). A gyro assembly, on the other hand, is a much more complex device and warrants a higher figure of 4000 fits, which is equivalent to one failure in 250 000 hours, or 28.5 years. The table illustrates the relative probabilities of failure for a number of typical spacecraft components.

Table A11.1 Example failure rates for spacecraft equipment.

	Fits (10^{-9} failures per hour)
Communications payload:	
TWT	1000
Diplexer	100
Filters	1 per section
Coaxial switch	10
Waveguide switch (mechanical)	50
Waveguide switch (ferrite)	20
Waveguide	1
Waveguide flange	0.1
Stripline	1
Fixed attenuator	20
Variable attenuator	80
Propulsion:	
Tank (bladder diaphragm)	310
Tank (surface tension)	40
Latching valve	160
Thruster (hot gas—failure to operate)	450†
Thruster (hot gas—failure to close)	60†
Thruster (cold gas—failure to operate)	300†
Thruster (cold gas—failure to close)	60†
AOCS:	
Motor	100
Reaction/momentum wheel	100
Solar array drive mechanism (SADM)	10
Gyro assembly	4000
Bolometer (IR sensor)	100
Thermopile	160
Power:	
Solar cell	1
Battery cell (NiCd)	100
Battery cell (NiH_2)	100
Thermal:	
Heater (high temperature—thruster)	30
Heater (low temperature—line/tank)	10
Thermostat	50
Pyrotechnic:	
Explosive nuts/bolts	300 000

† Failures in 10^9 cycles.

11.2.2 Life-limiting factors

If we assume that a satellite's systems will be operating to specification by the time it reaches orbit, there are four remaining factors which may curtail its operational life: radiation damage, random (or accidental) failures, wear-out mechanisms and the exhaustion of consumables (typically station-keeping propellant).

Natural radiation, from the Sun and from the Earth's radiation belts, causes the degradation of solar arrays and thermal coatings [see §§A5.4 and A6.4 respectively]. The decrease in cell efficiency is taken into account in the sizing of the solar array [see box A5.2] and should have no unforeseen effects on the operational lifetime. However, the disruption of individual semiconductor components, known as single-event effects or upsets (SEE or SEU), do have an effect and must be considered random failures.

Two terms commonly found in the literature relating to failure statistics are mean time to failure (MTTF) and mean time between failures (MTBF). These terms are largely self-explanatory, but it is worth pointing out that the failure rate is the inverse of the MTBF ($\lambda = 1/\text{MTBF}$). The MTBF for the various components of a subsystem will invariably differ, meaning that some components will fail before others. As far as random failures are concerned, it is necessary to design components with MTTFs greater than the desired lifetime of the total system (preferably by several decades). This places total system failure due to random failure of a critical component well outside the nominal lifetime, as figure A11.2 shows. Where this is impossible or impractical, or extra insurance against failure is desired, the provision of spare or *redundant* components is the solution.

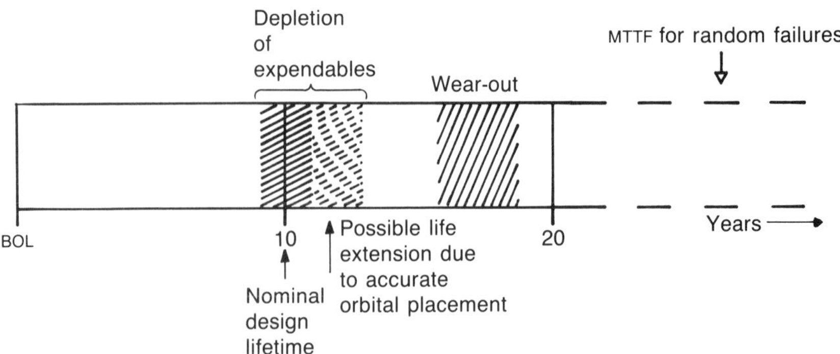

Figure A11.2 Satellite life-limiting factors.

As the satellite nears the end of its life, the overall probability of failure of a 'mission-critical' component increases, as components, such as bearings, solenoid-operated valves and TWT cathodes, begin to wear out. Wear-out mechanisms are non-random and are governed by a mathematically predictable 'normal' or Gaussian distribution—a probability distribution where the probability of failure, $P(t)$, increases to a maximum at a specified time, t_0, and decreases again thereafter [see figure A11.3]. A system is normally designed so that the mean wear-out times are long compared with the design lifetime. This provision and the placement of the MTTF for random failures makes for a reasonable system availability, which is important both to the user of a service and to the provider, for planning the number of satellite launches necessary to keep a system operating.

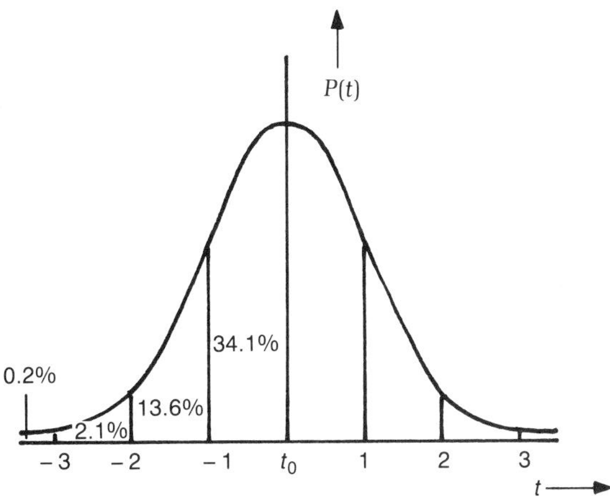

Figure A11.3 Normal or Gaussian probability distribution. Ordinate: relative frequency; abscissa: standard deviations (from the mean); figures within curve: percentage of total area under curve (% of total population).

With these provisions made in the basic design of a satellite, the factor which marks the end of its life is the depletion of its propellant resources. This is why the accurate positioning of the satellite by the launch vehicle is so important, since even an extra 2 years of station-keeping life will not bring the other life-limiting factors into play.

11.2.3 Product assurance—reliability improvement

Despite the difficulties associated with the provision of a satellite, there is much that can be done to improve the overall reliability of the system.

Most of the solutions can be grouped under the heading of *product assurance*.

Product assurance embodies a variety of related subject areas including reliability itself, quality assurance (QA), safety, the procurement and qualification of parts and the evaluation of materials and processes. Most industrial undertakings have a QA function to ensure the dependability and good name of their product, but for the reasons outlined earlier the PA function is particularly important when the product is a satellite destined for geostationary orbit.

Product assurance is, in essence, the 'policing of the design'. Despite the inherent wish of the good design engineer to produce a dependable product, the complex process of building a satellite requires a team of overseers who can devote themselves not to designing the most efficient 'widget', but to ensuring that all the widgets actually work, and, moreover, work in harmony with all the other widgets, subwidgets and gizmos!

Box A11.1 gives a summary of the factors that contribute to the doctrine of risk reduction which pervades the entire process of designing and manufacturing a spacecraft. The following sections describe the industrial standards, design philosophies and test procedures which combine to ensure that a satellite will be as reliable as possible.

BOX A11.1: REDUCTION OF FAILURE PROBABILITY: A SUMMARY OF DESIGN AND MANUFACTURING FACTORS

(a) Experienced contractors (prime and sub) with highly trained workforce;
(b) relatively labour-intensive procedures (i.e. not mass production);
(c) in general, materials, components, manufacturing procedures, etc, restricted to those already proven;
(d) high tolerances (mechanical and electrical);
(e) constant checking—quality assurance (QA);
(f) conservative design philosophy;
(g) redundancy provisions (reduction of potential single-point failures (SPFs));
(h) systems analysis and testing;
(i) configuration control;
(j) constant management of design (progress meetings, design reviews, etc);
(k) continual reliability studies (predicting and checking component and subsystem reliability);
(l) continual technical assurance (overseeing subcontractors to ensure that final component or subsystem will meet specification).

11.3 MANUFACTURERS AND COMPONENTS

As far as the purchaser of a product or service is concerned, one of the basic methods of product assurance involves choosing the right product or service provider. This activity begins at a very early stage—in the selection of the winning proposal. The proposal for a satellite system is the document in which the satellite manufacturer ‘sells’ itself by describing the key elements of its own design for the system. To ensure that the proposal meets the high standards required of the industry, it is managed by an experienced engineer with a team of experienced specialist engineers beneath him. The various subsystems of the spacecraft are addressed by individual specialist departments and the complete design is overseen by a number of overall systems engineers who check on its general feasibility. This is particularly important for the interfaces between the subsystems as previous chapters have shown. Of course the satellite manufacturer is more correctly termed the prime contractor. As with any other major industrial project, a satellite project involves a very large number of separate manufacturers and suppliers. The prime contractor manages the project and oversees the work performed by the various subcontractors. These subcontractors will in turn rely on teams of sub-subcontractors.

Although ideally members of a team are chosen on their merits, the situation for ESA contracts in Europe is complicated by the *juste retour* principle, which states that the value of contracts awarded to a member state should reflect, as closely as possible, the relative percentage of its subscription to ESA. Originally, the policy was introduced to encourage the growth of the European space industry, which it has succeeded in doing, but its management is becoming increasingly complex (and therefore costly) and it may one day be abolished.

Historically, companies which now manufacture space hardware have their roots in the aircraft industry which engenders similarly exacting engineering standards: examples include British Aerospace, Boeing, Matra, McDonnell Douglas, Saab, etc. Generally, they have grown gradually into space technology as opportunities in the field have expanded and usually have dedicated space divisions which are already experienced suppliers. The smaller subcontractors, who have a more individual speciality (e.g. solar cells, bearings, motors, etc), have also had to prove themselves. They have had to show, when they first embarked into manufacture for space, that they could meet the required standard and thence learn from further experience.

On the other hand, component manufacturers and material suppliers at the lowest level of subcontractor (e.g. for resistors, capacitors, wire, etc) may not produce components specifically for the space industry. However, in choosing a supplier the subcontractor can refer to a 'qualified parts list', developed by one of the space agencies (NASA, ESA, CNES, etc), from which a 'preferred components list' can be derived for the particular project. Components can be custom built, but this is expensive and not necessarily consistent with improved reliability. In general, off-the-shelf components of a military or industrial standard are preferred, since the high-volume production gives a good statistical basis for reliability determinations, and subsequent testing weeds out the poorer components. In some cases the aerospace industry takes the top few per cent of a batch of tested components, the rest being sold to less demanding industries (e.g. consumer electronics).

Despite the choice of experienced contractors and advanced facilities, the spectre of human error must still be faced. Most examples of this go unreported, but a fire in the Magellan Venus-mapping spacecraft while it was under test in 1988 was one of the exceptions. A technician, forced to insert an electrical connector without being able to see the socket and assuming it would be impossible to make a wrong connection, caused a short circuit. The resulting fire destroyed a battery and an electrical harness and cost an estimated $87 000 [AWST 1988b]. Another example was the case of the stumbling technician who 'wrote off' the carbon–carbon nozzle cone of an inertial upper stage when he tripped on his lab-coat, lost his balance and kicked it [AWST 1988c]. The moral to the story must be that where there are humans, there can be errors!

11.4 HERITAGE

Ensuring high reliability starts early in the design process, with the design concept of the satellite and all its subsystems. The overall design philosophy is generally conservative in that the number of entirely new developments is kept to a minimum. A proposed spacecraft is therefore invariably based on an existing design, incorporating flight-proven subsystems to reduce the probability of failure. This is especially true for the *bus* or platform which is usually intended to be used with a variety of different communications payloads.

An example of conservatism in the initial concept phase is the wish to keep deployable items, specifically communications antennas and solar arrays, to a minimum. However, the increased complexity of communications systems generally makes this a very difficult rule to adhere to. As the requirement for smaller spot beams grows, spacecraft antennas must increase in size, from the solid deployable type to the large unfurlable antennas described in chapter A7. Equally, the requirement for higher onboard power requires larger solar arrays. Despite this, the basic maxim prevails: the spacecraft design shall not include a deployable item where a fixed item will do.

Nowadays prospective customers are becoming increasingly wary of the satellite producer; they seem ever fearful that they might be the guinea pig for some new improved system or facility. Heritage tables, which show where major subsystems components have been used before, and their level of success, are therefore common in satellite proposal documents.

To the design engineer this conservatism can be frustrating, since it retards the acceptance of technical improvements, the very lifeblood of technology. Examples of such advances include the gradual expansion in operating frequencies from Ku-band to the higher frequency Ka-band; the increase in the RF output power available from the travelling wave tube; the use of combined bipropellant propulsion systems; and the move from silicon solar cells to those manufactured from gallium arsenide. Sometimes the customer will specifically exclude an improvement. The policy, as far as the national space agencies are concerned, has been to develop technology demonstration satellites to prove components and systems in space. Examples are ESA's Orbital Test Satellite (OTS) and Olympus, NASDA's Engineering Test Satellites (ETS), and NASA's series of Applications Technology Satellites (ATS). This policy is, however, fast becoming an unattainable ideal and, commercial pressures being what they are now, it is very rarely practicable to wait for the culmination of an in-orbit test programme before ordering a satellite. Most technology is years out of date before it reaches orbit as it is, without additional delays.

Thus the major satellite users, particularly Intelsat which orders relatively large numbers of any one specific generation, have become the standard bearers of technological innovation by specifying larger, more powerful, more capable satellites with each new generation. The US military authorities provide another example: without the prerequisite for commercial viability and with other economic considerations on the sidelines, military satellites sport untried sensors, huge unfurlable antennas and communications equipment of a complexity way beyond that of the lowly commercial satellite.

Despite the technophobia in the commercial arena, some advanced systems are beginning to appear on more ordinary commercial satellites (e.g. high-power TWTs on European DBS satellites). There are several contributory reasons for this. First, some satellite programmes are joint national undertakings which allow the risk to be shared among satellites for more than one country (e.g. France's TDF-1 and West Germany's TV-SAT). Secondly, based on their previous experience, most satellite manufacturers have developed a 'family' of different satellite types, thus tending to replace the one-off image with an off-the-shelf concept. Thirdly, in becoming wise to the needs of the market, component and subsystem manufacturers are highlighting the technical similarities in their evolving technologies, thus limiting the impression that an item with a new model number is a totally different device. Of course, some companies are better at the art of technical marketing than others, but then that is the nature of marketing departments, irrespective of the technology.

As a first step, therefore, it is usual to use proven designs where available and designs evolved from the proven designs where an advance is required or allowed. Specific advances can always be included in the proposal as options, but as far as possible the system design should be simple, with few of the 'bells and whistles' traditionally attributed to the practising engineer. This does not mean satellite design has been reduced to the process of choosing last year's subsystems from a component catalogue (although some of it *is* like that); it simply reflects the need for design solutions which are elegant rather than revolutionary.

11.5 REDUNDANCY

However simple and well tried a design, components can still fail. Although maintenance is impossible, spares can be carried in case of failure. This is the concept of redundancy.

If a number of components are linked together in series, the failure of any one will render the whole chain inoperative. This means that every item in the chain represents a *single-point failure*, or SPF. In a policy which keeps SPFs to a minimum, the most critical components, or those most prone to failure, are duplicated and connected in parallel so that one or other of them can be selected by a switch [see figure A7.6, for example]. If one spare unit is provided for every operational unit, it is termed 2-for-1 redundancy; if any two of three redundant units can be selected, it is 3-for-2 redundancy, etc. In a communications payload, for example, where a number of RF switches enable any of several transmit chains to be connected to any channel, the arrangement is known as ring redundancy.

In terms of the operating status of the equipment, three types of redundancy can be identified: parallel, active stand-by and passive stand-by. With parallel redundancy each item takes part in its assigned function throughout its lifetime and can continue even when the other has failed. This is the case for parallelled travelling wave tubes [see §A7.3.5]. Active stand-by is where one item is held in reserve in a powered condition but without contributing to the assigned function, and passive stand-by is the same except that the redundant unit is unpowered. These latter two states are sometimes referred to as hot and cold stand-by, respectively.

In the limit, this notion of redundancy can be extended to the whole satellite. There are cases where two satellites in similar orbital positions simultaneously provide the same service to a potentially large number of users. If one of the satellites was to fail, the other could continue, but the overall service (number of channels) would be reduced. Equally, a second satellite, termed an in-orbit spare, can be held in reserve in various states of readiness depending on the level of urgency for service continuity. Of course, spare satellites may also exist on the ground, but this is stretching the concept of redundancy to its limits!

No matter how many spares *are* incorporated within a satellite payload or subsystem, SPFs can never be entirely avoided. It is instructive to analyse a typical satellite mission to identify the potential SPFs and the associated redundancy arrangements.

11.5.1 The delivery system

The launch vehicle can be considered an SPF, since any failure which results in the destruction of the vehicle also terminates the mission of its payload. Launch vehicle reliability is a subject in its own right but beyond the purview of the present volume.

The other parts of the delivery system, apogee and perigee kick motors, are also SPFs and the success of a satellite's mission is totally reliant on the flawless operation of these propulsion systems in injecting their payloads into the correct orbit. This was shown only too well by the failure of the PAM (payload assist module) perigee motors attached to two communications satellites deployed on the tenth mission of the Space Shuttle in February 1984. One of the satellites, Westar VI, was destined for an elliptical transfer orbit with a perigee of around 300 km and an apogee of 35 800 km. When the errant spacecraft was finally located, it appeared to be in a 1190 × 250 km orbit. The other, Palapa B2, was later released into a similar, equally useless, 1220 × 310 km orbit. Following extensive analysis by the manufacturer, the reason for the failures was reported as the delamination of the exit cones as a result of density variations in the composite material. The two PAM solid motors were part of a lot of five and the failures were due to a lack of quality control on that particular batch [Williamson 1984]. The example shows succinctly the ramifications of a SPF. Unfortunately it is not possible to have a redundant exit cone on a perigee kick motor! The two satellites were later retrieved by the Space Shuttle, returned to Earth and refurbished for later use. Following several years of negotiation, they were finally resold and booked for a relaunch.

11.5.2 Deployable items

Once the satellite is in orbit, the deployable items, chiefly solar arrays and antennas, are the first sources of potential failure. Deployment mechanisms can fail for several reasons: uneven heating or cooling may distort and jam a moving part; a mechanism can become frozen because it has been allowed to become colder than intended; or cold welding may occur [see §A2.4]. These problems apart, array deployment mechanisms usually incorporate some redundancy measures, such as redundant springs and pyrotechnics for the three-axis satellites and redundant motors for the spinners. This makes them quite reliable, but an incident concerning the TV-SAT 1 spacecraft shows what can happen if quality control is deficient. When TV-SAT 1 reached orbit, it was found that one of the solar array panels had failed to deploy and that this had prevented the deployment of the receive antenna. It is thought that a number of latches, used to hold the array against the side of the satellite prior to launch, may have caused the panel to stick. Various attempts to free the panel failed and the spacecraft was declared a write-off [AWST 1988a]. An array panel on Arabsat 1, launched in 1985, also became stuck, but the problem was rectified by firing thrusters to overcome the friction which delayed deployment and 'shake the array out'. More complex deployable arrays (e.g. comprising an array *blanket* stored in a canister and extended by means of a motorized mast [see figure A5.2]) may prove even less reliable. Only time will tell.

The other main deployable item is the antenna reflector. Although the failure of an antenna to deploy can represent an SPF in itself, since part of the communication system would be incapable of operation, the deployment mechanism incorporates redundancy in the same way as the panel array mechanisms and failures are extremely rare.

The deployable omnidirectional TT&C antenna is another important SPF. On three-axis spacecraft the antenna is rarely deployed, since it is sited either on the satellite's feedhorn tower or on the antenna platform itself, but since contemporary spinners invariably have to deploy their entire antenna subsystem, the TT&C antenna must also be deployed. In the case of Syncom IV-3 (Leasat), launched by the Space Shuttle in 1985, the failure of the timer circuits designed to deploy the TT&C antenna after release made communication with the spacecraft impossible: commands could not be received and telemetry could not be transmitted. (The satellite was repaired on a later Shuttle mission [see figure A12.1].)

The general trend towards ever larger spacecraft antennas, to provide greater gain levels and smaller spot beams on the ground, results in an increasing requirement for deployable antennas. This is the case for three-axis satellites as well as spinners because of the finite size of their antenna platforms. Thus the future will see even larger antennas of the deployable and unfurlable types (like the ATS and TDRS umbrella-type antennas [see §A7.5.2]) whose mechanisms are even more complex. No one can currently say how reliable these reflectors are going to be, but a manufacturing or packing error could conceivably result in a nice tangle of antennas!

11.5.3 Service module

Although the early days of a satellite's life are fraught with uncertainties, once it has been working reliably for a number of months it is more likely than not to continue that way, at least for the main portion of its life.

The service module as a whole is generally fairly reliable, but this really depends on the individual subsystem design: some redundancy plans are better than others and, generally, the more flexible a system is, the better. Box A11.2 gives a summary of redundancy arrangements for the major spacecraft subsystems. Although it is largely self-explanatory, two items in the power subsystem are worthy of mention: BAPTAs and batteries.

Bearing and power transfer assemblies, by their nature as pivot points between two contra-rotating sections, represent SPFs. It is perhaps surprising that BAPTA failures are not observed, especially due to their mechanical nature, but very fine tolerances and careful manufacturing and handling techniques ensure a high degree of reliability.

Batteries have been known to experience problems, to the extent that they gained a name for unreliability in their early space applications, but modern batteries have a much better reliability record. Although redundancy arrangements usually maintain the required DC power levels for a satellite's operating lifetime, batteries are relatively heavy components and there is a limit to the number which can be carried.

11.5.4 Communications payload

Although deployable antennas were discussed above, the antenna subsystem is usually considered to be part of the communications payload. It may comprise a number of antennas mounted on the spacecraft's antenna platform, a feedhorn tower and lengths of interconnecting waveguide and coaxial cable. Whilst it is generally true that antennas are not duplicated, any failure is extremely unlikely since antennas, simple feedhorns and waveguides contain no moving or electronic parts. Coaxial cables and their associated connectors are, however, less rugged and therefore less reliable than waveguides and, although problems are rare, a number of satellites (e.g. Satcoms IR and V) have experienced failures of their TT&C antennas as a result of faulty coaxial connectors.

Generally, the more components an electronic unit has the less reliable it is likely to be, so more complicated payloads, for instance those which use digital signal processing techniques, have a greater risk of failure. The repeater itself always incorporates redundancy measures, often by the duplication of whole chains or subchains of equipment, which can be switched in by ground command depending on the fault. The switches are tested over thousands of cycles before payload integration to ensure reliability in orbit, and very rarely fail. The degree of redundancy depends entirely on the design of the payload and the requirements of the mission.

BOX A11.2: REDUNDANCY MEASURES IN SPACECRAFT SUBSYSTEMS

Propulsion: Solid AKMs—no redundancy.

Liquid bipropellant propulsion systems [see §A3.9]—more than one exit cone (and associated propellant feed hardware) is possible, but not universal.

AOCS: Reaction thrusters (gas jets)—invariably redundant. Often mounted in redundant pairs with redundant pipes and valves to allow one or other to be switched into operation. No SPF.

Reaction/momentum wheels—invariably redundant. Redundancy options depend on the specific wheel arrangement; several different systems are used [see §A4.5.3]. No SPF. (Note: on three-axis-stabilised satellites, wheels can be used with thrusters to orient the spacecraft.)

Sensors—invariably supplied in redundant pairs. Each array panel on a three-axis spacecraft has its own redundant pair of sun sensors, so that each array is individually controlled. No SPF.

Power: Solar arrays—essentially non-redundant since they are sized for a specified end-of-life power, but one panel on a three-axis-stabilised spacecraft could provide a reduced service.

Array deployment mechanisms do, however, incorporate redundancy: two pyrotechnic charges are installed with the release cable, either of which is sufficient to sever it. The 'drop-skirt' array on a spinner has redundant motors, but again if it failed to deploy the satellite could operate on about half-power.

BAPTA—no redundancy.

Batteries, power supplies, etc—mainly redundant.

Thermal: Heaters—mainly redundant.

Most of the subsystem comprises passive equipment and redundancy is not required.

Payload: Active components (e.g. amplifiers)—usually redundant.

Passive components (e.g. filters)—usually not redundant.

Antennas—usually not redundant (deployment mechanisms—same as arrays).

TT&C:—as with payload.

Travelling wave tube amplifiers are both the most critical and the most complex element of the spacecraft transponder, so a number of spares are always provided. Although the probability of failure of these devices is high relative to most other spacecraft components, it would be wrong to consider them as 'unreliable'. However, since the TWTA is such a crucial part of a communications satellite, a failure is, to say the least, noteworthy.

The failures experienced by the Japanese BS2a direct broadcast satellite in March and May 1984 provide a good example. Within 5 months of launch two of its three 100 W TWTs had apparently overheated and switched themselves off, leaving the satellite to operate on half its design capacity with the one remaining tube. Officially, the root-cause of the failures is unknown, since the limited amount of telemetry available from the satellite allows only a limited analysis of the problem. However, the deduction that the tubes were operating at a higher temperature than specified suggests that the EPCs, which provide the various voltages to the TWT, may have been overstressing the tubes. Not only is this instructive in providing a warning that the most obvious cause of a failure is not always the most accurate, it also shows the importance of the interfaces between equipment, especially when they are produced by different manufacturers as in this case.

In fact, TWTs have generally proved to be quite reliable devices, largely because of the rigorous test programmes they undergo, typically including some 2000 h of *burn-in* even before customer acceptance tests begin. The amount of test documentation delivered with each tube indicates its importance: it is reckoned that some 10 kg of paper accompanies every 600 g of tube hardware!

In addition to the test programme for flight tubes, a number of similar units are subjected to life testing to indicate their potential lifetime. For example, Thomson CSF began life testing a number of 20 W, 12 GHz tubes in 1974 when similar tubes were being prepared for ESA's Orbital Test Satellite (OTS). As of the beginning of 1987, 58 tubes had amassed about 4.3 million hours of operation between them: two or three had failed before their seven year design life was up, 26 survived more than 10 years and five for more than 12 years. It is now generally believed that tubes will last for at least 20 years, which is much longer than current satellite design lifetimes.

The OTS 2 spacecraft, launched in May 1978, survived its tenth anniversary with its tubes in working condition—more than three times its originally intended design life—although it was only used to a limited extent in the latter years. The survival of the tubes on its predecessor OTS 1 is, however, even more remarkable considering that its launch vehicle was destroyed at a height of over 9 km. Two of the tubes, attached to an aluminium honeycomb panel, were recovered later from a Florida beach and returned to ESA for analysis. Incredibly, they were found to be in working condition—even attaining acceptance test standards—and the one with its connectors intact was used thereafter as a laboratory RF source [see figure A11.4].

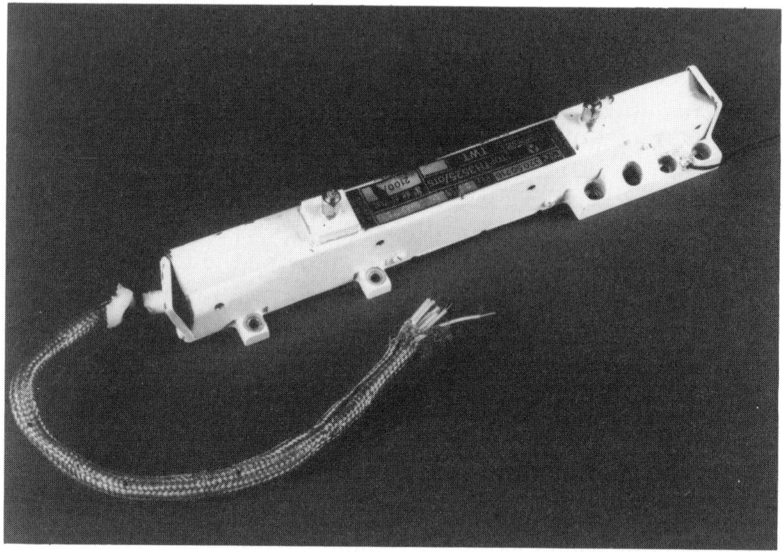

Figure A11.4 One of two travelling wave tubes recovered from the wreckage of the OTS-1 satellite after its launch vehicle was destroyed, which were later found to be in working condition. [Thomson CSF]

11.6 FAILURE MODES AND SYSTEMS ANALYSIS

Apart from a knowledge of what may fail, it is also essential to understand just how it will fail, that is its failure modes. Electronic devices can fail by becoming short-circuited (perhaps as a result of a wire filament joining two tracks on a circuit board), or open circuit (where a contact breaks), or they may become defective because of a drift in some parameter which should remain stable. Mechanical devices generally fail due to wear, corrosion, crack propagation and other time-dependent phenomena. Alternatively, some failures are ‘cycle dependent’ (e.g. the condition of a switch depends on the number of on/off cycles it has performed). It may be that identical spacecraft systems fail one after the other, due for example to a common manufacturing or design fault: these are called common-mode failures.

One step back from failure modes are the root-causes of the failures themselves, the *failure mechanisms*: the failure modes depict how an item fails, the failure mechanisms show why. They are the processes which occur at the atomic or molecular level but are observed by their effects at component or system level. Perhaps the best way to illustrate the difference between modes and mechanisms is by example—see table A11.2.

Table A11.2 Failure modes and mechanisms.

Component	Failure mode	Failure mechanism
Resistor	Open circuit	Wire corrosion due to contamination
Capacitor	Short circuit and leakage	Dielectric breakdown
Integrated circuit	Open circuit	Contamination of plating solutions

The reliability engineer regiments the information available on modes of failure and their relative importance into a failure mode effects and criticality analysis (FMECA). This analysis is conducted at all design stages to predict the effect of failure modes at component, subsystem and system level. Using these techniques the design can be evaluated before production begins as well as throughout the construction process.

Systems analysis, as a discipline, is a precursor to testing the actual hardware. It enables a theoretical, largely mathematical, model of the spacecraft and all its subsystems to be constructed, with which later test results are compared. For reliability purposes, it is common to consider a spacecraft as a number of independent subsystems—power, propulsion, AOCS and TT&C being the main ones. In a more detailed analysis the thermal control, structural, electrical distribution and pyrotechnics subsystems may also be specified. The communications payload is analysed separately from the bus subsystems and, if there are several different payloads, these are themselves treated as separate entities. Moreover, critical deployment systems, such as antennas and solar arrays, are the subjects of their own reliability analyses.

In terms of system modelling, the overall system reliability is a 'series model' where the composite probability of success is a product of the individual subsystem reliabilities. So if the reliability of three series elements is, respectively, $R1$, $R2$ and $R3$, the overall system reliability will be $R1 \times R2 \times R3$. For example, if all three elements have a reliability of 0.99, their combined reliability will be 0.97. This illustrates the fact that if a number of 'slightly unreliable' components are brought together and operated as a system, the overall reliability of the device is somewhat lower than the individual values: if one of the components is much less reliable than the others (say $R = 0.7$), it will bring the overall reliability of the system right down (0.69). Equally, a large number of components with fairly reasonable reliabilities (0.99) placed in series can make for a very unreliable system: a hundred such components would give a system reliability of 0.366! This is hardly surprising since the failure of any one element causes the failure of the whole system: they each represent an SPF.

This illustrates both the need for very high reliability of individual components and the need for redundancy. Redundancy alters the form of the series model to a network of series and parallel paths depicted in its simplest form by figure A11.5. Here, the reliability of two items in parallel is given by the sum minus the product, so the total reliability is $(R1 + R2 - R1 \times R2) \times R3 \times R4$. Real systems are, naturally, far more complex and computer techniques must be utilised.

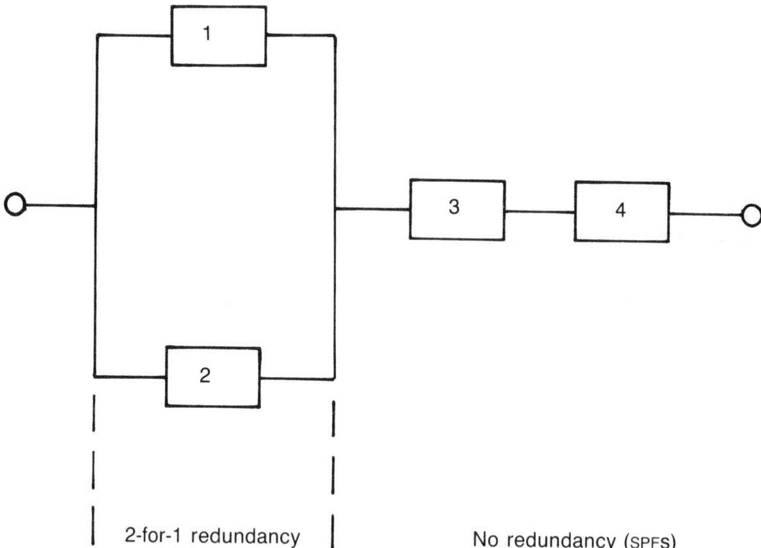

Figure A11.5 Basic reliability model showing series and parallel paths.

11.7 TESTING

At some point in the analysis process, theoretical data give way to actual test results as components and subsystems begin to arrive from the supplier. At the most basic level, tests are conducted purely to establish the reliability of a device or component, perhaps by a manufacturer wishing to quote a reliability value in his parts catalogue. Reliability testing is an expensive endeavour, mainly because of the long duration of the tests (for space equipment which must operate unattended for a decade or so) and the need to maintain the test *conditions* over a long period. The equipment may be committed to real-time tests of many years, but it is equally common to conduct *accelerated life tests* by increasing the value of a parameter, typically temperature, which is predominant in the aging process. This is, in effect, the opposite of *derating* [see §A11.2.1].

The usual objective of a reliability test is to measure the failure rate or MTBF. For example, a component claimed to have an MTBF of 1 million hours (about 114 years) at the 60% confidence level would have to have been tested for 916 000 hours with no failures, or more than 2 million hours with one failure [Meaker 1986]. An MTBF of 1 million hours is equivalent to 1000 failures in 10^9 hours (1000 fits), which is a relatively high failure rate as shown by table A11.1. Something like a solar cell or a length of waveguide, with a failure rate of 1 fit (once every 114 millenia!) is in a different league. As one might expect, these figures are the results of analysis and extrapolation rather than real-time tests.

Testing in general occurs throughout the manufacturing, integration and even pre-launch periods. Owing to the great expense of the major testing facilities, they are owned and operated by different companies and, particularly in Europe, in different countries. When a spacecraft has completed its final integration, it embarks on a tour of these facilities which include RF anechoic chambers [see figure A11.6], solar simulation facilities, thermal-vacuum chambers and vibration facilities. This enables all aspects of the satellite's operation to be checked before it is dispatched to the launch site. Theoretically, when the satellite has completed its tour of the test facilities, all of its component parts, both individually and as integrated entities, will have been tested to very high levels of confidence. The probability of failure should, therefore, be very small.

11.8 SPACE QUALIFICATION

The testing of components, devices and subsystems is all part of a process known as space qualification, the demonstration and certification that an item is fit for space. Items from single components to whole spacecraft are space qualified using a combination of previously acquired operational data and test experience. As part of this process it is traditional to build and test a number of spacecraft not intended for flight. This may include a mechanical or *structural model* to confirm that the structure and equipment will stand up to the shocks and vibrations of launch; a *thermal model* to verify the design of the thermal control system and confirm predictions of the thermal environment experienced by the various equipment; and an electrical or *engineering model* which tests the complete electrical/RF performance of payload and bus under vacuum and throughout a specified temperature range.

Figure A11.6 Antenna subsystem undergoing RF performance tests in an anechoic chamber. [Marconi Space Systems]

In this way the design is demonstrated to provide performance margins, such that deviations from the normal operating conditions will not render the satellite unusable—it is assumed that the spacecraft will be subjected only to random failures once in orbit. Qualification test values for ESA spacecraft are typically 1.5 times the expected levels for vibration and ±10°C on the predicted in-orbit temperatures; American qualification values usually have smaller margins. One of the many advantages is that faults can be introduced to check the analysis results without jeopardising the spacecraft's mission.

Separate *flight model* spacecraft are then built and tested, but only to predicted vibration levels and ±5 °C on temperature so as not to exacerbate failures. This is known as acceptance testing, since it provides a basis for acceptance of the spacecraft by the customer. It helps to ensure that no manufacturing errors have been introduced (to the already-proven design) during the production of the flight models and allows the spacecraft to experience the closest possible approximation to the space environment before it actually gets there.

Two distinct approaches to qualification can be identified: *prototype* and *protoflight*. The prototype approach, as expounded above, is intended for entirely new designs and involves testing a complete spacecraft well beyond the expected operating conditions. The spacecraft is then considered unfit for flight and may even be subject to destructive analysis. This is expensive, but since there are likely to be a number of similar spacecraft built, say for Intelsat or Eutelsat, the cost can be amortised over the production run. In some cases an engineering model is maintained in working condition throughout the life of the system to allow the simulation of problems which may occur and to demonstrate remedial actions before commands are sent to the orbiting spacecraft.

Where experience of the technology has been gained, it is becoming common to use the alternative protoflight approach, whereby it is argued that, since equipment which has flown successfully on a number of other spacecraft is being used, the design can be considered space qualified. The manufacturer using this approach is therefore obliged to demonstrate a good qualification heritage and to defend his claims in any dispute. This is not to say that the flight spacecraft are not tested: the equipment and the fully integrated spacecraft still undergo a qualification test programme, but the durations are halved [Meaker 1986]. There is, as yet, too little experience of the protoflight approach to state whether these spacecraft are any less reliable than those qualified under the prototype approach, but since protoflight is both faster and cheaper than prototype, its use is becoming more widespread.

11.9 CONFIGURATION CONTROL

A part of product assurance not yet considered but crucial to spacecraft reliability is configuration control. It involves monitoring the detailed design or configuration of the satellite, updating the documentation as necessary and making sure that everyone is working to the same edition of the design. It is a mammoth paper-based exercise in which everything in the design is reduced to a number, a long list of which constitutes the baseline design at any particular moment. A typical communications satellite may contain more than 300 separate items of equipment, designed, manufactured and tested by some 30 different companies over a period of about 3 years. Throughout this time, there can be more than 500 design, hardware, test and delivery reviews, and more than 2000 changes to the design may have to be made [Meaker 1986]. Although the design is eventually 'frozen' at the critical design review (CDR), after which flight hardware can be manufactured, the product assurance engineers must be on top of the whole process. The problem in synchronising and controlling all relevant documents and their seemingly endless updates is seen by many as a greater endeavour than actually building the satellite!

Despite the best efforts of the configuration control department, communication between individual members of the many different companies typically involved in a satellite design is not always as good as it might be. It has been known for parts from different manufacturers to be brought together at the integration stage, only to find that they do not fit! The importance of configuration control and the updating of configuration status documents is shown by the case of Solar Max, a low Earth orbit astronomical satellite launched in 1980. After only 9 months in orbit, four of its six telescopes became incapable of accurate pointing because a number of fuses in the attitude control equipment had blown. The fuses failed because they were of an incorrect rating for a circuit modification which had been made during the design process. Solar Max was, however, repaired in 1984 as part of the eleventh Space Shuttle mission, marking the first in-orbit satellite repair.

11.10 CONCLUSION

The reliability of a communications satellite is of paramount importance simply because of its enforced isolation. At best, in the case of a failure in low Earth orbit, it is very expensive to retrieve and repair a defective satellite; at worst, it is impossible. This chapter has attempted to give an overview of the principles of reliability and the relevance of reliability improvement to all aspects of satellite design and manufacture. The requirement to produce a satellite that is as failure-proof as possible affects the approach and performance of engineers in every speciality.

REFERENCES

AWST 1988a *Aviat. Week Space Technol.* 7 Mar. 57

—— 1988b *Aviat. Week Space Technol.* 14 Nov. 35

—— 1988c *Aviat. Week Space Technol.* 12 Dec. 30

Maral G and Bousquet M 1986 *Satellite Communications Systems* (Chichester: Wiley) ch. 9

Meaker T 1986 *Spacecraft Engineering Course Notes* (University of Southampton) ch. 17

Perrotta G and Somma R 1983 *Space Commun. Broadcast.* **1** 189–98

Williamson M 1984 *Phys. Technol.* **15** 284–90

Wilson M 1983 *Ocean Voice* **3** no. 4 15–18

A12 Space Insurance

12.1 INTRODUCTION

The previous chapter discussed the methods by which the reliability of a satellite can be improved. However, failures do occur and, since the operation of a communications satellite is usually a commercial venture, a mechanism for recouping the potentially high financial loss is required. This is where insurance comes in. In general, the insurance of a product of technology is of little interest to design engineers. They realise the need to produce a reliable product and, in most cases, strive to ensure this through an inherent technical diligence and professional pride (there is little kudos in designing a poor product!), but the requirements of the insurance industry are not high on their list of priorities. There is, however, a practical commercial reason for an appreciation of space insurance by the practising satellite engineer.

In a nutshell, the commercial entities that purchase communications satellites require substantial bank loans to finance their investment. Since the intention is to repay the loan from income generated by the operation of the satellite, the operator is reliant on the successful orbital placement and subsequent performance of the satellite. It is the risk involved in this which must be insured against if the bank is to provide a loan. If a large number of failures occur, insurance claims are high and the insurance industry makes a loss. If the industry sees no chance of an eventual profit, insurance will become unavailable. In summary: no insurance, no loan; no loan, no satellite. It is therefore encumbent upon the spacecraft manufacturers and launch agencies to make their product as reliable as possible. This is not to say that the insurance community *needs* to insure satellites: it is as yet but a small part of their business and, as the events of the 1980s have shown, their interest in this new field can wane [see §A12.5]. In fact, space insurance is confined almost entirely to commercial communications satellites. Military spacecraft and scientific and other satellites developed by space agencies (such as NASA's TDRS) are rarely insured.

This chapter describes the fundamentals of space insurance, its historical development and the likely prospects for the space industry in future.

12.2 INSURANCE POLICIES

12.2.1 Types of insurance

It is traditional to insure an item against specified risks for a set period of time. A satellite is, however, insured according to a sequence of events, some of which are unique to this type of product. The four main phases of a satellite's life in insurance terms are manufacturing, pre-launch, launch and in-orbit operation.

The *manufacturing phase* contains potential hazards which are well known in other types of insurance and can be covered in a similar way. It includes activities concerned with the assembly and test of space hardware to the point where it is ready for shipment to the launch site and is usually arranged by the spacecraft prime contractor (although it will of course be paid for in the price of the satellite).

Pre-launch covers transportation, temporary storage, integration with the launch vehicle and other preparations for launch, including the potentially dangerous loading of station-keeping fuel. Cover terminates at intentional engine ignition or upon release of the hold-down clamps, but can be reinstated in the case of a launch pad abort.

Launch insurance, in toto, provides coverage from launch vehicle ignition to the confirmation of correct placement and performance in geostationary orbit. It typically extends to 180 days from launch to allow time for pre-operational tests and initial operation, and to include at least one period of eclipse. In some cases, this phase is further subdivided into a launch phase and a positioning phase in an attempt to separate failure caused by the delivery system and the satellite itself. The majority of launch vehicles deliver their communications satellite to a geostationary transfer orbit [see chapter A10], so correct delivery to GTO is a convenient breakpoint.

The final *in-orbit phase,* when the satellite has reached operational status, can continue until the spacecraft is decommissioned at the end of its life, although the policy typically has to be renewed on a periodic, usually annual, basis to allow a reassessment of the risk which may arise from a change in the satellite's performance.

The most obvious failure in the launch phase is the total destruction of the launch vehicle and its payload which would be described as a total loss, but claims for a partial loss are also possible. For example, if the launch system fails to inject a satellite into GTO, it may be possible to boost its orbit using onboard station-keeping propellant. This was the case for the uninsured TDRS-A spacecraft launched from the Space Shuttle in 1983. The operators of an insured satellite that had its potential lifetime reduced by a similar use of station-keeping propellant would have grounds for an insurance claim for partial loss [see box A12.1]. On the other hand, if a satellite is injected into an orbit which cannot be corrected, as was the case for the Intelsat V-F9 spacecraft in 1984, it would be a total loss. In this case the orbit was so low and eccentric that the satellite had to be commanded to re-enter to reduce the hazard to other spacecraft and avoid a later uncontrolled re-entry.

During the positioning phase it is possible that the satellite's apogee motor could explode or that its solar arrays could fail to deploy fully, for example. These eventualities would constitute total and partial losses respectively.

BOX A12.1: PARTIAL LOSS FORMULAE

The example partial loss calculations given below are based on an actual insurance policy. Propellant depletion, transponder failures and available power are considered in separate clauses.

Partial loss (propellant)

In the event that the available station-keeping propellant is 50 kg or less, the satellite shall be deemed a total loss. In the event that the available propellant is less than 97.5 kg but more than 50 kg, the spacecraft shall be deemed a partial loss in accordance with the following formula:

$$PL_{prop} = \frac{97.5 - M}{97.5 - 50.0} \times 100$$

where PL_{prop} is the percentage of the insured sum payable for a partial loss (propellant) and M is the actual mass of propellant remaining.

Partial loss (transponder)

In the event that the satellite suffers transponder failures and is not declared a total loss, the amounts payable for each transponder failure will be US $1 million, but no claim shall be payable for a C-band transponder failure until the failure of the sixth transponder. In the event that two of the three TWTAs in the S-band payload fail, the payload shall be deemed a failure. The insurance value shall be the equivalent of five C-band transponders.

Partial loss (power)

In the event that the available electric power is 1000 W the satellite shall be deemed a total loss. In the event that the available power is less than 1495 W but more than 1000 W, the satellite shall be deemed a partial loss in accordance with the following formula:

$$PL_{power} = \frac{1495 - P}{1495 - 1000} \times 100$$

where PL_{power} is the percentage of the insured sum payable for a partial loss (power) and P is the actual power available.

Failures during the operating phase can be total, for example as a result of an uncontrollable pointing error, but are more likely to be partial. Although failure or degradation of satellite equipment can reduce the service life, other factors—like damage due to natural radiation or micrometeorites, and the human error of spacecraft controllers—must also be allowed for in a 'fully comprehensive' policy.

Launching spacecraft requires another type of insurance that sounds familiar to those who have motor insurance policies: third-party liability. This is intended to provide cover against damage suffered by third parties due to a launch accident or spacecraft re-entry, as demanded by the UN Outer Space Treaty of 1967 and the Liability Convention of 1972. The space industry has very little experience of third-party liability and there have been no claims of any note against such policies. This is largely due to range safety provisions, which stipulate that launches should be conducted from a coastal site in a seaward direction and that a range safety officer should have the ability to destroy a launch vehicle at any time in its flight. Although there is no record of an 'injury due to falling spacecraft', there have been incidents which could have been much more serious, for instance a Saturn V second stage fell into the Atlantic east of the Azores in 1975; fragments of the American Skylab space station fell on the west coast of Australia in 1979; and the Soviet Cosmos 954 satellite, including its nuclear reactor, re-entered over north-west Canada in 1978.

A variety of other insurances can be taken out by the spacecraft manufacturer and the purchaser. One way to buy a satellite is for the purchaser to make an initial down payment to the manufacturer and then to make *incentive payments* at agreed intervals if the specifications are met during operation in orbit. The manufacturer can insure against the loss of these payments, which would occur in the event of a failure. For the user, there is loss-of-revenue insurance, but claims can be difficult because of the problems involved in proving that a given revenue would have been gained had a failure not occurred (e.g. a satellite may not have been used to its full capacity).

12.2.2 Arranging insurance

As with most other types of insurance, space insurance is arranged for the client by a broker. The broker negotiates with the client, is responsible for the details of the contract and arranges coverage with the underwriters. The underwriter assumes the actual financial liability for the insurance policy and receives the insurance premium for his/her trouble; the broker, acting as intermediary between client and underwriter, is allowed a percentage of the premium for the business he/she arranges.

At present there are about half a dozen large American brokers involved in space insurance and a similar number of European ones. Since a significant percentage of the business has historically been placed with underwriters at Lloyds of London, most of the American brokers have developed a working relationship with a certified Lloyds broker in the UK. Lloyds underwriters are grouped into syndicates, a few of which are considered Lead Underwriters for space, in recognition of their pioneering spirit in accepting the risks of space insurance. The financial risk itself is spread across a large number of underwriting syndicates, and then spread further by reinsuring with other underwriters and reinsurance companies throughout Europe and the rest of the world.

Although some underwriters have a specific interest in space insurance, it is only a small part of their total portfolio and on any given day they are as likely to be providing insurance for aircraft, ships or buildings as for space launches. Some underwriters have their own technical advisers, but to many the brokers are the 'technical people': they are expert in policy wordings and take an interest in the engineering aspects of space technology. However, in general, neither brokers nor underwriters possess an engineering background and rely on independent technical advice when dealing with specifications and reliability estimates and assessing the significance of failures. One of the most important criteria for the assessment of risk is the reliability of a system derived by the manufacturer who is, of course, not entirely unbiased! The risk of receiving incorrect information, or of suffering the effects of deliberate overinsurance by the client, is known by the insurance community as a moral hazard.

As the insurance community becomes more involved in space insurance, the need for a comprehensive understanding of satellite and launch vehicle technology becomes more and more evident. Losses incurred by the insurance industry in the last decade or so have made it clear that the risks in space technology are quite different from those found in terrestrial fields. This is reflected by the wording of space insurance policies, which have grown to look more like the design specification for a satellite than an insurance document.

12.2.3 Wording and cover

As is often the case in understanding insurance policies, it is as useful to read the list of exclusions as it is to read what the policy *does* cover. Like the typical household policy, the space insurance policy excludes loss or damage caused by war, strikes, riots and other commotions and losses due to nuclear radiation, contamination, etc (except where it occurs naturally in the space environment). It is in deference to the destructive potential of space technology that current policies also exclude damage due to anti-satellite devices or any device employing nuclear fission or fusion, lasers or other directed energy beams. Apart from other general exclusions, there may be very specific exclusions based on previous failures which the insurance community would prefer not to cover again (e.g. the electrostatic discharge (ESD) problems experienced by the French Telecom 1A satellite in 1984 [see box A2.1]).

In defining the circumstances under which a claim will be justified, the policy document makes a number of technical definitions, including the previously mentioned concepts of total and partial loss. This is where the policy reads like a technical specification, quoting the allowable transponder failures in terms of usable bandwidth, the amount of propellant and power that should be available at the start of the satellite's operational life, and the attitude control pointing error. Any partial loss is calculated by considering the number of operational transponders and/or the life expectancy due to the station-keeping propellant and power available [see box A12.1].

Another nuance of the insurance policy is the 'deductible', a specified, otherwise allowable, type of loss for which the insurer makes no payment when it occurs. For example, if a series of satellites is insured together, 'one loss deductible' means that the first loss of a satellite will not be paid for, but any subsequent losses will. On a smaller scale, a policy could specify 'one transponder deductible', that is two transponders will have to fail before any insurance payment is made. This is analogous to the domestic policy which carries an 'excess' (i.e. claims are paid except for the first £15 or whatever).

12.3 SPACE INSURANCE RECORD

In the formative years of the space age space risks were uninsurable: launch vehicles were unreliable and most of the payloads were experimental in one way or another. It was not until 1965 that the first space insurance policy was arranged. It was attached to the first commercial geostationary communications satellite, Intelsat 1, otherwise known as Early Bird. However, the policy covered only third-party liability and pre-launch insurance. Launch and in-orbit cover were not provided, quite understandably, because of the lack of knowledge and experience of these phases at the time. The policy was placed with the aviation insurance community since the technology in that field was closer to space technology than all the others.

Of the first six Intelsat launches (Early Bird, four Intelsat IIs and an Intelsat III) only four successfully reached geostationary orbit. This led Comsat, the operator of the Intelsat system, to seek launch insurance for the next seven Intelsat IIIs. They succeeded in doing so, but only on the basis of 'one failure deductible' in each of two series of launches.

Unfortunately for Comsat there was a launch failure in the first series and an AKM malfunction in the second, so the insurance industry did not have to pay out. Despite the AKM problem, Intelsat III F-7 achieved geostationary orbit using station-keeping propellant, but suffered a reduction in its life as a result.

Again Comsat wanted to broaden the scope of insurance cover by including orbital positioning in the definition of launch insurance: such a policy was negotiated for the Intelsat IV series [Hill 1983]. However, policies were still being issued on a one failure deductible basis and the failure of one of the eight Intelsat IVs to reach orbit was, once again, not covered by insurance. In the mid 1970s the reliability of the Atlas and Delta launch vehicles looked good, and following the launch of the Intelsat IVs and a number of Westar and Anik commercial satellites, the insurance industry agreed (in 1975) to waive the deductible for the first time. The first satellite to benefit was the Marisat maritime communications satellite launched in 1976. Another insurance innovation initiated in 1975 was coverage for in-orbit failures, which paved the way for the comprehensive insurance of space risks. The first satellite life insurance was negotiated by RCA for its Satcom series [Hill 1983]. It was based on 10 years experience of geostationary satellites, which had given the insurance community sufficient encouragement to provide cover for the operational phase and allowed a sensible partial failure clause to be written.

However, no sooner had the space insurance market settled down to provide the modern form of cover in the four phases (manufacturing, pre-launch, launch and life) than it was obliged to make the first big pay-out. On 13 September 1977 ESA's OTS-1 (Orbital Test Satellite) was destroyed when its Delta launch vehicle malfunctioned. The insurance industry was still recovering from this failure when, in December 1979, Satcom III failed to reach orbit due to an AKM failure. The claims for these two losses, amounting to some $105 million, exhausted the entire premium income from the first decade and a half of space insurance and devoured the prospective premiums for years to come. In only 2 years, and as a result of only two claims, the space insurance industry's reasonable profits had turned into a substantial loss.

Table A12.1 is a summary of the major insurance payments made on space policies since the OTS-1 failure. Following the Satcom III claim, the failures of NASDA's ECS-II, the Indian Insat 1A and Inmarsat's Marecs B produced a similar amount in claims to that of the late 1970s, further depressing the insurance community. The situation could, however, have been much worse. But for the one loss deductible basis upon which ESA insured the Marecs programme, the Marecs B loss could have resulted in a claim for closer to $100 million than $20 million [Goudge 1985]. Marecs A and B together were insured for $96 million

[Danforth *et al* 1985], but when the Marecs A launch was successful this left Marecs B uninsured, so ESA decided to arrange an additional policy. The Italian Sirio II spacecraft was lost with Marecs B but was not insured. If it had been, the claim could have added another $50 million or so to underwriters' losses.

Table A12.1 Satellite failures and insurance payments.

Satellite	Insurance claim ($million)	Date of failure	Details of failure
OTS-1	29	13 Sep. 77	Delta failure
Satcom III	76	7 Dec. 79	AKM failure
ECSII (Ayame 2)	17	22 Feb. 80	AKM failure
Insat 1A	70	4 Sep. 82	In-orbit failure
Marecs B	20	10 Sep. 82	Ariane LO5 third stage
Satcom II	9	Apr. 83	Transponder failure (in seventh year of service)
Oscar 10	0.25	16 June 83	Ariane LO6 third-stage collision
Westar VI	105	3 Feb. 84	PAM-D failure (STS 41-B)
Palapa B2	75	3 Feb. 84	PAM-D failure (STS 41-B)
Intelsat V-F9	102	9 June 84	Atlas–Centaur second stage (orbit too low)
Telesat-H (Anik D2)	4.3	8 Mar. 85	Spin up due to ESD (life reduced by propellant use)
Arabsat 1A	6.7	16 Mar. 85	In-orbit malfunction
Syncom IV-3 (Leasat 3)	85	13 Apr. 85	TT&C antenna failed to deploy (STS 51-D)†
Arabsat 1B	3.8	Mid 85	In-orbit malfunction
Syncom IV-4 (Leasat 4)	85	6 Sep. 85	Communications failure after reaching GEO (STS 51-I)
GTE Spacenet F3	85	12 Sep. 85	Ariane V15 third stage
ECS3 (Eutelsat F3)	65	12 Sep. 85	Ariane V15 third stage
G-Star 2	2	28 Mar. 86	Transponder failure
Intelsat VA-F14	92.2	30 May 86	Ariane V18 third stage
TV-SAT 1	57	23 Nov. 86	Solar panel deployment failure
Intelsat V-F6 & F7	2.5		Minor in-orbit failures
Total claims:	991.75		(as of Feb. 1989)

† Repaired by STS 51-I Sep. 85 [see figure A12.1].

Figure A12.1 Space Shuttle recovery of the Syncom IV-3 communications satellite which was stranded in low Earth orbit when its TT&C antenna failed to deploy. [NASA]

Throughout 1983 losses were relatively small and the market regained some of its stability, but then came 1984 with its record claims exceeding $250 million. Two satellites released from the Space Shuttle on mission STS 41-B, Westar VI and Palapa B2, both suffered failures due to faulty nozzles on their PAM-D perigee motors [see figure A12.2]. The Shuttle was seen as more reliable than expendable launch vehicles because of extra safety measures associated with its nature as a manned spacecraft. What was not fully appreciated was that this reliability extended only as far as low Earth orbit: the PKM required to inject the payload into transfer orbit had reliability values in proportion to its expendable nature.

An equally bad year was 1985, which culminated in the loss of two insured satellites on the Ariane V15 mission. Even so, few were prepared for the period that would be labelled a 'launch vehicle crisis'. It began with the destruction of *Challenger*, which was bad enough, but continued with failures of an Ariane, carrying the insured Intelsat VA-F14 satellite, and a number of American expendable launch vehicles. There followed a period when none of the western world's major launch vehicles was considered operational and space insurance slipped further into the doldrums, suffering under a loss ratio of over 200%.

(a)

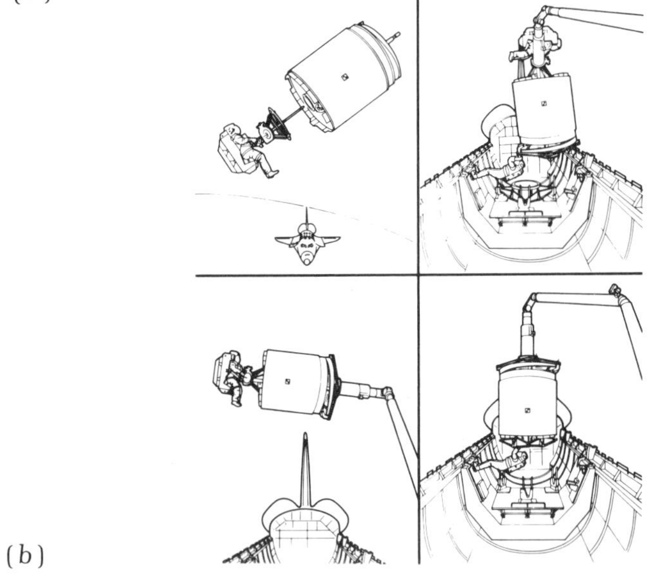

(b)

Figure A12.2 (*a*) Space Shuttle recovery of the Westar VI communications satellite which failed to reach orbit because of a faulty nozzle on its PAM-D perigee motor [NASA]. (*b*) The sequence of events designed to recover Westar VI and Palapa B2. In fact the aluminium truss, designed to be attached over the antenna (top right) to allow the Shuttle's manipulator arm to manoeuvre the satellite, could not be fitted, and one astronaut had to hold the satellite while another fitted a payload bay adaptor (lower right). The revised procedure was repeated for the second satellite.

12.4 ANALYSIS

With the benefit of hindsight, it is easy to see the reasons why the losses sustained by the insurance community were so huge. The roots of the problem lie in the small population of insured satellites and their high individual value compared with most other types of insured risk. The individual value of an insured Boeing 747, for instance, is also quite high, but the premium-paying population is large and the (fairly unlikely) destruction of an individual aircraft will not decimate the market.

This was, in itself, obvious to all, as was the lack of knowledge of space technology within the insurance industry, the reason for the initial caution. But it seemed that there was money to be made from insuring satellites and few worried over their ignorance of the technology: satellites were expensive so premiums were high and the short-term gains looked good on the balance sheets. After all, a satellite insured for $50 million at a rate of 10% would generate an income of $5 million. It is now widely agreed that the initial rates for space insurance cover were set too low, considering the known reliability of the Atlas and Delta vehicles. When the failure rate of the launch vehicles was around 10%, the insurance rates should have been set to reflect this *and* provide a margin for profit. As it was, early Atlas–Centaur launches secured rates of around 6% and, even between the OTS-1 and OTS-2 Delta launches, rates rose from only 7.2% to 7.9% [Danforth *et al* 1985].

Unfortunately, launch vehicle failure rates were based more on opinion than historical data. Just before the 1984 Westar and Palapa failures, rates for launch coverage varied between about 5% for the Shuttle and 10% for Ariane, the former being seen as a relatively reliable satellite launcher and the latter relatively unproven. By February 1985, however, Brazilsat and Arabsat reportedly paid record rates of 20% for an Ariane launch *and* found it difficult to secure even a minimum ($25 million) cover [Danforth *et al* 1985]. In fact rates were to climb further to about 25% before slowly reducing again. Table A12.2(*a*) compares the failure rates of four types of expendable launch vehicle currently available for commercial communications satellite launches.

Table A12.2 (*a*) Launch vehicle failure rates (for launches to GEO) to 26 January 1989.

Launch vehicle	Number launched	Number failed	Failure rate (%)	Success rate (%)
Atlas–Centaur	34	5	14.7	85.3
Delta	83	5	6.0	94.0
Titan 3C/34D	39	6	15.4	84.6
Ariane	24	3	12.5	87.5

Table A12.2 (*b*) Details of failures (for launches to GEO) to 26 January 1989.

Launch vehicle	Satellite	Date	Failure
Atlas–Centaur	ATS-4	10 Aug. 68	Second stage
	Intelsat IV-F6	20 Feb. 75	First stage
	Intelsat IVA-F5	29 Sep. 77	First stage
	Intelsat V-F9†	9 June 84	Centaur
	Fltsatcom 6	26 Mar. 87	Lightning strike
Delta	Intelsat III-F1	18 Sep. 68	First stage
	Intelsat III-F5	26 July 69	Third stage
	Skynet 2A	19 Jan. 74	Second stage
	GEOS 1	20 Apr. 77	Stage 2/3 spin up
	OTS 1†	13 Sep. 77	First stage
Titan 3C/34D	IDSCS 2	26 Aug. 66	First stage
	DSCS 5/6	20 May 75	Guidance failure
	DSCS 9/10	25 Mar. 78	Second stage
(3 other military launches failed: Aug. 85; 1986; Sep. 88)			
Ariane	Marecs B†/Sirio 2	10 Sep. 82	Third stage (LO5)
	Spacenet F3†/ECS 3†	12 Sep. 85	Third stage (V15)
	Intelsat VA-F14†	30 May 86	Third stage (V18)

Notes: upper stage failures not included in above statistics. Of Delta-launched payloads, Intelsat II-F1, Intelsat III-F8, Skynet 1B and RCA Satcom III (insured) suffered AKM failures. Of Titan-launched payloads, OV-2 and IMEWS 1 suffered Transtage apogee burn failures.

† Insured payloads [see table A12.1].

It can be seen from the table that the success rate of the Delta is slightly better than the other vehicles, but this is spread over approximately twice as many launches as the other vehicles. One could therefore conclude that twice the experience has resulted in an improved reliability. The figures should, however, be treated with great caution since they will be out of date well before this book is published. For example, if the Ariane launch vehicle following the last included in the table is a failure, the failure rate would jump to 16%. This indicates one of the problems of the small statistical population.

It is extremely important, and often very difficult, to compare like for like in such cases. The table includes only those launches intended to place satellites in geostationary orbit; the vehicles listed have, between them, launched many other spacecraft to low Earth orbits and interplanetary trajectories. This restriction is important since the launch sequence includes the crucial injection into geostationary transfer orbit.

Even more important, the failures quoted are due to malfunctions in the launch vehicle system, not the satellite's AKM, and certainly do not contain failures sustained by the satellite once in orbit.

Table A12.1 lists the causes of the failures which brought about the industry's major claims. A cursory analysis of the causes shows that, counting the loss of two satellites on Ariane V15 as one failure, 11 were due to propulsion or delivery systems of one kind or another and nine to the satellites themselves (although in one case the satellite *was* 7 years old!). Of the 11 propulsion failures, seven were due to the launch vehicle itself, while two were PKM (PAM-D) failures and two were due to the satellite's AKM. This sort of analysis may help to apportion the blame, but it does not immediately indicate the extent to which one category of insurance policy involves more risk than another. Although failures of the delivery system obviously fall within the launch insurance category [as defined in §A12.2.1], several of the 'satellite failures' occurred within 180 days of launch and thus fell into the same category. This shows that the majority of all claims are made against launch policies as opposed to in-orbit policies, reaffirming the observations made in the previous chapter regarding reliability: once a system has been shown to operate successfully over a period of time, it is more likely than not to continue in that fashion. Indeed, this is reflected by insurance rates for in-orbit coverage which are typically 2.5–3.5% of the satellite's value.

12.5 INSURANCE RESTRICTIONS

One of the major factors restricting space insurance is the lack of capacity in the market, that is the maximum dollar value of cover that underwriters are willing to attach to an individual launch. Even before the *Challenger* disaster, space insurance capacity was less than half what it had been 2 years earlier—the Westar and Palapa failures had seen to that. Capacity in 1988 was estimated to be between $80 and $110 million for launch insurance [Plochinger 1988], a marked improvement on the situation during the launch crisis of 1986. It did, however, pose a potential problem for dual or triple launches, since a single large communications satellite could easily take $100 million of the market's capacity. By early 1989, capacity was estimated to have grown to more than $200 million and, barring another launch crisis, seems likely to improve further.

Although provoking comparatively little interest compared with launch insurance, third-party liability insurance for space risks has also been affected by a reduction in capacity. The US Government sought to offset its liability, under international law, by demanding that users took out insurance policies naming the US Government and its contractors as beneficiaries, a strategy which, incidentally, led ESA to insure the OTS-1 launch [Hill 1983]. Thus NASA stipulated that coverage of $500 million be provided, but since rates were relatively low this could be arranged for a premium of only a few hundred thousand dollars. However, between January 1985 and November 1986, the rates were increased by a factor of 3 and were expected to rise further to bring the space industry into line with other high-risk industries [Greenberg and Gaelick 1986]. Why they were placed so low in the first place is unclear, since a rate of only 0.002% (2×10^{-3} dollars for every dollar of cover) would equate to a premium of $1 million for the required $500 million coverage. This would require the premiums from 500 launches to cover a single liability claim at the maximum allowed for, and at a typical launch rate of 20 per year it would take 25 years to collect sufficient premiums. One can only assume that the probability of a claim is very low (i.e. less than once every 25 years or so), but this is not to say that a claim will not be made next year.

Whatever the rate, because of the general reduction in space insurance capacity, in 1986 it became difficult to obtain more than $300 million of cover, which made NASA's figure of $500 million purely notional. In November 1988, however, President Reagan signed a launch insurance bill which established a shared liability between the US Government and commercial launch vehicle operators for third-party damage (up to $500 million).

Apart from a general lack of capacity, a number of other changes in the insurance industry have affected prospective policy holders. The common practice of insuring launches in groups all but ceased in the wake of the launch crisis, since underwriters wanted to see the result of the last launch before setting a rate for the next, or even agreeing to provide cover at all. Rates were originally quoted as valid for up to 3 years, which was useful in cases of a launch delay, but in 1985 this was reduced to 3 months. Although recent years have seen a partial return to group insurance, it is possible that future clients will suffer an increase in deductibles and a return to insurance cover limited to the more reliable mission phases.

Having originally been 'sold' the space business as an exciting and potentially profitable, tried-and-tested, low-risk venture, many underwriters reached a point where they began to feel they were funding launch vehicle research and development and satellite in-orbit test programmes. Naturally they should not be expected to do this, but equally the client should not be expected to fund the insurers' losses, incurred as a result of bad luck or an inability to set the correct rate at the right time.

12.6 ALTERNATIVES

In general, when insurance becomes too expensive, one of the common solutions that arises is self-insurance, which usually involves making or buying a back-up to the prime item, whatever that may be. From time to time, self-insurance is mooted for the space industry. Of course, providing an extra satellite and its launch is a very expensive option, but if a four-satellite series is under consideration and insurance rates are at 25%, the insurance premium is equivalent to the cost of one satellite. In this case, extending the production run from four to five satellites may be an alternative to insurance (as long as one is willing to bet that no more than one launch in five will be a failure). Not surprisingly, within the insurance industry this option is seen as unviable.

An extension of this is the self-insurance pool, which could be formed by a number of individual satellite owners. For instance, each of eight owners might provide a letter of credit amounting to 12.5% of the average value of the eight satellites. Then if one of the satellites fails, its replacement is paid for. The argument goes that, if the probability of one failure is 10% (based on the average launch success rate), the probability of two is $(0.1)^2$ or 1%, and that of three is $(0.1)^3$ or 0.1%. The assumption is that since a second failure has much lower odds than the first the insurance for subsequent failures will be that much cheaper. This does, however, assume that insurance would be available on this basis. Probability assumptions only work over large populations and launching satellites is not like throwing a die—there are many more variables.

Apart from this, insurance is a business and requires, at least eventually, a profit. The underwriters' argument is that self-insurance makes no provision for the long-term future: once a failure has occurred the client reverts to the insurance market, but the market has had its funds depleted due to a lack of premiums and can no longer afford a loss. Moreover, the 'pool' concept has been attempted several times by airlines and has met with little success, one of the problems being that airlines with high reliability records failed to see why they should pay for the lower standards of others.

Another alternative for the prospective satellite owner is 'delivery in orbit', an option pioneered by Hughes Aircraft Company for the contract to provide the British Satellite Broadcasting (BSB) DBS system. The contract stated that, in the case of a launch failure, Hughes would provide a cash refund or build and launch a replacement satellite. Title to the satellites, two of which were planned for launch in 1989 and 1990, would be transferred to BSB only if they passed their in-orbit acceptance tests, thereby relieving the customer of any launch risk. At the time of writing, although delivery in orbit has also been specified for the Australian Aussat B series, it remains to be seen whether this will become a common practice. Transferring the risk from the client to the manufacturer is seen by many as a purely short-term measure, intended only as a ploy to win contracts, since the manufacturer is likely to require insurance to cover his risks. We are back to the insurance industry.

12.7 CONCLUSION

There is little doubt that the time is nigh for a reassessment of the needs of the space insurance industry and its potential clients. There is little hope for the commercial expansion of space without the possibility of insurance—the risks are simply too great.

As far as launch vehicles are concerned, it is necessary to establish a true reliability record and adjust insurance rates in proportion to that record. It is recognised that a Shuttle launch is different from an ELV launch, but as figure A10.13 shows, payloads can be deployed from the Shuttle in several different ways: the risks associated with the PAM and IUS upper stages, the Leasat-type frisbee deployment, and RMS deployment are all different. If commercial payloads are to be carried again by the Space Shuttle, insurance analysts will have to take this into account. Equally, the record of the various ELVs must be open to further analysis: stage additions or modifications and strap-on boosters alter the character of a vehicle.

Satellites present even greater problems of comparison since they comprise a large number of subsystem components produced by different manufacturers. However, it is usually possible to identify the cause of a failure—at least to subsystem level—and compile statistics on this basis. Unfortunately, this has historically been done with too little regard to the true significance of the failure. Travelling wave tube failures, for example, have suffered from a degree of attention in the technical media out of all proportion to their significance. They have also suffered from uninformed 'blanket assumptions'. Little note has been made of the failure rate of TWTs from one manufacturer compared with another and TWTs have taken the blame when the fault has lain with their EPC (electronic power conditioner), perhaps supplied by a different manufacturer [see §A11.5.4].

If insurance policies were more closely related to the actual hardware being insured, spacecraft manufacturers would be obliged to enhance their reliability record by using only the most reliable subsystem suppliers. There must, however, be a balance if insurance cover is not to be totally weighed against advances in technology. This would force manufacturers to be even more conservative, an aspect which is already built into the satellite procurement process [see §A11.4]. The only way ahead for the two parties is through an understanding of each other's needs.

Future trends in spacecraft technology will make this even more necessary. The change from solid propellant apogee motors to combined liquid bipropellant systems, the progression from silicon to gallium arsenide solar cells and the inclusion of a communications payload operating at 20 GHz rather than 11 GHz are all aspects of subsystem technology whose significance requires special consideration. The next chapter discusses these and other developments.

REFERENCES

Danforth J C *et al* 1985 *Insurance and the Commercialisation of Space* (US Senate Committee on Commerce, Science and Transportation)

Goudge B 1985 *Satell. Technol.* Apr. 14–18

Greenberg J S and Gaelick C 1986 *Space Policy* **2** no. 4 307–21

Hill S M 1983 *Space Commun. Broadcast.* **1** 393–403

Plochinger L 1988 *ESA Bull.* no. 53 84–7

A13 Future Trends in Satellite Design

13.1 INTRODUCTION

It is never easy to predict the future, but in matters pertaining to space technology it seems, if anything, more difficult. Whereas with other technologies it is often the case that developments could not have been dreamed about 10 years earlier, it invariably seems that proponents of space exploration and commercialisation have a knack for dreaming and are often over-optimistic.

This optimism is in part due to the well publicised initial success of the American space programme. To contemplate a man walking on the Moon only 8 years after Yuri Gagarin made an orbit of the Earth, and only 7 years after the first American to do so, would have been classed as science fiction at the end of the 1950s. At the time, the USA was having trouble lifting its rockets off the pad, let alone to the Moon! Such was the euphoria of the early Apollo successes that pundits were optimistically predicting permanently manned space stations in Earth orbit by 1980 and a manned mission to Mars by 1990. The delays associated with the first launch of the Space Shuttle and the enervating effect of the *Challenger* disaster have brought a greater sense of realism to predictions of progress in space technology, but there is still the danger that those making the predictions are not fully aware of the need for conservative and reliable design, which takes both time and money.

Although the communications satellite sits on the sidelines of space technology, as far as media and public interest is concerned, its development is bound by the same constraints as any other, perhaps more newsworthy, space venture. Technical improvements require time and effort and can often consume copious amounts of money. There are, moreover, other factors which constrain advances in communications satellite design. One of these is the performance of the launch vehicle. Great advances in communications satellite engineering are easy to propose, as the concepts of the early 1980s for huge space platforms show, but if a launch vehicle is not available to lift the mass of the 'super satellite' into orbit, the design will probably remain on the drawing board.

Equally limiting are the needs of the satellite user. Satellite communications is the first big commercial application of space and advances are constrained by the need to provide a capability that the market can accommodate, at a price it can afford and with a level of reliability that allows a profit to be made. As a result, a balance is struck between the technocrats who propose the advanced concepts and the satellite buyer who prefers the more conservative design. The potential for the technological growth of the communications satellite is bounded only by the imagination of the designer, but the need for it to be a commercially viable product limits any advances to what is strictly necessary for the planned development of the communications market. Although this reduces the possibility of creating a 'service looking for a subscriber', it retards the growth of innovative satellite-based services like communications with mobiles and accurate position determination or 'global positioning' amongst others.

Despite the constraints, the number of commercial communications satellites in geostationary orbit has grown from 1 in 1965 to around 200 in 1989 (although they are not all operational [see appendix A]). The improvement in satellite technology over a similar period can be gauged from the number of transponders a satellite carries. Taking the 'measuring stick' to be the number of equivalent 36 MHz transponders, the average satellite capacity increased from 1 in 1965 to about 36 in 1985 [Hudson and Gartrell 1987]. In terms of the traffic these bandwidths could carry, this was equivalent to 240 two-way telephone circuits in 1965 and 15 000 circuits plus two TV channels in 1985 (Intelsat VA). In comparison, the Intelsat VI generation has taken advantage of advanced techniques to provide 120 000 voice circuits and three TV channels in only one and a half times the total bandwidth.

The Intelsat VI series is an example of satellite technology of the 1980s; in 1965 it would have been difficult to predict the capacity of the Intelsat V system let alone the Intelsat VI series. However, now that the subject is at least relatively mature, a knowledge of the relevant payload and subsystem technologies can be used to identify the likely development of the communications satellite in the near future. This chapter discusses the near-term prospects for both payload and subsystem technology and the possibilities for communications satellite systems in general. Since the accuracy of any long-term prediction is questionable, this analysis will be confined to the final decade of the twentieth century.

13.2 SATELLITE SIZE AND LIFETIME TRENDS

13.2.1 Size

The Intelsat example in the previous section illustrates the overriding trend in satellite communications in the last two and a half decades: satellites have become larger, heavier and able to carry an increasing amount of communications traffic [see figure A13.1]. The natural tendency would be for this trend to continue. As launch vehicles develop greater lifting capabilities and employ shrouds of greater volume, satellites can be designed to carry more payload, generate higher powers from larger arrays and radiate more heat from larger surface areas. It is an extrapolation of this trend which has produced plans for the large space platforms proposed for future communications. Although it is not a communications platform, the Columbus Polar Platform, shown in figure A13.2, is representative of the type of spacecraft under discussion.

Figure A13.1 Intelsat VI communications satellite compared with the original prototype of the first geostationary communications satellite, Syncom, which would fit inside one of the Intelsat VI propellant tanks. The masses of the two satellites are 35 kg and 4240 kg respectively. [Hughes Aircraft Company]

Figure A13.2 Artist's impression of the Columbus polar platform, part of the European Space Agency's contribution to the US/International Space Station programme.

There are, however, a number of arguments against the viability of large satellites, at least in the period considered here. Authorities on the commercial aspects of satellite telecommunications have argued continually over the past few years that there is a worldwide 'transponder glut', whereby the supply of transponders in orbit far exceeds demand. The decrease in newly orbited transponders due to the preponderance of launch vehicle failures in 1986 caused a temporary 'blip' in the upward trend, but by the end of 1988 the launch rate, particularly that of Ariane, was back to normal levels. It remains to be seen whether the resultant increase in the number of transponders in orbit, especially those with coverage of western Europe, will renew the cries of oversupply. As the majority of the new satellites with European coverage are designed to transmit television, this rather depends on how many channels the consumer is willing to take. Since larger satellites would increase transponder capacity still further, sales of those satellites (like Olympus and Intelsat VI) will depend on the trends in satellite usage in the next few years.

Satellites in the medium-size class have so far been the best sellers for developing countries and other first-time users. This is largely because the customers are impressed by the length of the production line (which numbers more than 30 for the Hughes HS376), the breadth of in-orbit experience and the competition between rival manufacturers, which helps to keep the price down [Renner 1983].

From a technical point of view, an increase in satellite size has both positive and negative aspects. As an alternative to simply cramming more active transponders into a large satellite, the redundancy of the payload elements could be increased. This would improve the in-orbit reliability of the satellite, a factor which is of increasing concern, not least with regard to obtaining insurance [see chapter A12]. Since adding redundant elements decreases the number of active revenue-earning transponders, the options must be very carefully considered. The loss of three or four active transponders could amount to about half the capacity of a small satellite, but only one-tenth for a large satellite. A 10% loss in potential revenue might be worth the extra reliability, but a 50% loss would not. On the other hand, increasing the overall size of a satellite causes its volume and mass to increase faster than its external surface area. This could cause problems in the radiation of excess thermal energy and the provision of places to mount the increasingly numerous and complex antennas which might be required.

13.2.2 Lifetime

One of the results of larger satellites and increased redundancy is longer potential lifetimes: redundancy means that a greater number of units can be allowed to fail before the satellite becomes uneconomic to operate, and its larger size means that more station-keeping fuel can be carried. However, if north–south station-keeping requirements can be waived, a great deal of propellant can be saved and the satellite's life can be extended. This option has been dubbed 'the Comsat manoeuvre', after the Comsat organisation which claims propellant savings of up to 90%. Details of the manoeuvre are the subject of a patent and difficult to ascertain, but when applied to an in-orbit Comstar satellite, annual consumption was reported to decrease from 16.8 kg to 1.4 kg, a significant saving. The main disadvantage of this manoeuvre is the daily motion of the satellite above and below the equatorial plane, which means that large-diameter/narrow-beamwidth earth stations have to be constantly repointed. Despite this, there are times when the manoeuvre could prove useful, for example when an aging satellite is awaiting a replacement that has been delayed by the unavailability of a launch vehicle.

A significant increase in the intended design lifetime of the communications satellite seems unlikely in the near term, however. In the first decade of the commercial communications satellite era (1965–75), the average design lifetime grew from 2 or 3 years to about 7, but the next 10 years saw a reduction in the growth rate as lifetimes approached 10 years [Hudson and Gartrell 1987]. Many satellites being launched at the end of the 1980s have nominal 10 year lifetimes. This places their demise in the early years of the twenty-first century, by which time the technology on board the satellites (which may date from as early as 1985, when the spacecraft was designed) will be well out of date.

Of course there is nothing wrong with out-of-date technology as long as it works, but one factor likely to support the replacement of such satellites by the year 2000 is the overcrowding of geostationary orbit. While a satellite is occupying an orbital position, another operating at the same frequency cannot take its place because of the potential for interference [see chapter B2]. This places a limitation on how efficiently the available frequency band can be used and restricts the number of voice circuits or TV channels that can be provided. The potential waste in boosting a perfectly good 10 year old satellite into the graveyard orbit is an eventuality which may have to be faced in the next decade.

The following section describes some of the improvements in payload technology which may lead to this competition for orbital and frequency space.

13.3 PAYLOAD TECHNOLOGY

Over the next 10 years, an increasing number of satellites using the 20/30 GHz communications band (Ka-band) are expected to join the C-band and Ku-band satellites currently in orbit. Expansion of satellite operations into this band obviously makes more bandwidth available and reduces overcrowding in the lower frequency bands. It may also allow the use of smaller earth terminals (because of the beamwidth relationship), but this advantage will tend to be cancelled out by the increased atmospheric attenuation at higher frequencies [see chapter B4]. Only a handful of satellites launched in the 1980s were designed to carry Ka-band payloads but, just as frequencies grew from C-band to Ku-band in the 1980s, there will be a steady growth of Ka-band systems in the 1990s.

The hardware design of the communications payload can be expected to develop to further complexity in the coming decade. As larger diameter unfurlable antennas become available, and are proved reliable, footprints comprising a pattern of multiple small beams will be more frequently specified. Complex beam patterns provide a greater degree of frequency reuse [see §A7.5.3] and bandwidth compression techniques allow an increased use of the limited frequency spectrum. These and other techniques have the potential to expand the types of communications service on offer. Thus communications with mobiles, down to the level of the private motor car and possibly to hand-held units, is likely to be an important market by the end of the century. Inmarsat's Standard C shoe-box-sized earth terminal development in the late 1980s has shown the great potential of mobile communications. It is not beyond the bounds of reason that some consumer electronics entrepreneur will see a market for the hand-held satellite communicator within the next 10 years. Although it is unlikely that many of the growing number of people who like to get away to the wilds of the countryside will want to carry a telephone, the facility to determine their position and height to within a few tens of metres may be a greater attraction, especially in the remoter areas.

To date, the great majority of satellite transponders have been of the *transparent* type, designed only to receive, amplify and retransmit a radio-frequency carrier and not to perform complex operations on the signal itself. The buzz words of the 1980s have included digital communications, satellite switching and onboard regeneration [see §A7.2.5], but there are unlikely to be many satellites in orbit by the end of the century which actually use these techniques. A few are under construction which demonstrate SSTDMA (satellite-switched time division multiple access), the technique which has likened the future satellite to a 'telephone switchboard in the sky', but some authorities consider it unlikely that we shall see anything but transparent transponders before the year 2000 [Bartholomé 1987].

In much the same position is the concept of the reconfigurable payload. So far, payloads have been designed for a unique application, which makes most satellites inflexible should there be a need for coverage of a different area. Reconfigurable payloads are designed to provide a new type of flexibility, but only within limits: although steerable antennas can usually be provided, transponders that operate on every possible frequency cannot. The frequencies used by two prospective users of the satellite would have to be the same, or at least similar, before such a system could work. One of the most likely scenarios for such a scheme would be for two or more countries with similar telecommunications needs each to purchase a satellite for their own use and jointly purchase an in-orbit spare with a reconfigurable payload to replace either of the prime spacecraft in the event of a failure. However, such an application is led by the needs of the market and there is no indication that a fully reconfigurable payload will fly in the next 10 years or so.

The final potential development in communications satellite payloads discussed here is the inter-satellite link or ISL. The satellites described in this book have communications links only with the Earth—an uplink and a downlink. Although the TDRS spacecraft system allows a link to be made between a spacecraft in low Earth orbit and the ground, no two satellites in geostationary orbit have yet been interconnected. To be able to do this would increase the flexibility of a satellite system. At present, a satellite communications link between two parties on opposite sides of the world is made either using the double-hop technique, which uses two satellites, or using a single-hop and a terrestrial link to cut down the delay. With an ISL, a signal could be uplinked to one satellite, beamed across space to another and downlinked to the receiving earth station [see figure A13.3]. Although for the largest cross-orbit paths this does little to reduce the signal delay, it allows a more efficient use of the available channels in a communications system: signals can be switched through whichever satellites have free channels, which varies with the time of day at a satellite's longitude.

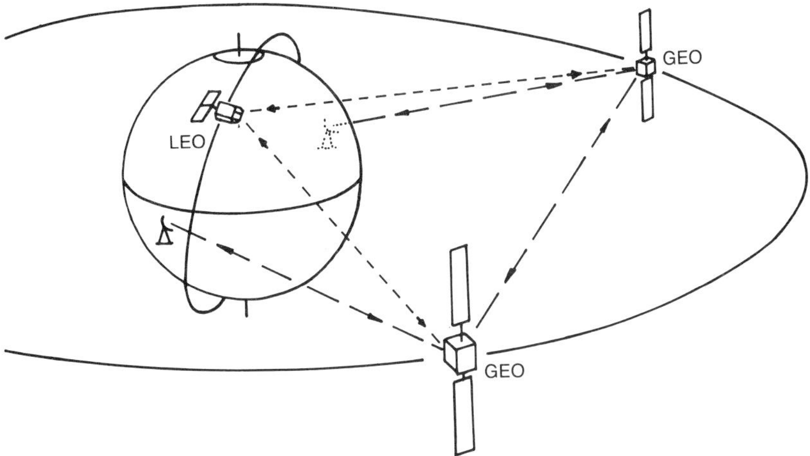

Figure A13.3 Inter-satellite links [see text of §A13.3].

The most likely candidate technology for ISLs is not radio-frequency but laser based. Optical links have several advantages: they avoid frequency allocation and congestion problems, offer high-rate data links and suffer virtually no interference with terrestrial systems [Lutz 1986]. A possible baseline ISL design, considered since 1977 for applications such as the European data relay satellite (DRS) system (the equivalent of NASA's TDRSS), featured a two-way ISL with a channel capacity of 1 Gbit per second, a very high rate. It would be suitable for GEO-to-GEO or LEO-to-GEO communications links as illustrated by figure A13.3. Although optical ISLs are as yet at the experimental stage and, at the time of writing, have yet to reach orbit, by the end of the century they are bound to be featured in industry proposals for future communications satellite systems.

13.4 SUBSYSTEM TECHNOLOGY

Some discussion of future trends in each of the satellite subsystems is included in the relevant preceding chapters. Here, an attempt is made to draw the prospective developments together to gain an overall picture of how the design of the communications satellite is likely to evolve.

Satellite *structures* are unlikely to experience significant change in the near term. The main change in the outward appearance of the communications satellite will come only when it evolves into the orbital platform or 'antenna farm', which seems unlikely to occur until the twenty-first century. A possible exception to this may be the large unfurlable antennas proposed for mobile communications applications. Inside, the structural designs may be altered by the requirements of other subsystems, as has already occurred as a result of the use of combined (or unified) propulsion systems which have tended to do away with the central thrust tube in three-axis-stabilised satellites [see §A2.3]. The other subsystems generally require only minor modifications to a basic structure and are unique to a particular bus design.

As for *materials* used in the construction of satellites, the field is one of continual change as an increasing variety of materials is developed. The next 10 years will see an increasing use of composites of all types, since they allow that eternal quest for minimal structural mass to continue.

The most significant development in spacecraft *propulsion systems* is likely to be the change from solid apogee kick motors to the liquid bipropellant systems which combine the apogee boost and station-keeping functions in a single system [see §A3.9]. Although integral satellite perigee motors have been suggested as an extension of the combined propulsion system, there is little evidence to suggest that this option will become widely available in the next decade. This is largely because the majority of launch vehicles are designed to inject their payload directly into geostationary transfer orbit, thereby obviating the separate perigee motor or *upper stage*. It is also due to the conservative attitude towards delivery systems in general. The expendable launch vehicles which use the upper stages designed originally for use with the Space Shuttle (namely PAM and IUS) have a proven reliability record with these stages and potential satellite purchasers know this. Additionally, the manufacturers of the upper stages have a vested interest in maintaining the status quo. Together these factors seem likely to militate against the integral perigee motor for at least the forseeable future.

A development in propulsion systems more likely to be applied to communications satellites in the near term is electric or ion propulsion [see §A3.10]. At first, it may feature only as a back-up to a standard system or take the place of only a part of the station-keeping subsystem, but by the end of the century the technology is likely to be sufficiently advanced for a better assessment of its capabilities. For the moment, the contest is between the combined bipropellant systems and the tried-and-tested hydrazine systems.

As for spacecraft *stabilisation* in general, the two main designs—spin and three-axis stabilisation—will remain, with the larger spacecraft tending towards the latter system because of its greater flexibility in providing greater onboard powers. Despite this there may be other contenders which show features of either or both of these types.

One of these is the *sun-pointing satellite*, which differs from the three-axis design mainly in that its solar array panels are mounted rigidly to the service module, instead of being connected by means of rotating bearings, and its payload module, topped with an array of antennas, rotates relative to the service module. Figure A13.4 illustrates a British Aerospace design concept for this type of spacecraft with a power supply capability of up to 15 kW, which compares favourably with the maximum of 7.7 kW for Olympus. Another advantage is that the spacecraft body, with its associated radiators, is permanently in the shade, thereby offering greatly increased capacity for heat rejection. The ability to dissipate more heat offers the possibility of a larger payload for a given spacecraft volume, and this leads to a more efficient and cost-effective use of orbital space and launch resources.

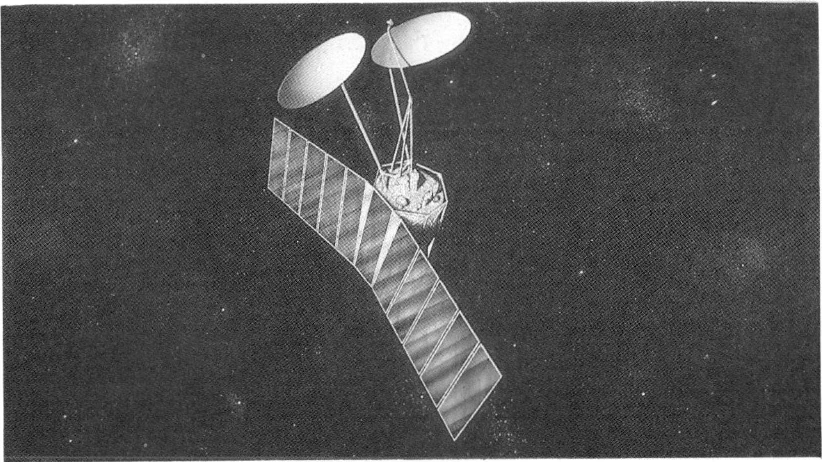

Figure A13.4 Artist's impression of a design concept for a sun-pointing satellite called 'Big Communicator'. [British Aerospace]

Another proposed spacecraft design is the Hughes HS394, which is a definite cross-breed between the two 'standard types' [see figure A13.5]. It comprises a spinning service module, a de-spun payload module and a dual flat panel array which rotates once every 24 h like the three-axis satellite array. It remains to be seen which, if either, of these novel designs will be first into orbit.

As for the *attitude and orbital control system* (AOCS), pointing control is likely to become ever more critical as spot beams become smaller, so the accuracy of sensors and actuators (including antenna-pointing mechanisms) must improve to suit requirements. In common with other subsystems the use of onboard control systems should increase as microprocessors of enhanced capability become space qualified. The development of onboard control will be especially important for the cluster-satellite concept discussed below.

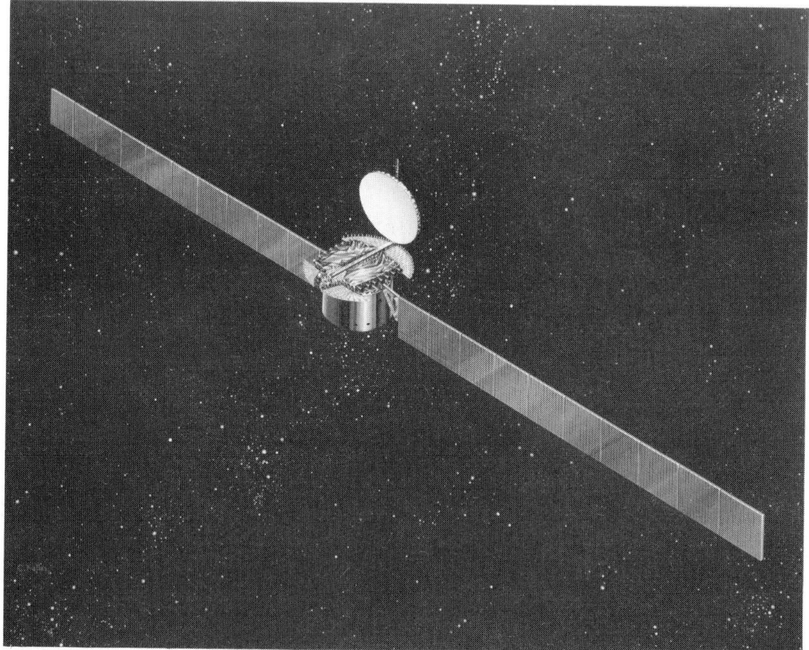

Figure A13.5 Artist's impression of the Hughes HS394 concept. [Hughes Aircraft Company]

For the *power subsystem*, apart from general improvements in efficiency, the most important development is likely to be in solar array design. If satellites become generally larger over the next decade, their power requirements will increase and new types of arrays will have to be flown. Most promising is the flexible concertina design of the type carried by the Olympus spacecraft [see figure A5.2], but for rigid panel arrays improvements are also at hand with the gallium arsenide solar cell, which is likely to be of equal importance to the silicon cell by the turn of the century.

The development of the *thermal subsystem* depends to a large extent on the growth in spacecraft size, since there will only be a need for radical changes in design if the power consumption shows a significant increase over present levels. It is only for the larger type of spacecraft or orbiting communications platform that one can foresee the inclusion of deployable radiators and pumped loop cooling systems. Nevertheless, improvements in the thermal capacity of heat pipes and the use of different-sized versions at both spacecraft-panel and circuit-board level seem likely to ensure the continual evolution of the thermal subsystem.

Most of this section has been concerned with aspects of satellite design which are direct extrapolations of current technology and largely involve increasing the capability and/or the efficiency of the respective subsystem. The next two sections cover aspects which exhibit much more of a departure from present-day thinking on communications satellite design.

13.5 SATELLITE CLUSTERS

An alternative to the large communications platform is the cluster concept, whereby a group of contemporary-sized satellites would be stationed around the same nominal orbital position and controlled as a single entity. The individual payloads would be joined together using inter-satellite links and, as far as the user was concerned, the cluster of satellites would appear as one. This would, of course, require all the satellites to remain within the beamwidth of a single earth station and may require tighter station-keeping control than is currently practised (e.g. $\pm 0.05°$ compared with the current $\pm 0.1°$). Several types of cluster formation are foreseen: in one the satellites all have the same mean longitude; in another they are placed side by side along the geostationary arc. A further possibility is a hybrid of the two, called a 'necklace cluster', whereby the satellites are arranged in a tilted circle about the mean position to obviate any inter-satellite line of sight problems experienced with other arrangements. The constituent satellites could be of equal importance in the operation of the cluster or one could be designated the master satellite in a master/slave relationship. The latter option would enable the master satellite to carry out TT&C functions for the whole group, for instance. Redundancy could be provided by an in-orbit spare which could replace any of the other satellites as necessary, although special provisions would have to be made for a master/slave system. One of the chief advantages of the satellite cluster over the space platform is the possibility of implementing the system in stages, thus matching a gradual growth in demand.

13.6 ALTERNATIVE ORBITS

Most communications satellites occupy positions in geostationary orbit, and the great majority of future ones will too, but there is no rule that says that a communications satellite must occupy a position in GEO. It is simply that GEO has proved the most convenient orbit for satellites covering the areas of the Earth upon which most of the western world resides [see appendix A]. The main disadvantage of GEO is its inability to provide coverage of the poles and high-latitude sites. Even in the far northern latitudes of Europe and Canada, mountains, buildings and other obstructions can screen the satellite's signal, an aspect which will become increasingly important as mobile–satellite communication grows.

The solution used in the USSR for many years is the Molniya orbit, a highly elliptical orbit with an inclination of 63.4°, which provides good coverage of the northern latitudes and polar regions as well as the southern regions of the USSR. Over the years, many studies have been undertaken in Europe concerning the potential of high inclination orbits, especially for mobile applications. One of these, undertaken for the Deutsche Bundespost in the mid 1980s [Nauck *et al* 1987], resulted in a satellite concept which goes under what must be one of the most contrived acronyms in history—LOOPUS (for geosynchronous loops in orbit occupied permanently by unstationary satellites)!

The LOOPUS system comprises three satellites in Molniya-type orbits (63.4° inclination, 1200 km perigee and 39 000 km apogee) spaced at angles of 120° around the Earth. The satellites are phased in their orbits so that, as the Earth turns below them, each describes the same ground track 8 hours behind the one in front, thus providing 24 hour coverage [see figure A13.6]. The range of movement of the sub-satellite point over the 8 hour period is between about 45°N and 63.4°N. Since the sub-satellite point moves very slowly in the north–south and hardly at all in the east–west direction, the satellite would be almost directly overhead at high-latitude sites for the majority of the coverage period. In fact, an earth station elevation angle greater than about 75° can be expected throughout the 8 hour coverage, resulting in a freedom from shadowing and multipath effects. A large family of Molniya-type orbits could be established about the Earth, each with its apogee over a different longitude and therefore serving a different area. Similar inverted orbits could be utilised in the southern hemisphere.

If such an orbit was to be utilised for communications by the western nations, the limited resources of orbital space and frequency spectrum would be greatly expanded. It would offer an effective system of frequency reuse in that frequencies already used in geostationary orbit could be reused in the high inclination orbit. The directivity of the ground station antennas would be sufficient to prevent interference, since they would be pointing in a completely different direction to antennas serving satellites in geostationary orbit. Indeed, the Molniya/LOOPUS type of orbit has been proposed for Inmarsat's third generation system [Nauck *et al* 1987]. However, it seems unlikely that (apart from Soviet spacecraft) we shall see anything but experimental satellites in these orbits before the turn of the century.

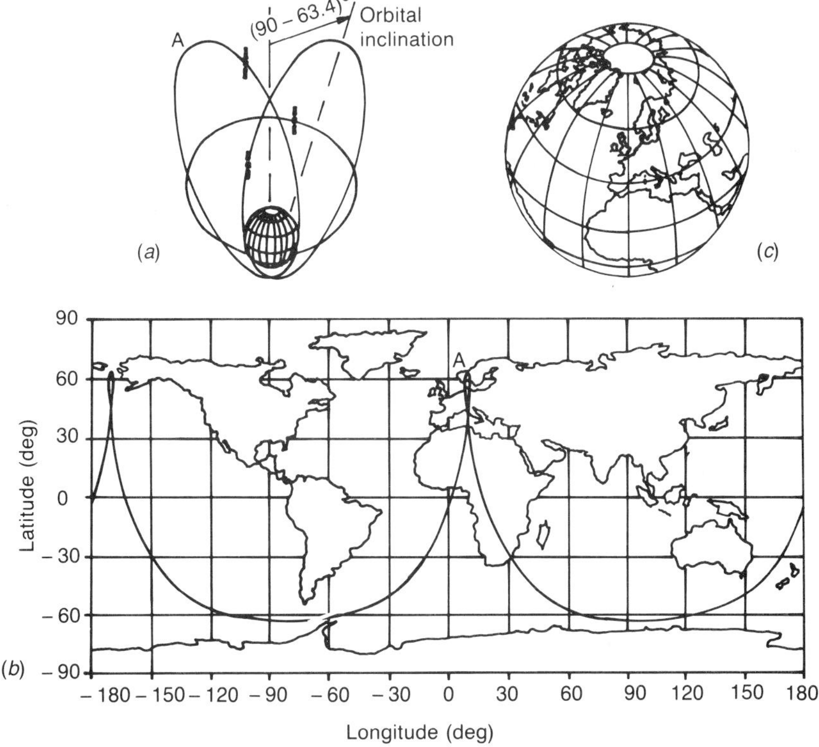

Figure A13.6 LOOPUS concept for satellites in Molniya-type orbits: (*a*) three-satellite LOOPUS/Molniya system (as seen by a fixed observer); (*b*) ground track of three satellites; (*c*) view from apogee (A).

13.7 CONCLUSION

To the engineer involved in the design of the communications satellite, the day-to-day problems of working in an industrial and commercial environment can often conspire to give the impression that a satellite is just another product of industry. But few, when forced to speak about their job to someone bound by the constraints of a purely terrestrial view, will fail to show a spark of enthusiasm for their subject. To design a complex collection of equipment that has to operate unattended for up to 10 years some 36 000 km out in space must rank alongside the best of mankind's technical challenges. However, the communications satellite is still a commercial product and a customer will buy it with the expectation that it will 'still work when he gets it home'. Few allowances will be made for the technical challenge that the satellite represents and few excuses will be acceptable in the face of a failure.

As a product, it is unlikely that the communications satellite will ever be truly recognised by the general public. After all, it has been around since 1965 and the closest approach to lay recognition has been the 'live via satellite' caption on our TV screens. Other products of space technology have a more glamorous image—the manned spacecraft, the deep-space probes, etc—but as far as the communications satellite is concerned, the interest of the media is confined to the failures and in-orbit rescues.

The only way the satellite will be noticed in the future is through the services it brings to the populace at large. Direct broadcasting of television into the home (DBS) is perhaps the first example of a service which will change the public conception of the satellite, but only then if the service is provided by an individual antenna on the subscriber's house. Satellite-derived TV brought to the home by cable will be indistinguishable from cable TV. But then the technology which brings the service to the home is of little interest to the average subscriber, unless of course it affects the quality. The arrival of the hand-held satellite-based mobile telephone, perhaps in the style of the '*Star Trek* communicator', may force a few to ponder on the usefulness of the communications satellite, but the appreciation will fade after a few weeks.

In the final analysis, the most successful service attracts the least attention, and the unquestioning acceptance of a technology is a sign of its maturity. Few people reading these words will marvel at the perfection of the typeface or ponder on the technology required to print and bind this volume. If the communications satellite evokes the same degree of interest in its users, then it bears the mark of a mature technology, and those involved in its design will know that the problems have been solved and the challenge has been met.

REFERENCES

Bartholomé P 1987 *Satellite Communication Systems* ed. B G Evans (London: Peter Peregrinus) ch. 22

Hudson W R and Gartrell C F 1987 *Int. Astronautical Fed. Congr.* IAF-87-07

Lutz H 1986 *ESA Bull.* no. 45 74–80

Nauck J, Gunther H J and Plate K 1987 *Int. Astronautical Fed. Congr.* IAF-87-481

Renner U 1983 *Space Commun. Broadcast.* **1** 145–54

Part B

Communications System Design Case Study: Direct Broadcasting by Satellite (DBS)

Introduction

All spacecraft—communications satellites, remote sensing satellites, deep-space probes or any form of manned vehicle—have one very basic and overriding requirement: the need to communicate with a ground station back on Earth. The main subject of this part of the book is the design of the communications link that makes this possible and the technical constraints which govern its operation.

By way of example, we shall investigate the requirements for a direct broadcasting by satellite (DBS) system, intended to broadcast television programmes direct to the home. Not only is this application important in the context of the satellite technology of the 1980s and 1990s, it is also relatively straightforward in its utilisation of the available resources—the complexities of multiplexing, multiple access and digital networks, prevalent in telecommunications systems, are unnecessary complications. The selection of DBS also allows the following chapters to give an account of the regulatory process which governs our communications networks and may provide an insight into the problems involved with the initiation of any new telecommunications service.

This section, therefore, is a case study of the development of a system for DBS, based largely on a plan to use high-power satellites to provide coverage of the countries of Europe. The rest of the world is discussed in general terms, but the case study concentrates on Europe since it is within Europe that satellites which actually adhere to the DBS plan have been built (e.g. TDF, TV-SAT and Tele-X). The techniques are applicable to any satellite communications link and adaptable to the links required for other orbiting spacecraft.

Chapter B1 introduces the subject with a discussion of the regulatory aspects of providing a new satellite service. Chapter B2 describes the two main natural resources for a communications satellite system: a position in geostationary orbit and a band of frequencies on which to broadcast. Chapter B3 describes the workings of the plan: the potential problems of interference on both the uplink and the downlink and the techniques used to overcome them. Finally, chapter B4 concludes the analysis with the compilation of a link budget for a DBS system and an explanation of its technical characteristics.

B1 The DBS Plan

1.1 DIRECT BROADCASTING BY SATELLITE

1.1.1 Definitions

Since DBS became technically feasible, the term direct broadcasting by satellite has become increasingly vague. It now seems that almost any satellite system which delivers a TV signal to any antenna, up to the 3 m diameter dishes on hotel roofs say, is described as DBS. In light of this, it is useful to re-examine the reasoning behind the term *direct broadcasting*.

Direct was intended to mean *direct to the home* and not to imply direct to the cable head-end, the point of reception and distribution in a cable system. TV *broadcast* refers to the wide dissemination of a TV service to the consumer and should be distinguished from TV distribution which is not intended to involve the consumer. TV distribution is the transfer of TV programming from an originating source, such as a TV company, to a point of distribution into the public network. The transmissions are not intended to be received directly by the public, as they are with direct broadcast, although they are often intercepted by radio and TV 'hams' or 'pirates'. What we may term *true* DBS is what was intended by the delegates of the World Administrative Radio Conference of 1977 (WARC-77) when they developed the plan for DBS in Europe. In a nutshell, this postulated satellites transmitting five Ku-band channels, each of 27 MHz bandwidth, at EIRP (equivalent isotropic radiated power) levels of 65 dBW, requiring perhaps 200 or 300 W per channel of transponder power. The WARC suggestion that individual reception should be possible using 90 cm dishes drove the requirement for a new type of high-power satellite, using high-power travelling wave tube amplifiers (TWTAs). This distinguished DBS satellites from other 'low-power' spacecraft used for general telecommunications which, at the time, typically had transponders with only about 10 W or 20 W of output power, which required earth stations with 20 or 30 m antennas to collect sufficient signal power.

This high-power DBS is now seen by many as an extremely wasteful and expensive way to broadcast TV, a view which has led to the appearance of what is called variously low-power DBS, medium-power DBS or quasi-DBS. Although definitions can suffer from subjectivity, the output power of the satellite travelling wave tube is probably the best guide in defining the service. Most contemporary true-DBS spacecraft have been designed to use tube powers greater than 200 W (e.g. TDF-1, TV-SAT, Tele-X, and one of the four payloads on Olympus 1). Satellites using 100 W tubes, or thereabouts, can be referred to as medium-power DBS (e.g. the Japanese BS-2), and those operating on only about 50 W, which seems to be the average anticipated, are low-power DBS (e.g. Luxemburg's Astra). Regional telecommunications satellites, such as the European Eutelsat 1 and the French Telecom 1, transmit TV to domestic antennas as small as 1 m in diameter using only 20 W TWTAs. This type of satellite was, however, not designed for DBS and the service should be termed quasi-DBS.

The distinction between the various types of DBS is not truly fundamental. It is like trying to define hi-fi: when does it become medium- or low-fi? With DBS too, the important difference is *quality* and, although there are many technical factors to take into account, quality depends to a large extent on the signal power (or power density) received at the ground. Lower power satellites provide lower power densities; smaller domestic antennas collect less signal power. Combining the two may provide acceptable quality most of the time but, for a receiver towards the edge of the coverage area in the middle of a heavy rainstorm, reception may be decidedly substandard. In the same situation the high-power satellite would generally have the power margin to maintain good-quality reception.

1.1.2 TV transmission standards

Another factor affecting the quality of the DBS service is the transmission standard. The three world standards for colour TV (PAL, SECAM and NTSC), used today in a variety of subforms, were developed in the 1950s to be compatible with existing monochrome receivers. Because of this the designs were a compromise, but the advent of DBS offered the opportunity to develop a completely new system that would provide higher definition pictures and have the potential for further improvements. It was with this in mind that the UK Independent Broadcasting Authority (IBA) developed the system currently being implemented for European DBS: multiplexed analogue component (MAC).

A colour TV picture comprises two different sets of picture information, details of brightness (luminance) and of colour (chrominance), whereas the monochrome systems have only luminance. With the terrestrial colour systems mentioned above the chrominance and luminance information is interleaved within the frequency bandwidth of the signal in an ingenious, if complex, way. One of the drawbacks to this mixing of colour and brightness is that the two information sets can interfere. One effect of this is the 'colour-flashing' seen on clothing with a closely lined pattern or check (typically newsreaders' jackets). The MAC system virtually eliminates such problems by electronically compressing the luminance and chrominance information and transmitting them sequentially rather than simultaneously. The signals are expanded and used to create a picture in the receiver.

The main drawback is that MAC is not compatible with current TV receivers and requires the addition of an adaptor or converter. It was for this reason that the contending system proposed by the British Broadcasting Corporation (BBC), *extended PAL*, was made compatible with standard PAL receivers. This was, of course, the same philosophy used for the move from monochrome to colour and was criticised for being old fashioned and restrictive. In November 1982 the argument was settled in the UK by the report of the Part Committee which recommended MAC to the UK Government since it was considered to be the best technical solution and have the better chance for European unity of standards. At the time, the BBC had a valid counter-argument that the initiation of MAC would leave the UK isolated in Europe, since France and Germany, already building their DBS satellites, had entered into a bilateral agreement to use versions of SECAM and PAL which would severely restrict the market for UK manufacturers of DBS receiving equipment. However, Germany and France were later persuaded to adopt the MAC system, largely because patents governing the manufacture of receivers excluded Japanese imports and would give European manufacturers time to establish themselves. There would be no such barrier to the import of Japanese PAL and SECAM systems. But technical standards are rarely this clear cut: the UK had originally adopted a version of MAC called C-MAC but, after great deliberation, changed to D-MAC in 1987. This version is still, however, not entirely compatible with that of Germany and France who have chosen D2-MAC, a reduced bandwidth version that is compatible with the lower bandwidth cable TV links. The drawback here is that, whereas C-MAC has sufficient bandwidth for eight digital data or sound channels, D2-MAC has only four. This means that D2-MAC is far less suited to future developments in television such as the large-screen high-definition pictures planned for the 1990s which will have perhaps 1250 lines as opposed to the present 625 currently standard throughout Europe.

Agreement on technical standards is invariably fraught with problems, like the comparable field of domestic video which offered VHS, Betamax and V2000. or the more historic non-compatibility of railway gauges. Although not concerned with the details of transmission standards, the plan for DBS was devised in a similar atmosphere of mixed international opinions.

1.2 REGULATION

The allocation of communications frequencies for all satellites and space systems is a very complex task and the process of allocation for a major new service like that of DBS requires many months of meetings at an international level.

The regulating body for worldwide radio frequencies is the International Telecommunications Union (ITU) which holds, from time to time, World or Regional Administrative Radio Conferences, otherwise known as WARCs or RARCs, at which the frequency allocations are decided. This subject embraces a plethora of acronyms, so a list of the major ones has been provided in table B1.1.

Table B1.1 List of acronyms.

BSS	Broadcasting-satellite service
CCIR	International Radio Consultative Committee
CEPT	European Confederation for Posts and Telecommunications
CPM	Conference Preparatory Meeting
DBS	Direct broadcasting by satellite
EBU	European Broadcasting Union
ESA	European Space Agency
FAC	Federal Advisory Committee
FCC	Federal Communications Commission
FSS	Fixed-satellite service
IFRB	International Frequency Registration Board
ITU	International Telecommunications Union
RARC	Regional Administrative Radio Conference
WARC	World Administrative Radio Conference
WARC-BS	WARC for Broadcasting Satellites
WARC-ST	WARC for Space Telecommunications

For administrative convenience, the ITU divides the Earth into the three regions illustrated by figure B1.1: Region 1 includes Europe, Africa, the USSR and Mongolia; Region 2 the Americas and Greenland; and Region 3 Asia, Australasia and the Pacific. One cannot read about DBS for long without encountering the acronym WARC-77, the conference at which the plan for DBS in Regions 1 and 3 was devised; a similar plan for Region 2 was the subject of RARC-83 (discussed in §B1.4).

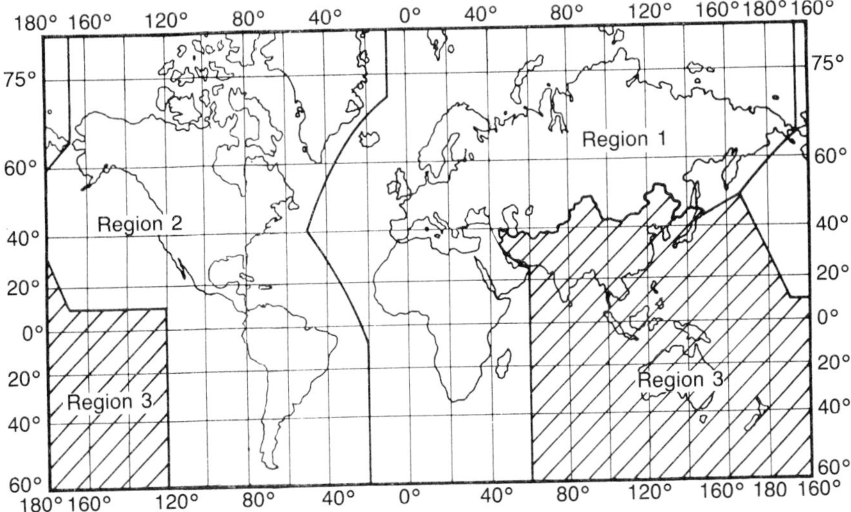

Figure B1.1 ITU Regions.

The origin of the two conferences dates back to 1971, when the WARC for Space Telecommunications (WARC-ST) made the first allocation of frequency bands to the broadcasting-satellite service [see box B1.1]. WARC-ST recognised that problems would arise because of the sharing of the bands with terrestrial and other satellite services and recommended the definition of a world or regional plan for the broadcasting-satellite service (BSS). It made no attempt, however, to formulate the plan, leaving it for WARC-77 and RARC-83.

1.3 WARC-77

The main goal of WARC-77, held in Geneva from 10 January to 13 February 1977, was the establishment of a plan for the BSS in the 12 GHz (downlink) band. The cast of WARC-77 was truly international: it was attended by more than 600 delegates from 111 countries, as well as officials from the ITU and observers from a number of international organisations, including the European Broadcasting Union (EBU) [Brown and Mertens 1977].

1.3.1 Attitudes to the plan

The broadcasting organisations which were EBU members were in favour of a plan from the start, since they felt this was the only fair way to initiate and develop satellite broadcasting. This was particularly important for the poorer nations, who would be prevented from establishing a DBS service at a later date if frequencies and orbital slots were to be handed out on a first-come first-served basis. In fact, the 12 GHz band was chosen for DBS downlink because it was the band for which contemporary technology could provide relatively inexpensive receivers for individual reception. If direct satellite TV was ever to be available to the less developed countries, this had to be the case.

The CEPT terrestrial telecommunications administrations were also in favour of a plan, mainly because they needed to know which frequencies would be usable for the terrestrial services they were planning to install. For similar reasons the other countries in Regions 1 and 3 were in agreement, particularly the developing nations who were anxious that their future requirements should be recognised. Opinions

BOX B1.1: REGIONAL FREQUENCY ALLOCATIONS FOR DBS

Downlink

The frequency bands allocated at WARC-ST for DBS downlinks were:

Region 1: 11.7 to 12.5 GHz (to be shared with terrestrial services).
Region 2: 11.7 to 12.2 GHz (to be shared with terrestrial services and the fixed-satellite service (FSS) [space-to-Earth]).
Region 3: 11.7 to 12.2 GHz (to be shared with terrestrial services).

The allocation for Region 2 was, however, altered significantly at WARC-79, with 11.7–12.1 GHz being allocated to the FSS and 12.3–12.7 GHz to the BSS. The 12.1–12.3 GHz band was allocated to both services at WARC-79, but divided further at RARC-83 so that the FSS had the 11.7–12.2 GHz band and the BSS had 12.2–12.7 GHz.

Uplink

The frequency bands allocated at WARC-79 for DBS uplinks (otherwise known as feeder links) were:

10.7–11.7 GHz: shared with space-to-Earth links of the FSS.
14.0–14.8 GHz: use restricted to countries outside Europe (14.0–14.5 GHz shared with Earth-to-space links of the FSS).
17.3–18.1 GHz: shared with space-to-Earth links of the FSS.

RARC-83 planned the uplinks for Region 2 in the band 17.3–17.8 GHz.

A feeder link plan for Regions 1 and 3 was only finally adopted at WARC-88 (a conference 'on the use of the geostationary satellite orbit and the planning of the space services using it'); the 17.3–18.1 GHz band was planned for general use and the 14.5–14.8 GHz band for use outside Europe.

in Region 2, however, were not so concordant. Some supported the idea of an immediate and detailed plan, but most, led by the United States, favoured an evolutionary plan where assignments would be made on an *ad hoc* basis. Their view was that services should be assigned frequencies as and when required, so that full advantage could be taken of state-of-the-art technology. This would be tempered, moreover, by a guaranteed access to geostationary orbit and the frequency spectrum, both of which were considered natural resources. Certain principles would also be adhered to, so that the band could be shared with the fixed-satellite service (FSS) without interference. The result of this difference of opinion was that WARC-77 included only regions 1 and 3 in its detailed deliberations.

1.3.2 Formulation of the plan

Although a colossal amount of work was done by many organisations, the technical preparations for the conference were entrusted to the Study Groups of the CCIR, which met throughout 1976. As might be expected from the above, most of the evolutionary planning originated from the United States while Europe provided the detailed planning. The essential technical definition of the system was carried out primarily by the EBU, firstly in collaboration with ESA and then with the members of the CEPT. In addition the ITU organised three seminars to assist the developing countries, at venues which could easily have been picked from a world tour brochure: Rio de Janeiro in August; Kyoto in September; and Khartoum in November.

The work of the actual conference was shared between eight Committees, most of it being undertaken by Committees 4 (Technical), 5 (Planning) and 6 (Procedures), the last of which devised the rules for the use of the plan. It was Committee 5 that formulated the plan itself. Since Region 2 was not in favour of a detailed plan, the debate was largely concerned with Regions 1 and 3. Even this, as it turned out, was no easy task: after all, who could expect 600 delegates from 111 different countries to agree on anything!

Naturally enough the large number of competing national requirements for broadcasting services and the predictable national sensitivities led to conflict. Unfortunately, few people had the technical background required to solve these problems by devising a plan to which everyone would adhere. Although others made contributions, the success or failure of WARC-77 fell on the shoulders of a few experts from the members of the EBU and the CEPT . . . and their computers. During the latter part of the conference, a series of computer programs was constantly in use, night and day, juggling channels, powers and all the other parameters in an attempt to provide all nations with an acceptable service.

Individual national requirements for frequencies (or channels) and orbital positions had been collected and published by the ITU prior to the conference. As negotiations progressed the requirements were amended and, practically every day, the IFRB issued a new set of updated requirements. Sometimes details would be altered because the delegates, who were learning more about the problems, simply changed their minds. Even before the conference began, it was obvious that some requirements were excessive and would lead to higher levels of interference than could be allowed. The main points of difficulty were: reducing the number of channels; persuading some nations to be more flexible in their preferences for orbital positions; and obtaining a more precise definition of the service areas.

Some delegations wanted more channels than could be accommodated within the limited bandwidth assigned to the service and some requirements for power flux density (a measure of the signal power on the ground) would have led to serious interference problems. With so many nations in Europe clustered closely together, their first preference was for the same orbital position which, in much of Europe's case, was 19°W [see chapter B2]. This meant that those delegations which argued less persistently were relegated to less favourable positions like the UK's allocation of 31°W. The drawbacks of this position include a reduced elevation of the satellite above the horizon and an increased obliqueness of the beam footprint on the Earth's surface [see chapter B3].

Each delegation was free to specify the limits of its service area, which could be chosen to correspond to its national boundary or to overspill deliberately into adjacent countries, providing the agreement of countries subject to overspill was obtained. Owing in part to a lack of technical understanding, confusion arose in relating the satellite antenna beam to the service area it was designed to cover. Had the definition of service area been more specific and more binding, some of this confusion might not have occurred.

After 2 weeks of discussion the conference decided to conduct two parallel planning exercises: one based strictly on the original excessive channel requirements and the other, a 'regular plan', allocating a maximum of five channels to each country. Perhaps it is indicative of the problems of international negotiation that it took 2 weeks to reach this stage. Alternatively it could be due to the initial lack of technical appreciation possessed by the majority of the delegates. With hindsight, at all times a wonderful thing, it is perhaps more difficult to see why it took another week for the obvious plan to be chosen. The regular plan, although not perfect, was eventually agreed to be significantly better than the other and the requirements were reduced.

1.3.3 Results

Towards the end of the conference, Committee 5 published a draft plan which included parameters under the 10 headings shown in table B1.2. The plan involved the definition and integration of a number of technical characteristics for each country. They included a geostationary orbital position for each satellite and the definition of frequencies, polarisations and transmission powers. It was also necessary to define the service area on the ground so that a satellite antenna beam could be designed to cover it. The synthesis of these and other items allowed an interference analysis [chapter B3] to be conducted, the result of which was a plan whereby each country would be able to receive interference-free satellite TV.

Table B1.2 Parameters defined in the DBS plan.

Parameter		Chapter reference number
1.	Country symbol and IFRB serial number	
2.	Nominal orbital position	[B2]
3.	Channel number	[B2]
4.	Boresight geographical coordinates	[B3]
5.	Antenna aperture (major and minor axes of elliptical beam at half-power points)	[B3 and B4]
6.	Ellipse orientation	[B3]
7.	Gain	[B4]
8.	Polarisation	[B2]
9.	Equivalent isotropic radiated power (EIRP)	[B4]
10.	Protection margin	[B4]

The plan, and the procedures designed to concur with the existing Radio Regulations, were incorporated into the bible of WARC known as Final Acts. The plan would be valid for at least 15 years from 1979 and until it was revised by, what the procedures called, 'a competent conference'. The month-long meeting of WARC-BS-1977 succeeded in formulating a plan for satellite broadcasting (for the downlink in Regions 1 and 3) which will be used, therefore, at least until 1994 and possibly into the next century.

An important limitation of WARC-77, shown by the frequency allocations of WARC-ST, was that it considered only the downlink (space-to Earth) frequencies for DBS; no allocations had been made for the uplink (Earth-to-space) path. The allocations, provided 2 years later by the plenary assembly of WARC-79, were better than some had feared [Block 1980], since they involved a compromise solution allocating three frequency bands instead of one [see box B1.1]. However, it was almost 10 years after WARC-79 ratified the WARC-77 plan for DBS downlinks that, at WARC-88, the formal plan for DBS uplinks for Regions 1 and 3 was finally adopted.

1.4 RARC-83

It was against the background of WARC-77 and WARC-79 that RARC-83 opened its doors to the 47 countries of Region 2. Prior to WARC-79, the United States had received a good deal of criticism that they were not prepared politically for the conference. Their unwillingness to accept an organised plan for Region 2 in 1977 may have engendered this view, but the situation for RARC-83 was quite different. Success for the US delegation meant leaving Geneva with 500 MHz of spectrum, full polarisation (both right- and left-hand circularly polarised channels on all allocated frequencies) and what they called 'the freedom to get DBS going' [Kachmar 1983a].

1.4.1 Preparation

Some countries of Region 2, however, suffered a more disorganised start: they were supposed to submit their orbital and channel requirements to the IFRB by June 1982, but 1 year later, with the conference underway, they had still failed to do so. Making use of what information *had* been submitted, the CCIR held a Conference Preparatory Meeting (CPM) in Geneva to recommend the planning methods and technical parameters for RARC-83. It was here that the first steps to the kind of flexible approach desired by the United States were taken.

The CPM emphasised a number of points which would be taken into account in devising a plan for Region 2: the high orbital capacity of the region (due to its wide east–west extent), the new technology becoming available, and the wide-ranging TV requirements of the region. The plan also had to be sufficiently detailed to identify potentially interfering services and allow protection to be arranged.

Another ITU preparatory group known as the PoE or, believe it or not, the Panel of Experts, considered that the Final Acts adopted at WARC-77 contained 'a rigid spectrum/orbit plan based on conservative technological assumptions'. The Federal Communications Commission (FCC) pointed out that it was very difficult to make modifications to the plan because of its rigidity, a rigidity which was seen on Europe's side of the Atlantic as the plan's cohesive strength. The 1977 parameters would, however, be updated because of advances in computer-modelling techniques and satellite technology generally. Although the US delegates admitted that really they would prefer to have no plan at all and introduce communications systems as they occurred, the rest of the world wanted a plan, so a plan there would be.

1.4.2 Technical requirements

Many groups advised and commented on the make-up of the American technical requirements, including a Federal Advisory Committee (FAC) which was open to the public. Predictably, the bulk of the contributions came from companies and organisations who thought that they would either gain or lose significantly from direct satellite broadcasting. The former category included the 'early-entry' DBS applicants, those companies which, considering DBS to be a money-making concern, jumped on the bandwagon even before the technical machinery had been oiled. Most of them have now pulled out as the economic gains begin to look less certain. Others in the group who stood to benefit were the major broadcast networks, AT&T (the national telephone company) and the engineering companies Rockwell and Harris. The latter category (the potential losers) included the National Association of Broadcasters [Kachmar 1983a].

The United States considered that the allocation of frequencies and orbital positions would be a relatively straightforward task for Region 2 compared with that for Region 1, whose countries had been obliged to reduce their initial requirements to give the production of a plan at least a fighting chance. Region 2, after all, has a relatively small number of countries and covers a comparatively wide range of longitudes (and therefore orbital slots). The opinion was, therefore, that it should be unnecessary to assess every possible combination of frequency and orbital position. The rather insular assumption inherent in this view was that the other countries in Region 2 would not request more than a handful of channels between them. In fact Paraguay and Uruguay requested 18 channels each!

Of course there is a question as to when these countries, and many others like them, will use their resource. The FAC studied this aspect and concluded that they would never implement a BSS, because it will always be cheaper for them to use terrestrial networks. This remains to be seen, but it does seem that no other country in Region 2, apart from the United States, has any immediate plans to launch a DBS system.

The United States stipulated that the 48 contiguous states should be divided into four service areas, corresponding approximately to the four time zones, and that a minimum of eight eclipse-protected orbital positions (two for each service area) should be allocated, thus providing a minimum of 72 video channels for each area. There should also be provision for additional channels, for use by the United States at a later date, which need not be eclipse-protected. The plan was to be based on *individual reception* using small antennas, which would also permit *community reception* with larger antennas and more complex receivers.

The United States also pioneered a policy they called 'block allotment', whereby each service area (US time zone, moderate-sized country, or group of smaller countries) would be allocated the entire 500 MHz of spectrum between 11.7 and 12.2 GHz and both polarisations for each orbital position. They wanted a sufficiently flexible plan to minimise coordination with other countries and one within which modifications could be made as the technology advanced. In other words, a large unchallenged allocation which could be expanded at any time. In preparing for RARC-83 the name of the plan was first changed to 'multichannel assignment' and then dropped altogether to mollify some of the other countries in the region, although their aims remained the same.

The Americans seemed to be perpetrating a scheme in which there were no specific channel allocations, a move that would make predictive interference calculations impossible until a forthcoming service was announced. This was precisely the argument that had impeded the formulation of a plan for Region 2 at WARC-77.

1.4.3 Results

When the US delegation arrived at the conference, they found that many administrations had greatly increased their channel requirements—the total number had doubled to approximately 2000 [Kachmar 1983b]. This level of competition must have surprised the delegation, but luckily the world of the ITU radio conference is a democratic one and every oversubscribed country had to reduce its requirements accordingly.

The results of RARC-83 show that, not surprisingly, the delegates had to compromise to get any sort of plan at all, just as the countries of Region 1 had in 1977. As far as orbital positions were concerned, of the eight requested positions the United States received five good and three not so good. One of the latter was at 61.5°W, which is east of the Eastern Time Zone and, in season, suffers a 1.5 h eclipse beginning at 9.30 pm. The two others are at 175° and 166° in the far west offering only low elevation angles for the Pacific Time Zone. Nor did the delegation get their requested 72 channels per service area, having to settle for 64 (32 on each polarisation).

1.5 CONCLUSION

In spite of the inevitable political wranglings that accompany most international agreements, a worldwide plan for DBS had been formulated. The WARC-77 plan for DBS, and its 1983 successor, are now part of the history of space telecommunications. It remains to be seen whether it will remain simply a paper exercise, but at least it offers an illustration of what could be achieved. The following chapters will describe the resources and technical attributes of the DBS plan, particularly for Europe, and show how they are combined to allow all nations a DBS service that is protected from interference.

REFERENCES

Block G F 1980 *ESA Bull.* no. 22 30–3
Brown A and Mertens H 1977 *EBU Rev.* Apr. 60–7
Kachmar M 1983a *Microwave. RF* **22** no. 6 28–37
——— 1983b *Microwave. RF* **22** no. 8 29

B2 Allocation of Resources

2.1 INTRODUCTION

The previous chapter introduced the two fundamental requirements for any satellite communications system which should be regarded as limited resources: a position in geostationary orbit and a band of radio frequencies on which to transmit and receive. This chapter describes how WARC-77 allocated these resources in the European section of ITU Region 1 and discusses how they combine to form the DBS 'Plan for Europe'.

2.2 GEOSTATIONARY ORBIT

There are an infinite number of paths an orbiting satellite can take around the planet, many of which are used for applications as varied as space astronomy and military surveillance. There are, however, very few orbits which can be conveniently utilised for communications [see appendix B]. The best known and most used is the geostationary orbit (GEO), because of its unique attribute of providing positions for satellites which remain essentially fixed from the point of view of small earth terminals, which therefore do not require tracking systems—an important commercial advantage for DBS.

2.2.1 The WARC plan for orbital positions

Table B2.1 shows the plan of orbital positions for European direct broadcast satellites formulated by WARC-77.

One obvious point in the table is that satellites of many countries share the same nominal orbital position. Though at first this seems inadvisable, it should be noted that GEO has a total circumference of 265 000 km (165 000 miles), which equates to some 736 km (457 miles) per degree of longitude, which is room enough for a family of satellites! The limiting factor is not physical space in the orbit but the accuracy with which an earth station can distinguish between two closely spaced satellites. If two satellites stationed at the same orbital position use the same frequencies, both satellites will receive each other's as well as their own signals, leading to interference. The closer the satellites are, the narrower the beamwidth of the ground-based antenna will need to be to discriminate between them.

Table B2.1 The WARC-77 'Plan for Europe': orbital positions.

Orbital position	Country
5°E	Cyprus
	Denmark
	Finland
	Greece
	Iceland
	Nordic countries†
	Norway
	Sweden
	Turkey
1°W	Bulgaria
	Czechoslovakia
	East Germany
	Poland
	Romania
7°W	Albania
	Yugoslavia
19°W	Austria
	Belgium
	France
	West Germany
	Italy
	Luxemburg
	The Netherlands
	Switzerland
23°W	USSR ‡
31°W	Ireland
	Portugal
	Spain
	UK
37°W	Andorra
	Liechtenstein
	Monaco
	San Marino
	Vatican State
44°W	USSR‡

† The five Nordic countries (Denmark, Finland, Iceland, Norway, Sweden) will be allowed to use some of their channels for a joint regional service.
‡ The USSR has two orbital slots because over a large part of the globe it is the only country that both requires satellite channels and has the sovereign right to use them.

Part of the solution is, of course, to allocate different frequencies to satellites sharing the same orbital position. However, given that radio frequencies have to be shared out as carefully for space communications as with traditional radio and TV channels here on Earth, there is a limit to the number of different channels which can be beamed to a particular orbital position, and this places a limit on the number of satellites sharing that position. The practice of grouping satellites around nominal orbital positions is known as *clustering*. Spacecraft are grouped according to their sensitivity to interference and their potential for generating interference, so that, for instance, spacecraft using the same channel frequencies are placed in different orbital groupings.

The plan for Regions 1 and 3 was based generally on a nominal orbital spacing of 6°, groups of satellites being stationed predominantly at 5°E, 1°W, 7°W, 19°W, 31°W and 37°W for countries in Region 1 for example. The USSR, in addition, occupies the orbital positions at 23°W and 44°W [see table B2.1]. Various preferences for orbital position were voiced by the delegates and, in general, were respected by WARC. Thus we find the eight countries of the European 'supergroup' (Austria, Belgium, France, West Germany, Italy, Luxemburg, the Netherlands and Switzerland) at longitude 19°W, the Nordic countries at 5°E, the Eastern European countries at 1°W, and so on. It has to be noted that the UK falls outside the supergroup, in which it might have been expected, by sharing 31°W with Ireland, Spain and Portugal. This position has the disadvantage that it reduces the elevation of the UK's satellite. In Spain and Portugal, elevation angles are much higher because of their southerly latitude.

An interesting point about the orbital positions in table B2.1 is that most of them have been chosen to be a number of degrees to the west of their respective country. This takes into account the effect of the *eclipse season* when, around midnight (in the country concerned) for a number of days, the satellite orbits into the Earth's shadow where its solar arrays can produce no power [see §A5.5]. If the satellite was placed in an orbital position with the same meridian as the country, the direct TV service would have to cease as midnight approached and the eclipse began. To counteract this, the satellite is positioned to the west so that it is eclipsed in the early morning, when a shutdown is less important. There is, therefore, a small advantage to the UK in the allocation of an orbital position so far west: with a satellite at 19°W the service would have to close down 1 hour and 16 minutes after midnight, whereas at 31°W it could be extended to 2 hours and 4 minutes past local midnight, although delaying the inevitable by 48 minutes may be a debatable advantage.

Naturally, a 24 hour service could be provided, but for high-power DBS systems it is considered uneconomic to carry the large mass of batteries sufficient to provide a high-power DBS beam. The argument here is in favour of quasi-DBS satellites [discussed in §B1.1] which, having lower power requirements, can maintain a service throughout an eclipse period.

The practice of positioning satellites to the west of their service area posed a problem with the sharing of frequency bands with the fixed-satellite service (FSS) (ordinary telecommunications) in Region 2 [Brown and Mertens 1977]. The problem was critical for the section of GEO above the Atlantic, where satellites stationed to the west of their service areas (to maintain extended services under eclipse conditions) would be juxtaposed with FSS satellites. The latter are able to carry sufficient batteries to provide a service in eclipse and can therefore be positioned at the same meridian or even to the east of the area they serve. This would mean that an FSS satellite could be placed quite close to a BSS satellite, which would force an FSS earth station to point in the general direction of the high-power DBS satellite. For example, the wide-coverage beams of the large African countries, from satellites stationed well to the west of Africa, posed a problem for the east of Brazil.

The Region 2 delegations, principally the USA and Brazil, insisted that Region 1 countries should keep the power flux density towards the edges of their beams, which would be likely to overspill into Region 2, to a very low level so as not to interfere with the FSS. After lengthy discussions on the subject, Regions 1 and 3 agreed to improve the satellite antenna radiation patterns, as far as contemporary technology would allow, to reduce the possibility of interference. It was also agreed that, wherever possible, channels in the band where sharing problems did not arise would be allocated to those countries most likely to cause interference. In addition, it was agreed to limit the range of orbital positions in case of future modifications to the plan:

(i) No broadcasting satellite for region 1 (using frequencies in the 11.7–12.2 GHz band) is allowed to occupy a position further west than 37°W or further east than 146°E.

(ii) No new orbital assignments or modifications to existing assignments, in the range 37°W to 10°E, are allowed to be coincident with or within 1° to the east of any of the existing nominal orbital positions. (This rule is intended to eliminate interference through overcrowding.)

2.2.2 Overcrowding of the orbit

Satellites have been launched into GEO since 1963 (Syncom I) and although many of those already launched have ceased to operate, and some have been boosted to the graveyard orbit, more and more satellites are being launched to replace them. The orbit has a limited capacity in each of the frequency bands and every allocation has to be carefully planned and closely adhered to. At the moment, GEO contains the majority of the world's communications satellites, a class which includes the Intelsat, Eutelsat, Arabsat, Inmarsat, Satcom and Westar networks and a host of other individual commercial satellites. There are also a number of military communications satellites in GEO, such as the Fleetsatcom series, and certain terrestrial applications satellites (e.g. Meteosat) which require a stationary orbit position. Appendix A gives a list of the satellites in GEO as of February 1989.

The limited amount of frequency space allocated to DBS services is a reflection of the severe problems that exist for every service requiring a radio frequency. An illustration of why frequency space, as well as geostationary orbit space, must be considered a conserved resource is given by the expansion of commercial GEO communications traffic. To extend the frequency space available to satellites, engineers are constantly striving to push the frequencies higher to give increased bandwidth for more communications services. The early communications satellites operated in the 4/6 GHz bands, the contemporary band is 11/14 GHz and an expansion into the 20/30 GHz region is beginning. This is, naturally, not without its difficulties and is an extremely expensive endeavour.

2.2.3 GEO sovereignty claims

Although WARC-77 was very largely a technical success, it would have been optimistic in the extreme not to expect some political disagreement in a forum of 111 participating countries. One fundamental question that was discussed at great length without the appearance of a clear solution was the sovereignty of GEO.

Several of the countries which lie about the Earth's equator developed a number of legal arguments (laid down originally in the 1976 Bogata Declaration) to assert their rights of control over the portion of GEO directly above their territory. They wished to apply national legislation to their 'geostationary orbit space' in the same manner as the already recognised airspace. This would mean that a broadcasting satellite in full conformity with the WARC plan could not be brought into service above an equatorial country without its express agreement.

Naturally the opposition sought to show that GEO was a part of outer space and thus escaped all claims of national sovereignty. It appears that the problem was one which could not be resolved within the framework of an ITU conference and no date or venue was set for a resurrection of the disagreement. The matter did, however, re-emerge at WARC-85, a major conference on the use of GEO [Martinez 1986]. The signatories of the Bogota Declaration, particularly Colombia and Ecuador, tried to persuade the conference to endorse their claim, but again it was concluded to be outside the purview of the ITU. In the end the conference referred the matter to the United Nations Committee on the Peaceful Uses of Outer Space [Lowndes 1985]. One can only hope that near-Earth space is not divided into national territories at some future meeting in a similar manner to the division of the continent of Antarctica.

2.3 FREQUENCIES

2.3.1 The WARC plan for downlink frequencies

From the deliberations of WARC-77, most countries in Region 1 (Europe and Africa) received five channels, while the countries of Region 3 (Asia and Australasia) received either four or five. The countries of Region 2 (the Americas), covered by RARC-83, received anywhere between four and 32 channels [Amero *et al* 1983]. Table B2.2 shows the channel allocations for the European sector of Region 1, which is by far the easiest to assimilate because of the limited number of both channels and countries, one of the reasons for using it as our case study.

With their characteristic flair for equality, the WARC committees made equivalent assignments to all countries, however small. For instance, Andorra, Monaco, Liechtenstein, San Marino and the Vatican State received allocations equal to those of France, West Germany, Switzerland and the UK. It is, at present, difficult to visualise five DBS channels emanating from San Marino, for example, but WARC-77 was a plan for the future—and which of us can foretell the future?

Perhaps surprisingly, the Vatican State already runs one of the most extensive communications systems in the world. Radio Vaticana, designed and installed in 1931 by Guglielmo Marconi himself, now broadcasts 24 hours a day in at least 34 languages worldwide, so they may well find a use for their DBS allocations long before some of the larger European nations.

Of course, it is highly unlikely that all these options will be taken up. Of the 33 countries listed in the plan, only nine (France, West Germany, Italy, Luxemburg, Norway, Sweden, Switzerland, the UK and Ireland) have shown any significant interest in operating a dedicated direct broadcasting satellite. Moreover, only a handful of these have actually committed money to building a satellite to provide the service, and then not for the full five channels. By the mid-1990s there will be only a few European DBS satellites in orbit and perhaps a few more under construction or on the drawing board.

Table B2.2 The WARC-77 'Plan for Europe': channel assignments.

Country	Channels	Polarisation†
Albania	22,26,30,34,38	L
Andorra	4, 8,12,16,20	L
Austria	4, 8,12,16,20	L
Belgium	21,25,29,33,37	R
Bulgaria	4, 8,12,16,20	R
Czechoslovakia	3, 7,11,15,19	L
Cyprus	21,25,29,33,37	R
Denmark	12,16,20,24,36	L
Finland	2, 6,10,22,26	L
France	1, 5, 9,13,17	R
West Germany	2, 6,10,14,18	L
East Germany	21,25,29,33,37	L
Greece	3, 7,11,15,19	R
Iceland	23,(27),31,(35),39	R‡
Ireland	2, 6,10,14,18	R
Italy	24,28,32,36,40	L
Liechtenstein	3, 7,11,15,19	R
Luxemburg	3, 7,11,15,19	R
Monaco	21,25,29,33,37	R
The Netherlands	23,27,31,35,39	R
Nordic countries	22,24,26,28,30,32,36,40	L§
Norway	14,18,28,32,38	L
Poland	1, 5, 9,13,17	L
Portugal	3, 7,11,15,19	L
Romania	2, 6,10,14,18	R
San Marino	1, 5, 9,13,17	R
Spain	23,27,31,35,39	L
Sweden	4, 8,30,34,40	L
Switzerland	22,26,30,34,38	L
Turkey	1, 5, 9,13,17	R
UK	4, 8,12,16,20	R
USSR	1,3–5,7–9,11–13,15–20, 22–28,(29),30–32,(33),34–36, (37),38–40	R&L‖
Vatican State	23,27,31,35,39	R
Yugoslavia	21,25,29,33,37	R

† Right-handed polarisation (R) is otherwise known as 'clockwise' or 'polarisation 1'. Left-handed polarisation (L) is 'anticlockwise' or '2'.
‡ Iceland's beams also cover the Azores and Greenland: channels 27 and 35 are part-assigned to Denmark for Greenland's use.
§ The five Nordic countries (Denmark, Finland, Iceland, Norway and Sweden) were allowed by WARC to use some of their channels to provide a joint regional service, namely Denmark 24,36; Finland 2,26; Norway 28,32; and Sweden 30,40.
‖ The USSR has a large number of channels because over a large part of the globe it is the only country that both requires satellite channels and has the sovereign right to use them.

2.3.2 Channel bandwidths

The frequencies assigned to DBS downlink for Region 1 occupy a total bandwidth of around 800 MHz, from about 11.7 to 12.5 GHz. The corresponding 800 MHz uplink allocation is between about 17.3 and 18.1 GHz. Figure B2.1 shows these bands in relation to the rest of the frequency spectrum. Incidentally, the uplink part of a satellite communications link is typically placed at a higher frequency than the downlink for two main reasons. First, the attenuation of radio frequencies by the atmosphere is greater at higher frequencies, so the responsibility for producing the higher power to overcome this is placed with the earth station, where higher powers are more readily available. Secondly, satellite hardware operating at higher frequencies tends to be more difficult to engineer and is therefore more expensive.

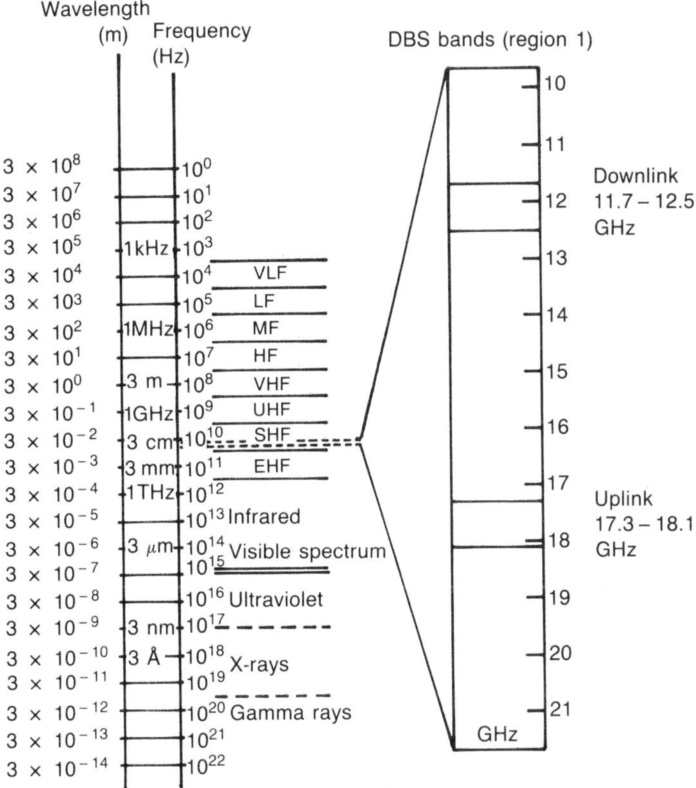

Figure B2.1 DBS frequency bands and the electromagnetic spectrum.

The downlink allocated bandwidth was divided into 40 separate channels: table B2.3 shows the exact frequencies against their respective channel numbers. The total allocated frequency band between the centre of channel 1 (11 727.48 MHz) and the centre of channel 40 (12 475.5 MHz) is 748.02 MHz, so each channel centre is separated by 748.02/39 = 19.18 MHz. Since the individual channel bandwidth required for an FM (frequency modulated) TV signal is 27 MHz, the channels must overlap. Figure B2.2, which depicts the channel separation, shows this overlap to be 7.82 MHz, giving the separation between 'semi-adjacent' channels (e.g. 3 and 5) as 11.36 MHz.

Table B2.3 The WARC-77 'Plan for Europe': correspondence between channel numbers and assigned frequencies.

Channel no.	Assigned frequency (MHz)	Channel no.	Assigned frequency (MHz)
1	11 727.48	21	12 111.08
2	11 746.66	22	12 130.26
3	11 765.84	23	12 149.44
4	11 785.02	24	12 168.62
5	11 804.20	25	12 187.80
6	11 823.38	26	12 206.98
7	11 842.56	27	12 226.16
8	11 861.74	28	12 245.34
9	11 880.92	29	12 264.52
10	11 900.10	30	12 283.70
11	11 919.28	31	12 302.88
12	11 938.46	32	12 322.06
13	11 957.64	33	12 341.24
14	11 976.82	34	12 360.42
15	11 996.00	35	12 379.60
16	12 015.18	36	12 398.78
17	12 034.36	37	12 417.96
18	12 053.54	38	12 437.14
19	12 072.72	39	12 456.32
20	12 091.90	40	12 475.50

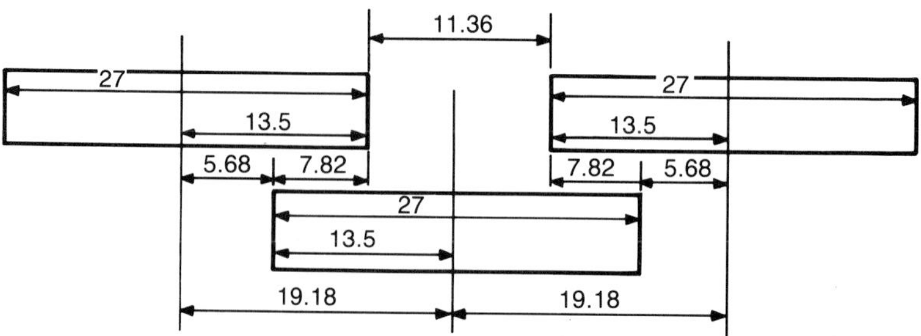

Figure B2.2 DBS channel separation and overlap (figures are in MHz).

In the scheme of channel allocation to the various countries, it is obvious that no one country would be offered adjacent channels since the signals would interfere. In fact, the separation is, in most cases, four channels as shown in table B2.2. Thus the UK has been allocated channels 4, 8, 12, 16 and 20. If all of the channels are transmitted through the same satellite and on the same polarisation, the only option for interference reduction is this channel separation.

Figure B2.3 shows a representative selection of channel allocations from the WARC plan, where each channel is depicted by its 27 MHz-wide passband. The channel separation diagram [figure B2.2] is mirrored in the way the passbands overlap along the direction of the frequency axis. A similar diagram, including all 33 countries in the WARC Plan for Europe, would indicate the level of complexity with which the frequency allocation committees had to deal.

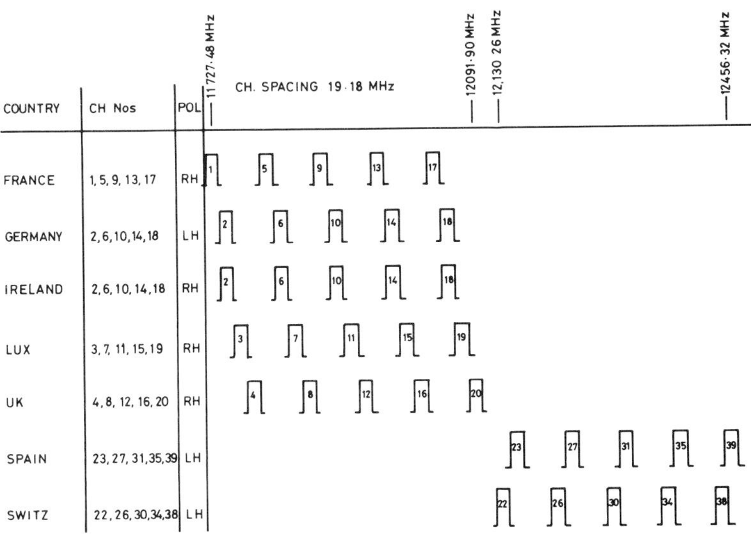

Figure B2.3 DBS channel allocations—European examples.

2.4 POLARISATION

The overall number of channels available to the planners was effectively doubled by using right-hand circular polarisation (RHCP) and left-hand circular polarisation (LHCP). Radio frequencies can be polarised either linearly or circularly. An unpolarised wave may vibrate in any direction, whereas linearly polarised radiation is confined to one plane. The polarisation of radiation can be represented by a waveform as shown in figure B2.4(*a*) for vertically and B2.4(*c*) for horizontally polarised waves, or alternatively by the direction of the electric field vectors

[B2.4(*b*) and (*d*)]. Figure B2.4(*e*) depicts an RHCP wave described by the rotation of its electric field vector. Circular polarisation is the superposition of two orthogonal linear polarisations with a 90° phase difference.

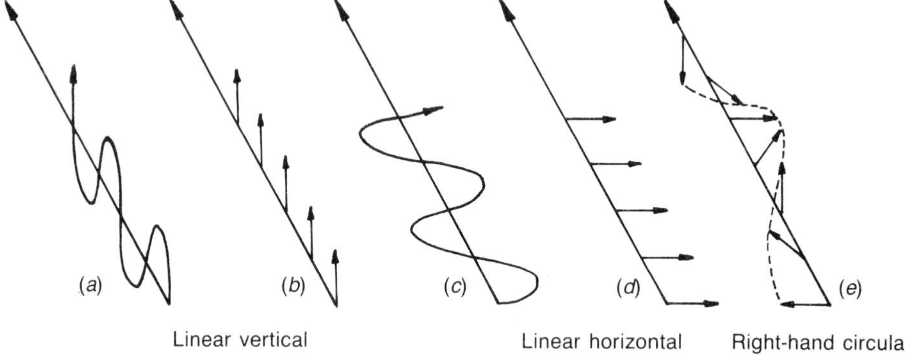

Figure B2.4 Polarised waves.

Satellite systems use both linear and circular polarisation. Using linear polarisation the transmit and receive antennas must be accurately aligned, as far as the polarisation axis is concerned, to ensure the maximum signal strength. Unfortunately, the atmosphere can cause the polarisation axis to rotate, and a system that tracks the polarisation is necessary to maintain the alignment. Circular polarisation was chosen for DBS at WARC-77 because polarisation orientation by the receiving terminal is unnecessary. However, the USA and Iran expressed reservations because this would make sharing with the fixed-satellite service (FSS), which uses linear polarisation, more difficult. Although circular polarisation was generally agreed upon, Iran stubbornly stated its intention to use linear polarisation, the only delegation to do so.

An antenna feed can be designed to cause the electric field vector to rotate clockwise or anticlockwise. Only a small amount of the radiation of one polarisation is receivable in a feed designed for the opposite hand, so polarisation represents an effective method of interference reduction. Indeed, opposite polarisations can give as much as 30 dB of isolation between channels on the same frequency.

In addition to the channel numbers, table B2.2 also lists the hand of circular polarisation allocated to each service. As with the orbital positions, national preferences for this parameter were met as far as possible. Thus an attempt was made to assign the same polarisation to countries in close geographical proximity to allow the simplification of receiving equipment for a particular area. So it is, for example, that

France, Belgium, Luxemburg and the Netherlands have right-hand polarisation and Norway, Sweden and Denmark left-hand. The channels for a given individual service area all have the same polarisation.

Table B2.2 shows how the allocation of opposite polarisations works in the DBS plan. For instance, the UK and Austria have identical channel numbers, meaning that their frequency separation is zero, but the channels are on opposite polarisations. In addition, the two countries have been allocated different orbital positions, 31°W and 19°W respectively, which will further reduce the possibility of interference on the uplink. The problems of interference and their solutions are discussed further in chapter B3.

2.5 CONCLUSION

Despite being faced with the difficult task of coordinating the sometimes excessive and conflicting demands of the delegates of WARC-77, the planning committee succeeded in devising a plan that allocated fairly the resources available to a DBS system. However, the allocation of resources is only part of the plan. The following chapter describes how its technical characteristics are integrated to ensure the future happy coexistence of a number of non-interfering DBS services.

REFERENCES

Amero R G, Bowen R R, Chambers J G and Forsey J 1983 *Space Commun. Broadcast.* **1** 339–49

Brown A and Mertens H 1977 *EBU Rev.* Apr. 60–7

Lowndes J C 1985 *Aviat. Week Space Technol.* 4 Nov. 69–73

Martinez L F 1986 *Space Policy* **2** no. 1 60–2

B3 The Architecture of the Plan

3.1 INTRODUCTION

The underlying principles of the DBS plan include two fundamental provisions: a broadcast TV signal direct to the home (individual reception) and national coverage for most participating countries. In pursuance of these principles each country was allocated an orbital position and a number of channels of a particular hand of polarisation. Every country in Europe was given the statutory right to initiate a DBS service, as long as they observe the technical guidelines laid down to ensure that no other service suffers any interference or degradation of their signal. This chapter describes the methods for interference reduction and the protection it offers potential users of DBS systems.

A satellite in orbit is well isolated from sources of terrestrial interference because of its distance from Earth. However, as the number of earth stations transmitting to satellites grows, the potential for interference grows too. There are three main factors in the design of a satellite communications system used to reduce the possibility of interference: the allocation of different transmit and receive frequencies, different polarisations, and the angular or spatial separation of the uplink and downlink beams. On the uplink spatial separation is realised by allocating widely spaced orbital positions, and on the downlink by the geographical separation of the countries.

3.2 THE UPLINK

Having been allocated a frequency band, it is up to the earth station transmit chain to confine the uplink frequencies to this passband. However, some unwanted frequencies may be radiated as a result of the combination of different frequencies to form mixing products. These and other spurious signals can sometimes cause significant interference problems. Despite this, the mechanism of frequency separation still reduces the likely degree of interference. Stationing the satellites in different orbital positions is all very well, but the beamwidth of the uplink earth station's antenna must be sufficiently narrow to confine its signals to just one satellite.

The energy radiated from an earth station is shaped into a beam, typically by an antenna reflector with a parabolic cross section [see chapter A9]. Naturally the beam widens as it leaves the proximity of the antenna and by the time the radiated signal reaches the geostationary arc it has spread out considerably. The measure of spreading is represented by the *beamwidth* of an antenna which can be defined as 'the width of the beam of radiation shaped by the parabolic surface and extending to the points where the radiated power is half of its peak value'. Figure B3.1 shows the antenna beamwidth graphically. The peak (or *boresight*) of the beam has been placed at 0 dB on the power axis to indicate that, at the half-power level (– 3 dB), the width of the beam is 2°. The half-power beamwidth (HPBW) of the antenna is therefore 2°.

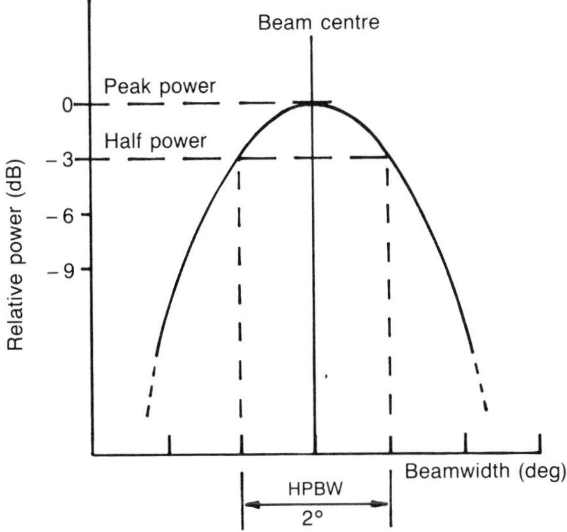

Figure B3.1 Illustration of antenna half-power beamwidth (HPBW).

Beamwidth is related to antenna diameter and frequency, such that at constant frequency the beamwidth decreases as the antenna diameter increases. This is described by the simple relationship

$$\text{HPBW} = 72\lambda/D \qquad \text{(B3.1)}$$

where HPBW is the half-power (3 dB) beamwidth in degrees, λ is the wavelength in metres, D is the earth station antenna diameter in metres and $\lambda = c/f$ where c is the speed of light (3×10^8 m s^{-1}) and f is the frequency measured in hertz (Hz). The constant, 72, in the above expression assumes a circular aperture with parabolic illumination; if the illumination is uniform across the aperture, the constant is 58.4. The same expression can be used for linear apertures, but the constants are 66 for parabolic illumination and 51 for uniform illumination.

As an example, let us take a DBS uplink station with a 7 m diameter antenna operating at a transmit frequency of 17 GHz, which is equivalent to a wavelength of 0.018 m. This would make its half-power beamwidth just less than 0.2°, which gives the beam a diameter of about 125 km at the satellite (from $r \sin \theta$, where r = 36 000 km).

Of course the HPBW is just a convenient way to define beamwidth: it is not a cut-off point and quite high levels of potentially interfering power are radiated over a much wider angle than this. Satellites sharing the same orbital position, although well spaced, would not necessarily be as much as 125 km apart, so satellites sharing the same nominal orbital position would not be allocated the same frequency and polarisation values. Put another way, if two DBS services are to be allocated closely spaced frequencies, the respective satellites themselves should occupy different orbital positions. This would give their earth station transmit beams sufficient spatial separation [compare tables B2.1 and B2.2 for examples].

3.3 THE DOWNLINK

3.3.1 The spacecraft antenna

One of the tasks of WARC-77 was to cover each European country with an elliptical beam. The ellipses in figure B3.2 represent the truncation by the Earth's surface of the beams of radiation transmitted from the respective satellites. All points on the ellipse represent an equivalent power density.

The beam is shaped by the spacecraft antenna, so the shape and orientation of the ellipse bears a direct relationship to the shape and orientation of the antenna. Although the beamwidth relationship quoted as equation (B3.1) was used to derive the beamwidth of a circular earth station antenna, it can also be used to determine the size of a spacecraft antenna required to produce a given elliptical beam. Rearranged to give $D = 72\lambda/\text{HPBW}$ (where D is the spacecraft antenna diameter in metres) the expression is applied separately to each of the orthogonal axes, which of course have different beamwidths. At a downlink frequency of 12 GHz the quantity 72λ is equal to 1.8, so the antenna required to give a $1° \times 2°$ beam, say, has a major axis of 1.8 m and a minor axis of 0.9 m. This illustrates a fundamental property of antenna reflectors: since the beamwidth in the expression is inversely proportional to the antenna diameter, the larger axis of the antenna produces the narrower beamwidth and vice versa. This means that the major axis of the spacecraft antenna must be orientated at right angles to the 'major axis' of the country to give the required beamshape for coverage.

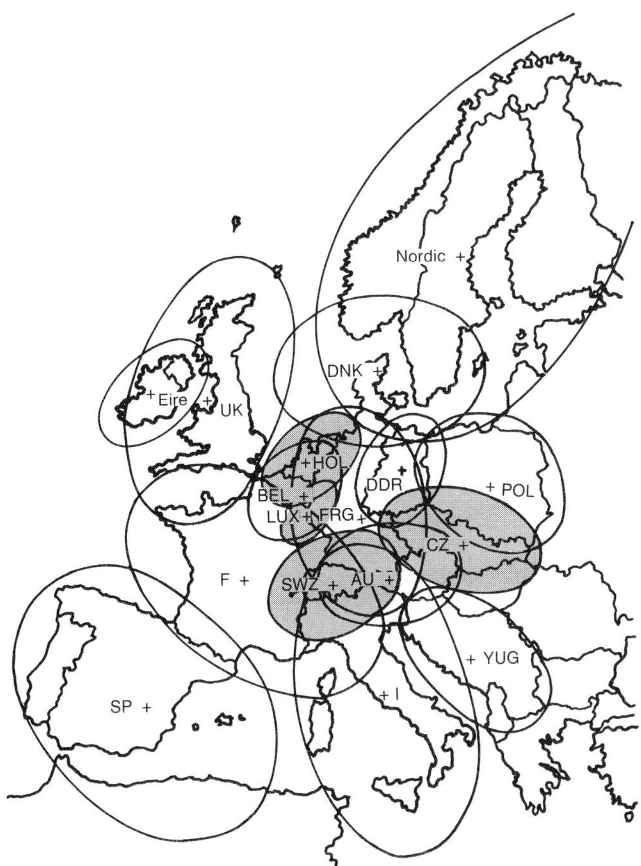

Figure B3.2 European DBS 'beam areas' (approximate): Austria, Belgium, Czechoslovakia, Denmark, Eire, France, East & West Germany (DDR & FRG), Italy, Luxemburg, the Netherlands (HOL), Nordic countries, Poland, Spain, Switzerland, UK, Yugoslavia (four ellipses are shaded for clarity).

Part of the antenna design process is the orientation of the coverage ellipse to provide efficient coverage of the service area. This is defined by the angle its major axis subtends to the east–west direction. To produce the correct beam, the spacecraft antenna must be pointed accurately at the point on the Earth designated as the beam centre. This is called *boresighting*. It is fixed at the design stage, and the spacecraft is built with its antenna aligned such that it will point in the correct direction when it is successfully on-station. Pointing and orientation are maintained by the satellite, but the shape of the beam is fixed by the size and shape of the antenna. Figure B3.3 shows all the information (the boresight, the orientation of the major axis and the size of the ellipse) needed to lay down any country's beam on a map, provided the scale of the map is known.

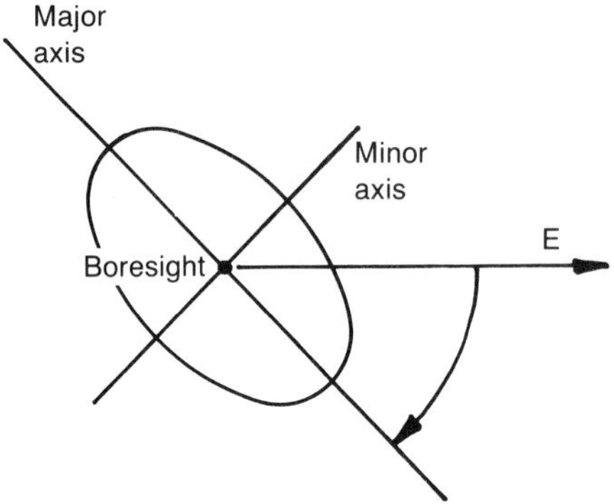

Figure B3.3 Beam orientation measured by the angle its major axis subtends to the easterly direction.

A common misconception is that signals cannot be received outside the area delineated by the simple ellipses in figure B3.2. In fact these ellipses usually represent the contour around which the power from the satellite is half what it is at the peak of the beam. The area within this contour (known as the − 3 dB contour or half-power contour) is termed the *beam area*, but of course the definition only applies if it really is a half-power contour—anyone can draw an ellipse on a map!

In an attempt to obviate confusion between terms which were being used interchangeably, WARC-77 made a further distinction between *service area* and *coverage area*. Figure B3.4 shows their relationship.

The *service area* is defined as the area within which the administration originating the service can demand protection against interference from other transmissions. A receiver in the service area should therefore be able to receive an interference-free signal from that country's satellite. It is assumed that the service area lies inside the political boundary of the country, but WARC is not specific on the point, recognising that some political boundaries are subject to dispute.

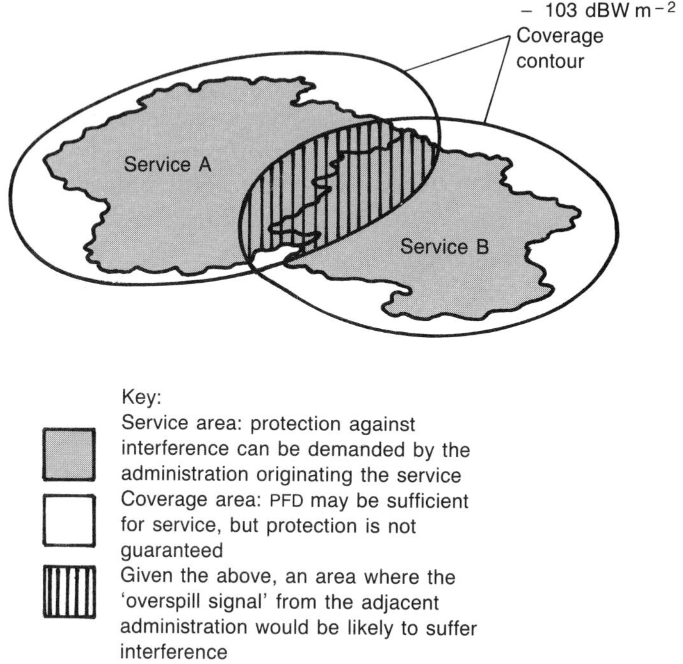

Figure B3.4 Relationship between the service areas and coverage areas of two adjacent countries. Note 'beam area' is that within the – 3 dB contour (or half-power beamwidth).

The *coverage area* is defined as the area over which the signal power level at the ground may be sufficient for a DBS service but in which there is no guaranteed protection. The signal power, in this case, would be sufficiently high to permit the desired quality of reception, but only in the absence of interference.

Although such definitions of area appear at first sight academic, it is to technicalities such as these that the legal profession may refer in future cases of dispute, where one country's services are found to be interfering with another's.

To define the coverage area completely, a value of power must be assigned to the contour ellipse which bounds the area. The received power on the ground is measured in terms of power flux density (PFD) and is quoted in dBW m^{-2}. Of course, the level of power received from a satellite 36 000 km away in geostationary orbit is exceedingly low. For a DBS service the PFD which defines the edge of the coverage area is – 103 dBW m^{-2} (i.e. 5×10^{-11} W m^{-2})!

The complete description of the way the satellite's radiated power is distributed on the surface of the Earth is contained within the contour diagram known as a footprint. Figure B3.5 shows the WARC footprint for the UK, which has been simplified to show only selected isoflux contours (contours of the same flux). The contours may be likened to those on an Ordnance Survey map depicting a hill, since the centre contour borders the area within which the highest power density is available and the others represent a progression to lower power densities. The centre of the beam (marked by a cross) is known as the peak, providing the *peak power* within the footprint (it is also called the beam centre or boresight). The coverage area within the -103 dBW m^{-2} contour is shaded. In general, a beam leaves the satellite antenna with an elliptical cross section but, since the Earth is spherical, its projection on the surface is distorted, especially to the higher latitudes as the surface curves away. The result is shown to effect in figure B3.5.

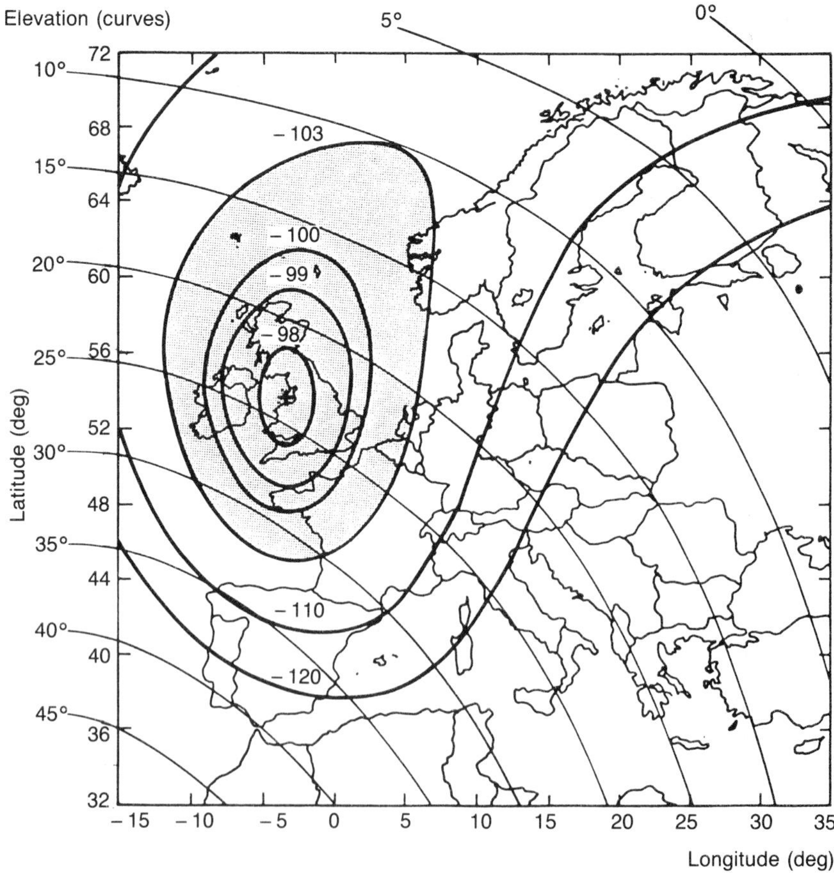

Figure B3.5 UK WARC footprint (power flux density at peak is -97.7 dBW m^{-2}). [British Aerospace]

The map is set in three axes: latitude, longitude and the elevation angle of the satellite as seen from the ground station. Localities in mainland UK have elevation angles between about 19° and 27°—in other words, the satellite is not very far above the horizon in northern latitudes. This is largely because the UK's geostationary orbit position is so far west. If the satellite was to be stationed at 19°W with most of the rest of north-west Europe's direct broadcast spacecraft instead of 31°W, the antenna elevation angles would range between about 22° and 31°, offering some improvement.

The elevation of the satellite is particularly important in the city environment where signal reflections from buildings and other solid objects can cause an effect known as *multipath*. Part of the received signal takes a fractionally longer path to the receiver via the reflecting surface and, since both signals travel at the speed of light, the reflected signal reaches the receiver fractionally after the main signal. This is manifested on the TV screen as a faint ghost image or images and, since the electron beam which builds up the picture from the coded signal scans from left to right across the tube, the 'ghosts' are offset just to the right of the main image. In built-up areas the solution is to use one larger community antenna on a building with a good line of sight to the satellite and distribute the signals through a local cable network.

The PFD must be sufficient for receivers in the service area to obtain a usable signal under the worst weather conditions (WARC recommends that a usable signal should be available for 99% of the worst month). This means that the PFD under clement conditions is more than enough for a good signal and explains why coverage areas are much larger than service areas. DBS consumers just outside the political boundaries of the originating country will be able to receive a good TV signal (using the same class of equipment) most of the time. And the -103 dBW m^{-2} contour is in no way a physical cut-off point, so a more sensitive receiver even further outside this contour could obtain a signal. For example, the UK's DBS signal would be receivable at or above the -110 dBW m^{-2} level through most of Norway and Sweden even though the satellite would appear very close to the horizon.

One of the provisions of the Radio Regulations states that the coverage area must be the smallest area which encompasses the service area. This is especially difficult with small countries, since the satellite antenna required to produce such a small footprint would be too large to fit inside the shroud of the launch vehicle. Many of the resultant footprints are therefore larger than actually required, as can be seen from figure B3.2. Of course no satellite beam can possibly adhere exactly to a country's political boundary, and overspill is inevitable.

The size of the service area affects the design of the satellite in another way. Since the service areas vary in size, the spacecraft designed to cover the larger countries will require higher power TWTAs than for the smaller countries to give an equivalent PFD (and therefore signal quality) at the domestic receiver. Covering the USA, Australia or the USSR with one beam would, however, require higher transponder powers than are presently available. For this reason, apart from the fact that one beam might not be desirable anyway, the larger countries have been divided into several service areas. For example, Australia has been divided into six areas, India into 12, the USSR into 21 and China into 35, all of which make the planning for Europe look trivial!

Within the plan a number of examples of legal overspill were recognised. Amongst them were a number of so-called 'superbeams' which cover a group of neighbouring countries or greatly exceed the national frontiers of one particular country. For example, a Nordic superbeam covers Denmark, Finland, Norway and Sweden with eight channels (two for each country). By way of compensation, the number of national channels for the Nordic countries has been reduced to three. The request for a similar superbeam for the German-speaking countries (West Germany, Austria and Switzerland) was denied, with the allocation being limited to the standard five national channels. The coverage areas of the beams were, however, slightly extended as a consolation.

Overspill does not necessarily mean interference, since services have been allocated different radio frequencies, perhaps on opposite polarisations, and the ground receiving antennas may be pointing at satellites in different orbital positions. But protection is all a matter of degree. It is obvious that where beams overlap the possibility of interference is higher.

3.3.2 The domestic antenna

Probably the most obvious component of a DBS system to the casual observer is the 'dish' antenna which collects the signal radiated from the satellite and channels it to the receiving equipment adjacent to or within the TV set. It also apparently represents an opportunity to proclaim one's individuality! [See figure B3.6.]

The size of the antenna specified in the WARC plan has greater ramifications for the system than are at first evident. The CCIR originally proposed that a 75 cm dish should be the standard, obviously having well in mind the cost and environmental impact of the antenna. The diameter was, however, increased to 90 cm by the WARC committee in response to recommendations from the US delegation, who pointed out the importance of improving the total capacity of the frequency band

and the geostationary orbit. Their argument rested on the relationship between the size of an antenna and its beamwidth. Although the domestic antenna does not transmit (hence the acronym TVRO, for television receive only), it does have a direction-dependent receive sensitivity analogous to the transmit beamwidth.

Figure B3.6 'What's up, Doc?' [Mark Williamson]

Increasing the antenna size brought about a slight increase in gain, but more importantly decreased the antenna's 3 dB beamwidth from 2.4° to 2°, which allowed the nominal orbital spacing of broadcasting satellites to be reduced from the 7.5° proposed in earlier planning exercises to the 6° now adopted. The resultant 25% increase in the number of available orbital positions was one of the factors which made the WARC plan possible in the first place. Table B3.1 shows, amongst other parameters, the half-power beamwidth of the receiving antennas for all regions, as specified by WARC-77 and RARC-83.

It has been suggested that 90 cm amounts to 'overkill', since all orbital positions for DBS will not be occupied in the foreseeable future, so 30 cm 'window-sill' antennas have been proposed. However, an additional characteristic of small antennas is that their gain is lower: they receive less of the satellite's radiated power because of their small area. In fact the 30 cm dish has a gain of only about a tenth of the 90 cm antenna, which implies that either the satellite would have to transmit 10 times the power, or the receiver electronics would have to be significantly improved to enable the same quality of TV picture to be received. Common sense shows that it would be far cheaper to increase the size of a mass-produced antenna rather than enhance the design of the domestic receiver or the satellite. Moreover, if low-power DBS becomes the norm, the larger, higher gain antennas will be required anyway.

Table B3.1 Summary of parameters from WARC-77 and RARC-83.

Parameter	WARC-77 (regions 1 and 3)	RARC-83 (region 2)
Orbital spacing	6°	>9°
RF channels	1: 40 in 800 MHz 3: 20 in 400 MHz	32 in 500 MHz
Channel bandwidth†	27 MHz	24 MHz
Channel spacing	19.18 MHz	14.58 MHz
Guardbands: lower	14 MHz	12 MHz
upper	11 MHz	9 MHz
Protection ratios‡	31 dB co-channel 15 dB 1st adjacent —	28 dB co-channel 13.6 dB 1st adjacent −9.9 dB 2nd adjacent
PFD§	−103 dBW m^{-2}	−107 dBW m^{-2}‖
C/N (99% worst month)	14 dB	14 dB
Receive terminal G/T	6 dB K^{-1}	10 dB K^{-1}
Rx terminal diameter	0.9 m	1.0 m
Rx terminal beamwidth	2°	1.8°
Polarisation	circular	circular
Type of modulation	FM	FM

† Difference in channel bandwidth reflects that required for NTSC video compared with PAL and SECAM.
‡ Against interference from other DBS (FM TV) carriers. See §B3.4.1 for terrestrial interferers.
§ At edge of coverage area for 99% of the worst month.
‖ But see discussion in §B3.4.2.

A development which may please the environmentalists who fear the prospect of antenna dishes on every rooftop is that of the flat-plate antenna. Much experimental work has been conducted on microwave antennas comprising a number of radiating elements mounted on a flat surface, which can be electrically coupled to form a beam. For DBS the plate could be mounted on a house wall or installed as part of a pitched roof. The beam could be steered electronically to point towards a number of satellites in different orbital positions. Much of the work so far has been associated with military projects and, until recently, flat-plate antennas could not be made economically enough for the consumer market, but, thanks to more recent research, the simpler fixed-beam versions are already available.

3.3.3 Antenna performance patterns

Figure B3.7(*a*) shows the curves which specify the electrical design of the receiving antenna: the co-polar and cross-polar reference patterns.

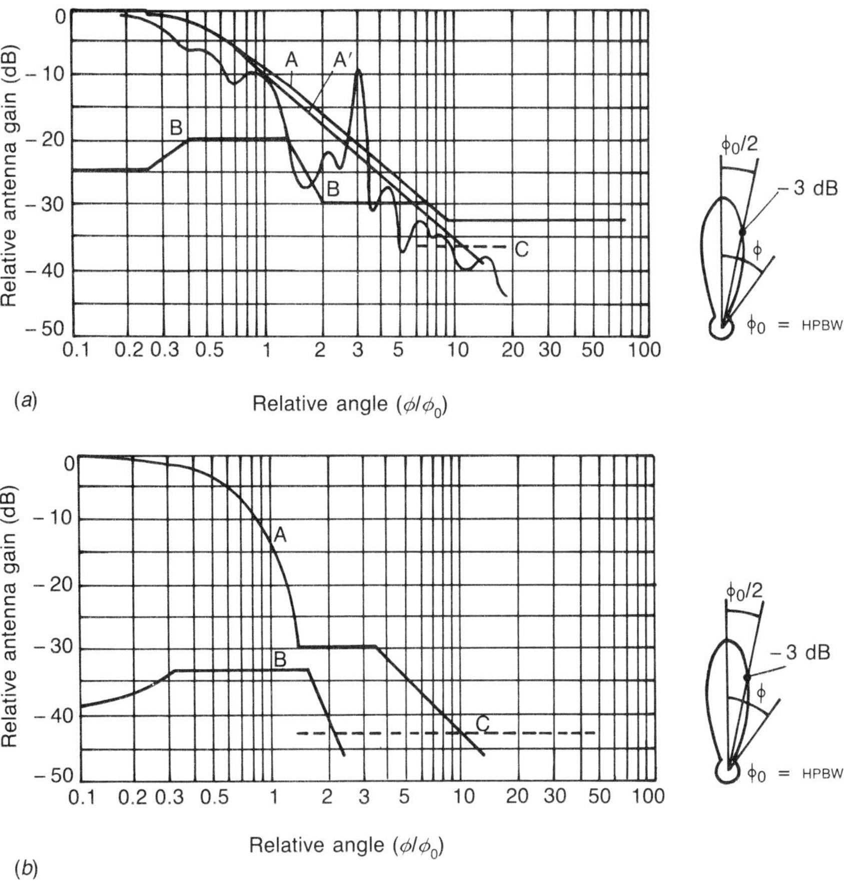

Figure B3.7 (*a*) Co-polar and cross-polar reference patterns for receiving antennas (with example pattern). Individual reception: co-polar—curve A for Regions 1 and 3 (curve for Region 2 is similar but flattens off at – 38 dB); cross-polar—curve B for all regions. Community reception: co-polar—curve A′ to intersection with curve C, then C for all regions; cross-polar—curve B for all regions. (*b*) Co-polar and cross-polar reference patterns for satellite transmitting antennas: co-polar—curve A to intersection with curve C, then C; cross-polar—curve B to intersection with C, then C.

Co-polar radiation is that received on the operational (wanted) polarisation; cross-polarised radiation is that of the opposite (unwanted) polarisation which, if received, could cause interference. The polar patterns defined by WARC are designed to make sure that interference is minimal.

The sketch to the right of figure B3.7(*a*) is a polar diagram of the sensitivity of an antenna. An antenna has a greater sensitivity in a given direction, that is a *directivity* and thus, by definition, a value of gain. The vertical line is the direction of greatest sensitivity (the boresight) of the antenna and the polar diagram is a plot of the antenna's sensitivity at all points of the compass. In this way we can define the antenna's beamshape and the 3 dB (half-power) beamwidth. The vertical axis gives a measure of how the antenna gain varies from the boresight ($\phi = 0$).

The antenna pattern for the actual receiving antenna must be contained below the relevant curves: the example pattern given does not meet the requirements at $\phi/\phi_0 = 3$, for instance. Whereas the idealised sketch shows a pattern with only a main lobe, our example shows a sidelobe outside the 3 dB beamwidth. Some of the sensitivity of the antenna has, in this case, been shifted outside the useful beam and unwanted signals from a different direction (a different satellite) may be received [see §A9.5.1]. The antenna's sensitivity to cross-polarised radiation is shown by a pattern which must be below curve B to afford protection from a service using the opposite polarisation.

The satellite antenna must also adhere to requirements for co- and cross-polar gain similar to those for the domestic receiving antennas. The reference patterns used in preparing the WARC plan are shown in figure B3.7(*b*). In the case of the satellite, there is an extra requirement for pointing accuracy. The satellite must keep its antenna pointing at the boresight on the Earth to within 0.1° in any direction, so that the beam does not drift unduly into other service areas. If the satellite control systems did not correct for them, the effects of gravity and the solar wind would perturb the satellite to such an extent that Luxemburg's beam, for instance, might end up over Brussels or Paris.

3.4 TECHNICAL CONSIDERATIONS

3.4.1 Guardbands and protection ratios

The division of the downlink band for Region 1 into 40 overlapping channels, as derived in chapter B2, allowed for a *guardband* of 14 MHz at the lower end of the band and 11 MHz at the upper end. Guardbands for all three regions are summarised in table B3.1. The guardbands may be thought of as the 'hard shoulders' of the DBS carriageway and are designed to protect services in adjacent frequency bands. They are necessary because even the best system will produce undesirable low-level signals (so called spurious emissions), as a result of the modulation process, which fall outside the 27 MHz channel bandwidth. The guardbands allow DBS systems to operate without interference to existing terrestrial services, so it is only fair that, in return, DBS should be protected from other services. This is the function of the protection ratio, which is the ratio of the power of the wanted signal to that of the unwanted (interfering) signal. For example, if the value of the protection ratio is 3 dB, the wanted signal is 3 dB higher than the unwanted signal.

Figure B3.8 shows a plot of protection ratio, R(dB), against the difference (in MHz) between the two carrier frequencies in question, Δf, for 'single-entry' interference (i.e. only one interferer). The single-entry protection ratio against all types of terrestrial transmissions (except amplitude modulated multichannel TV systems†) is 35 dB for differences in carrier frequency of up to ±10 MHz, as shown by the plateau in figure B3.8. Thus the wanted signal power is 3162 times higher than the interfering signal ($10\log_{10} 3162 = 35$ dB). The ratio decreases linearly from 35 to 0 dB for frequency differences between 10 and 35 MHz and is 0 dB for differences in excess of 35 MHz. This shows that frequency separation, in its own right, is sufficient protection against interference.

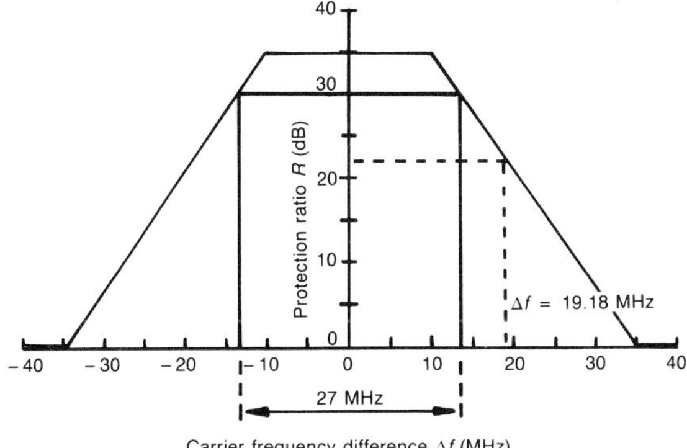

Figure B3.8 Protection ratio: truncated triangle describes the level of protection for a broadcasting satellite signal against a single terrestrial interferer; the other lines are explained in the text [§B3.4.1].

† AM multichannel TV systems produce peaks of high power flux density spread over the wide range of their channel bandwidth, so the protection ratio in this case is 35 dB irrespective of the difference in carrier frequency.

Figure B3.8 also shows that the protection ratio, or the interference ratio, should not be less than 30 dB for all the interference in a given 27 MHz bandwidth DBS channel. If there is more than one interfering signal the resultant ratio is simply the power of the wanted signal divided by the sum of the interfering powers, which can be performed for both co-channel and adjacent channel interferers as required.

Of course DBS systems can also interfere with each other. WARC-77 used values of 31 dB for co-channel interferers and 15 dB for adjacent channels to calculate the protection margins. These values are less than for their respective terrestrial interferers, that is co-channel (Δf = 0) at 35 dB and the equivalent adjacent channel spacing (Δf = 19.18 MHz) at about 22 dB [see figure B3.8], indicating that a greater protection margin is required against terrestrial interferers than for other BSS TV signals.

In addition to the guidelines on relative power levels, WARC-77 also ruled that the absolute EIRP level from satellites must never be more than 0.25 dB above the nominal value used in the plan to calculate interference levels. Obviously, if this was not observed the plan would cease to be valid.

The bottom line of the WARC plan is given in terms of the *protection margin*, that is 'by how many decibels a service is protected from interference'. In more technical terms, it is the difference (in dB) between: (i) the actual carrier-to-interference ratio (*C/I*), and (ii) the theoretical guideline, *C/I*, or *protection ratio*. Both (i) and (ii) are quoted in terms of the ratio of the wanted power to the sum of the interfering powers, but (ii) represents a recommended limit and (i) an actual value.

Negative protection margins correspond to interference which exceeds the limit considered acceptable, and therefore indicate defects in the plan. An analysis performed by the IFRB (International Frequency Registration Board) at the end of the WARC-77 conference showed certain negative protection ratios in some channels of certain service areas. The most negative in the EBU countries' area concerned Luxemburg, Jordan, Turkey and Yugoslavia. The proportion of negative protection ratios was, however, very small considering the fact that the plan contains a total of 907 assignments of channels for 247 service areas.

The conference preparatory meeting for RARC-83 had decided that similar technical characteristics to WARC-77 should be defined. Unfortunately the two nations most likely to initiate DBS systems, the USA and Canada, could not agree on the technical parameters. The two parameters which caused the most disagreement were the adjacent channel protection ratio and the domestic antenna receive pattern, which together largely determine the minimum orbital spacing required between satellites serving adjacent areas on the same frequency. The former stemmed from the difficulty in agreeing the amount of 'picture impairment' (due to interference) which the viewers would accept. The latter produced (from Canada) 2 or 3 dB less isolation from adjacent satellite interference and, according to an American spokesman, the two parameters combined would produce an orbital spacing for adjacent satellites as high as 18°! The USA had proposed offset-fed antennas of 0.9 m diameter, but the conference chose the Canadian 1 m symmetric-fed antennas with Canadian reference patterns and set the nominal orbital spacing for satellites servicing adjacent service areas (e.g. the Eastern and Central Time Zones of the USA) at 9 or 10°, compared with the 6° spacing chosen in Europe at WARC-77. Several administrations thought the offsets too expensive, despite the insistance of various American sources that they could manufacture them competitively given sufficient numbers.

3.4.2 Power flux density and earth station figure of merit

To make sure that a DBS service adheres to its protection ratios, the power flux density (PFD) from the satellite must be kept within limits by careful design of the communications link. But at the same time it must be high enough for acceptable pictures to be received by a certain quality of domestic receiver. The parameter which defines the quality of the receiving system (i.e. domestic antenna and electronics) is G/T or *figure of merit*. To ensure a workable system the PFD and G/T must be balanced in a downlink budget like that described in chapter B4.

WARC-77 found the PFD a difficult parameter to fix, since it has direct repercussions on both the satellite transponder power and the G/T of domestic receivers. The EBU, amongst others, was keen to minimise the cost of receiving equipment even at the expense of the satellite end of the link and strongly recommended to the CCIR that the G/T should be set at 4 dB K^{-1}. This would correspond to a carrier-to-noise ratio (C/N) of 14 dB and a minimum PFD of -101 dBW m^{-2} for 99% of the time. However, studies made by ESA and others showed that an improved G/T of 6 dB K^{-1} could be obtained with contemporary technology without increasing receiver costs. This 2 dB increase in G/T meant that the satellite power could be reduced by 2 dB to give a PFD of -103 dBW m^{-2}. This will serve to reduce space segment costs, since the satellite is required to produce less power, but will also lead to a slight increase in the price of receivers. Naturally, it is hoped that much of this increase will 'disappear into the noise' through the economies of mass production.

This value is the minimum PFD deemed to be sufficient for the reception of a good TV picture—the actual PFD may be somewhat higher.

The quoted figure holds for 99% of the worst month (in terms of weather conditions). Much of the time the signal strength will be higher than this and the coverage area will widen accordingly. In fact, planning for the UK beam set the boresight PFD at -97.7 dBW m^{-2}, an increase of 5.3 over -103 dBW m^{-2} [see figure B3.5]. The minimum value of -103 was finally adopted mainly as a result of discussions with the Soviet delegation, who argued that the lower PFD should be demanded of BSS systems to permit sharing with multichannel telephone radio relays. Of course, the reduction of PFD will also tend to reduce the overspill of western programming into Eastern bloc countries.

The decision did, however, raise doubts in the propagation model of the Earth's atmosphere adopted by the conference. A radio signal is attenuated, or weakened, by its passage through the atmosphere, the degree of attenuation depending on the atmospheric conditions. The conference accepted the CCIR theoretical atmospheric models giving atmospheric attenuation as a function of elevation angle for five hydrometeorological areas of the world, which correspond to various statistical distributions of rainfall [see figure B4.3]. For Europe, the CCIR model differed slightly due to results of experiments performed by ESA. There was, however, a total lack of experimental data for 12 GHz propagation in Africa. Many participants therefore, and particularly the delegates from African countries, expressed concern at the rather theoretical nature of the CCIR models.

The problem was not helped by the fact that the conference decided to calculate satellite transmission power purely to satisfy the condition for a carrier-to-noise ratio of 14 dB, for at least 99% of the worst month [see chapter B4]. The conference had the option of a 'threshold condition' of '99.9% of the time', but this would have required a higher transmission power, which the USSR argued against for fear of interference.

The resultant decrease in PFD was of particular concern to countries with a tropical climate, where the attenuation due to rainfall could be expected to be exceedingly high at certain times. It was pointed out that the effect of sand- and dust-storms would also have to be taken into account, but preliminary laboratory measurements on simulated sandstorms, submitted by Sudan, showed that the effect would not be very serious. Nevertheless, the CCIR was requested to make an urgent study of these problems.

The level of power flux density in the service area was also a point of contention at RARC-83. Canada and Brazil opted for -107 dBW m^{-2}, but the US delegation argued for the higher PFD of -105 dBW m^{-2} because they wanted the extra power for a future high-definition television service. The matter had to be settled by a secret vote, which produced a majority for the lower power. The result is thought to be largely due to worries about interference from and overspill of American TV transmissions into adjacent countries. However, when it came to signing the Final Acts, the USA and Venezuela reserved the right to implement systems providing -105 dBW m^{-2} and Mexico reserved the right to use any PFD value necessary to provide convenient coverage [Amero *et al* 1983].

The proclamations of the ITU are not legally binding and it has to rely on the common sense of its members, most of whom believe that international communications would be impossible without some form of regulation. On the other hand, it could be argued that the ITU overspill limitations contravene the European Convention on Human Rights [Hughes 1988]: Article 10 allows anyone the right of freedom of expression, which should include the programmer's freedom to transmit. Technical issues are rarely quite as clear cut as they might at first appear!

3.5 INTERFERENCE ANALYSIS

The complex synthesis of the parameters described above (and others too numerous to mention here) formed the technical basis for the non-interfering DBS Service Plan devised by WARC-77. When each country had been allocated an orbital position and a number of broadcast channels on a particular polarisation, the expected level of interference between each service and any other could be calculated. This analysis is a complex task and it should come as no surprise that the protection from interference afforded some services by the WARC plan is less than originally expected.

Two examples illustrate the workings of the DBS plan. France and Ireland have been allocated adjacent channel numbers on the same (right-hand circular) polarisation. Adjacent channel bandwidths overlap by 7.82 MHz, so on its own this allocation would lead to interference. Moreover, the French coverage area overlaps the service area of Ireland very slightly in the south-east. It is only the discrimination due to the pointing of the receive antenna towards a satellite at 31°W (as opposed to 19°W for the French satellite) which protects the Irish service against interference from the French system. For the second example, consider Italy and Switzerland, whose satellites will occupy the same orbital position (19°W) and transmit on the same polarisation (left-hand). The two countries share a border and the Swiss coverage area is almost completely within the larger Italian coverage area. The only saving grace is the frequency separation, but with semi-adjacent channels this is not great: the high-frequency end of channel 22 (Swiss) is only 11.36 MHz from the low end of channel 24 (Italian)! This example shows the plan in its true colours: it is often just one of the three interference reduction mechanisms (spatial separation, frequency separation and polarisation) which protects two services from interference, and sometimes the margin is narrow.

A country for which the protection from interference is marginal might be tempted to increase its downlink PFD to ensure that its signals get through, but WARC recommendations are designed to guard against this. Any communications service, whether terrestrial or satellite, must go through the complicated process of frequency coordination with its PTT to be granted a licence in the first place; therefore, interference due to non-compliance with the recommendations is not expected. However, if one service was to suffer interference from another, a possible solution would be to increase the *uplink* power to overcome the interference by brute force, so to speak. Perhaps surprisingly, there are no official restrictions on uplink power in the WARC plan. Increasing the uplink power is not necessarily a good solution, however. If the satellite transponder receives too powerful a signal, it can become oversaturated and automatically turn down the gain on the downlink path. This would mean that the power received by the domestic antenna would be too low to detect a usable signal. A problem of this nature, should it occur, would have to be resolved by the authorities.

In fact, a worldwide coordination plan may not be feasible in practice and coordination may have to be performed on an *ad hoc* basis as with communications satellite systems in the fixed-satellite service. It is, however, most unlikely that all the European options on a DBS service will be taken up and this alone, apart from improvements in technology, will reduce the possibilities of serious interference.

3.6 CONCLUSION

Radio transmissions do not observe the restrictions of international borders and satellites offer the ultimate potential for the broadcast of political, or simply cultural, propaganda. Historically, Luxemburg is one European country which broadcasts radio or TV programmes over a wide area. One result of this is the claim to a much larger advertising audience. France, Belgium, Holland and West Germany lie in Luxemburg's nominal coverage area, with a much larger international audience just beyond the -103 dBW m^{-2} contour. Soviet satellite TV is already easily received in parts of Europe; European programming is spilling into the countries of the Eastern bloc. Within Europe itself, this could lead to the reception of what one nation considers pornography from another which considers it merely a vehicle for advertising.

To implement the policy of non-interference, the beamshapes of both the satellite transmitting and domestic receiving antennas must be carefully controlled and all antennas should be pointed as accurately as possible, but the skills of the communications systems engineer really come into their own in the design of the link budget. It is all a question of signal level: the equipment on board the satellite must receive a sufficiently noise-free signal to retransmit to the ground receiver as an acceptable TV signal. Link budgets must be balanced correctly by varying, on the one hand, the satellite EIRP to give a specific PFD at the ground and, on the other, the receiver G/T and the size of the receiving antenna. The design of the communications link budget constitutes the key to the successful satellite communications system.

REFERENCES

Amero R G, Bowen R R, Chambers J G and Forsey J 1983 *Space Commun. Broadcast.* **1** 339–49

Hughes R 1988 *Satellite Broadcasting* ed. R Negrine (London: Routledge) ch. 2

B4 The Link Budget

4.1 INTRODUCTION

A link budget for a satellite communications system is in many ways like a financial budget. The *raison d'être* of a direct broadcast satellite service is the transmission of a signal from the ground up to a satellite in geostationary orbit and its return to ground-based receivers. The signal itself must be of sufficient quality for the receivers to decode it as a good TV picture. A link budget is the 'profit and loss account' of the transmitted signal and produces a communications link design adhering to these requirements. 'Profits' are accrued by the amplification of the signal, either in the satellite transponder or in the earth station's amplifier chain and from the inherent gain of the spacecraft and earth station antennas. 'Losses' arise from the signal's distance of travel on its radiated path through space, through Earth's atmosphere and in the communications hardware.

The link budget table constitutes a 'balance sheet' which combines these profits and losses; the resultant margin in signal power is the balance. If the account is in credit, all is well and the communications system is one of profitable design. If the link budget shows a debit, or negative margin, then the system is not viable and must be redesigned. Either way, the link budget has proved its worth.

The signal path from the earth station, through the satellite's communications payload and back to Earth is described in the relevant sections of chapters A7 and A9. This chapter describes how the power of the signal varies on the uplink and the downlink [Williamson 1984]. Figure B4.1 gives a visual indication of the signal power throughout the round-trip from the transmitting earth station to the domestic receive terminal.

4.2 THE UPLINK BUDGET

The uplink budget can be divided into 14 stages as shown in table B4.1.

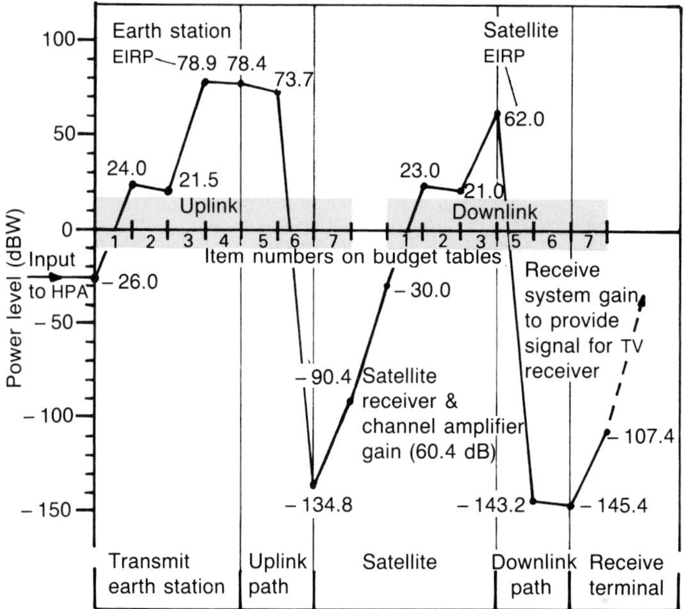

Figure B4.1 Relative power levels in a satellite communications link.

Uplink

1. Earth station HPA gain (50 dB)
2. Feed losses
3. Antenna gain
4. Pointing loss
5. Atmospheric/rain attenuation
6. Free-space loss
7. Satellite antenna gain

Downlink

1. Satellite TWTA gain (53 dB)
2. Output losses
3. Antenna gain

5. Free-space loss
6. Attenuation
7. Domestic antenna gain

[item numbers refer to link budget tables B4.1 and B4.2].

1. *HPA power.*

The signal fed into the transmitting earth station must, of course, be amplified for the uplink. The high-power amplifier (HPA) chosen to make this example link work is one with 250 W of radio-frequency output power. The level of RF power produced by the amplifying device, a travelling wave tube (TWT), is the important quantity, in contrast to the level of direct current (DC) power fed into the tube [see §A7.3]. Like any device used for the conversion of energy from one form to another, the TWT has a finite efficiency. For instance, the production of 250 W of RF power would require between 500 and 700 W of DC power depending on the conversion efficiency.

Table B4.1 Uplink budget. Frequency $f = 17.6$ GHz, wavelength $\lambda = 0.017$ m.

	Parameter	Value	Notes
1.	HPA O/P power (250 W TWT)	24.0 dBW	
2.	Feeder losses	2.5 dB	
3.	Earth station peak antenna gain	57.4 dB	†
4.	Pointing loss	0.5 dB	
5.	Atmospheric attenuation (including rain)	4.7 dB	‡
6.	Free-space loss	208.5 dB	§
7.	Satellite antenna gain (at half-power point)	44.4 dB	‖
8.	**Carrier (C)**: Received carrier power at satellite ($C = 1 + 3 + 7 - 2 - 4 - 5 - 6$)	−90.4 dBW	
9.	Boltzmann's constant	−228.6 dBW Hz^{-1} K^{-1}	
10.	Satellite noise temperature	31.2 dBK	¶
11.	**Noise (N_0)**: System noise power density at satellite ($N_0 = 9 + 10$)	−197.4 dBW Hz^{-1}	
12.	Carrier-to-noise power density ratio ($C/N_0 = 8 - 11$)	107.0 dBHz	
13.	Bandwidth (27 MHz)	74.3 dBHz	
14.	Carrier-to-noise ratio at satellite ($C/N = 12 - 13$)	32.7 dB	

† Peak gain: $G = 10\log_{10} \eta(\pi D/\lambda)^2$, $D = 5$ m, $\eta = 65\%$.
‡ Elevation angle 25°: oxygen and water vapour 0.7 dB; rain (6 mm hr^{-1}) 4.0 dB.
§ FSL $= 20\log_{10}(\lambda/4\pi R)$, $R = 36\,000$ km.
‖ Half-power gain: $G = 10\log_{10} \eta(27\,800/\Theta\phi)$, $\eta = 60\%$, $\Theta = 1.2°$, $\phi = 0.5°$.
¶ Earth at 290 K = 24.6 dBK. Satellite system noise figure: 4.6 K = 6.6 dBK.

For budget calculations, the value of 250 W is converted to its equivalent on the decibel scale. Since the value is a power, it is referenced to the watt—the unit dBW shows how many times greater than 1W the power level is, so 250 W $= 10\log_{10} 250 = 24$ dBW. A good guide for those unfamiliar with the decibel scale is that each 3 dB increase equals a doubling of the power: 3 dB is twice, 6 dB is four times, 9 dB is eight times, and so on [see appendix D]. Figure B4.1 shows the HPA to have a gain of 50 dB, bringing the power level of the carrier up to the 24 dBW level.

2. Feeder losses.
The amplified signal leaving the HPA must now be fed, through a length of waveguide, to the earth station antenna which will radiate the signal

to the satellite. The length of the waveguide run from the HPA to the antenna is minimised to reduce power losses [see box A7.3]. Directly following the HPA, and connected by a waveguide, is another loss-making device in the shape of an RF filter. This finally confines the frequency of the radiated signal to its appointed bandwidth [see box A7.2]. It also plays an important part in reducing interference with other services. The waveguide run continues to the feedhorn, a specially shaped open-ended termination to the waveguide, designed to illuminate the surface of the antenna with the radiated RF power.

The precise value of the losses along the feed system depends on several variables, such as the radio frequency and the length of waveguide. The 2.5 dB adopted for this link budget is a typical value.

3. *Earth station antenna gain.*

The size of the uplink earth station antenna depends on the beamwidth required to meet non-interference regulations and on the power multiplication factor or *gain* needed to deliver a sufficiently high signal level to the satellite receiver. In a nutshell, one would like the antenna beamwidth to be narrow and the gain to be high. Luckily the two factors are complementary: increasing the antenna diameter decreases the beamwidth, thereby reducing the possibility of interference to other satellites, and increases the gain. The obvious drawback is that larger antennas cost more.

A useful equation for the calculation of antenna gain, G, is

$$G = 10\log_{10} \eta \, (\pi D/\lambda)^2 \qquad \text{(B4.1)}$$

where η is the antenna efficiency, D is the antenna diameter and λ is the wavelength. At a typical frequency for a UK DBS uplink channel of 17.6 GHz, the wavelength is about 1.7 cm [see table B4.1]. Taking a 5 m diameter antenna with an efficiency of 65% (η = 0.65), which is sufficient for a DBS uplink station, the equation gives the peak gain as 57.4 dB.

Thus the carrier has been amplified in preparation for the long journey to the satellite where it can again be manipulated. Until then it can benefit from no further gain—all budgetary transactions will be losses.

4. *Pointing loss.*

The first of the losses is known as pointing loss. In an ideal world, the earth station antenna would point directly at the satellite and the satellite antenna would be centred exactly on the earth station. In reality, the satellite is not precisely stationary with respect to the earth station. It drifts both above and below the equatorial plane, describing a figure-of-eight pattern in the sky which is repeated daily. The size of the 'eight' can be controlled to confine the satellite to a 'box' which is typically set at 0.2° on a side [see figure A4.1].

Ideally, the beamwidth of the earth station antenna should be greater than this so that the satellite remains within its beam without the need for tracking. Using equation (B3.1),

$$\text{Half-power (3 dB) beamwidth} = 72\lambda/D$$

the beamwidth of a 5 m antenna is calculated as 0.24°, which is only a little larger than the 0.2° box. Because of this, the antenna would be required to follow the movement of the satellite to keep it at the peak of its beam [see §A9.7]. Since antenna tracking cannot be absolutely precise, a small pointing loss must be taken into account: even though the satellite is permanently within the 3 dB beamwidth of the earth station antenna it may spend much of its time a fraction of a decibel down from the beam centre. The budget is therefore debited by a nominal 0.5 dB.

5. *Atmospheric attenuation.*
The signal must now pass through a layer of radio-absorbant atmosphere. The effect of the atmosphere on propagating radio waves is a topic of scientific investigation in itself, but this section will be confined to the two main mechanisms of atmospheric attenuation which affect satellite communications: the attenuation due to oxygen and atmospheric water vapour; and the attenuation due to rainfall.

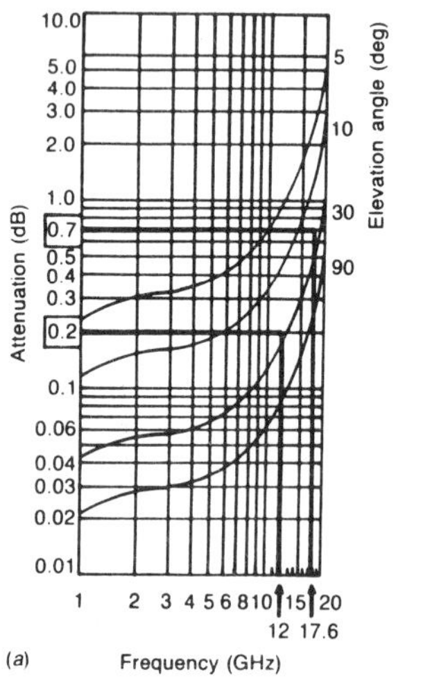

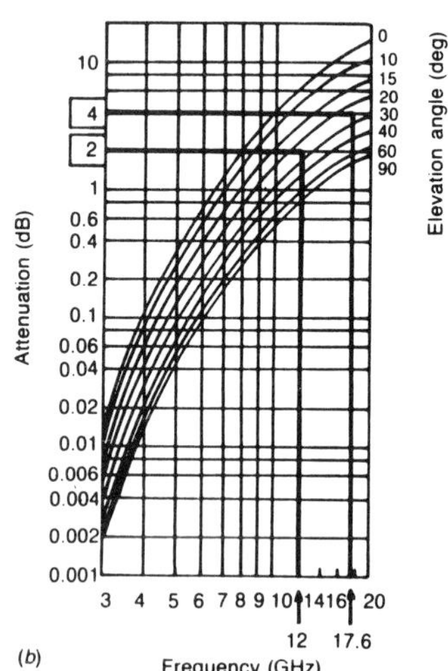

Figure B4.2 (*a*) Oxygen and water vapour attenuation against frequency. (*b*) Rainfall attenuation against frequency (temperate climate: precipitation layer 3 km, rate 6 mm h^{-1}, temperature 18 °C).

Attenuation due to oxygen and water vapour results from an interaction between the radiation and the atmospheric molecules at a resonant frequency, the most important of which for satellite communications is centred at 22.235 GHz [Pratt and Bostian 1986]. The degree of attenuation is dependent on the radio frequency and the elevation angle of the satellite from the earth station. The latter is due to the fact that the atmospheric path is longer the further the satellite is from the zenith (the same factor causes stars near the horizon to twinkle far more than those overhead). The UK's direct broadcast satellite located at 31°W, for example, will present an elevation angle of about 25° from southern England and Wales [see figure B3.5]. Figure B4.2(*a*) is a graph (based on CCIR recommendations) of attenuation against frequency, with regard to elevation angle. It shows the attenuation at 17.6 GHz (the uplink frequency) to be about 0.7 dB.

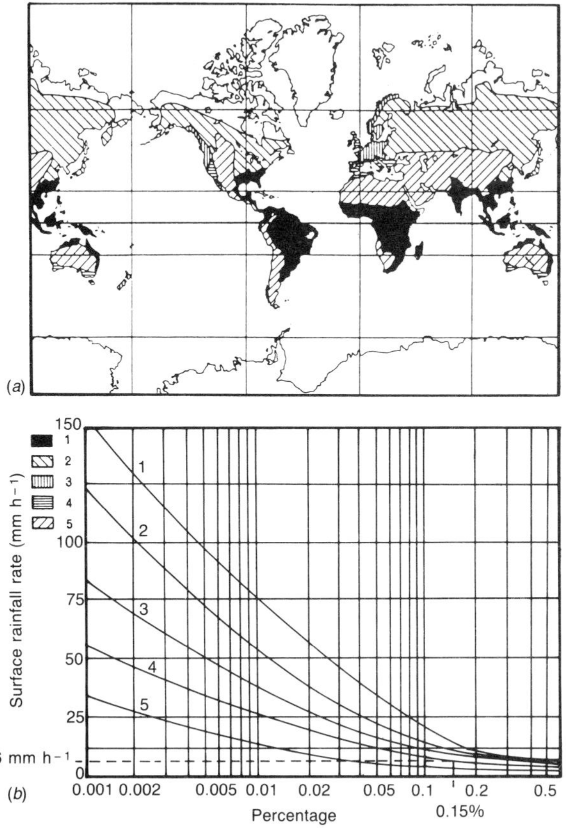

Figure B4.3 (*a*) CCIR rain climatic zones. (*b*) Percentage of an average year for which rainfall rate is exceeded in the five rain climates.

Attenuation due to rainfall occurs because individual raindrops absorb radio energy and because some energy is scattered out of the propagation path. Rain attenuation also causes depolarisation. Figure B4.2(*b*), which shows the frequency and elevation angle dependence of rain attenuation, is based on a precipitation layer of 3 km which, at an elevation of 25°, gives a path length through rain of about 7 km. The curves also assume a precipitation rate of 6 mm h^{-1} which, according to CCIR figures, is exceeded for about 0.15% of the average year [CCIR Recommendation 453, Report 563-1 pp83–4]. The CCIR divides the world into five regions corresponding to rainfall rate and defined according to the graph in figure B4.3. Since 0.15% of a year is equivalent to about half a day, even without recourse to meteorological records, it seems feasible that a rate of 6 mm h^{-1} could be exceeded for this time—an hour or so of really heavy rain now and then throughout the year. The curves can therefore be used for the UK with impunity, even if they represent a worst-case situation.

Indeed the worst case is just what is required for a DBS link budget, since WARC recommends that a usable signal should be available for 99% of the worst month. But the penalty for the budget is high: for a signal at 17.6 GHz and a path elevation of 25°, the attenuation is 4 dB. Coupled with the attenuation due to atmospheric oxygen and water vapour, this brings the total loss of signal power due to the atmosphere to 4.7 dB. Since this value is equivalent to a factor of 2.95, this loss is significant: 4.7 dB of atmospheric attenuation means that the power of the earth station HPA must be nearly three times what it would be if there was no atmosphere. However, numerically speaking, atmospheric losses should not be considered exceptional, especially when compared with the free-space loss.

6. *Free-space loss.*

The power radiated from the earth station is dissipated simply because of the distance between the earth station and the satellite. The power is spread, in accordance with the antenna beamwidth, like a huge searchlight beam which, in the case of a 0.24° beam, is over 150 km across when it reaches geostationary orbit. The power level at the satellite is reduced by the same ratio as that between the area of the beam at that height and the surface area of the ground antenna, which is around 10^{10}:1. In these terms space loss is otherwise known as *spreading loss*, but since beamwidth is frequency dependent, it is common (and convenient) to include a wavelength term in the calculation of free-space loss (FSL).

Thus the loss in signal power between the earth station and the satellite may be calculated using the following expression:

$$\text{FSL} = 20\log_{10}(\lambda/4\pi R) \qquad \text{(B4.2)}$$

where λ is the wavelength and R is the distance (both in metres) separating the satellite and the earth station (the depth of the atmosphere is usually ignored in this calculation but is taken into account in the determination of atmospheric attenuation). Thus FSL debits the account by a massive 208.5 dB, which is equivalent to saying that the transmitted signal power is 7×10^{20} times the received power. This major source of loss is inherent in a geostationary satellite communications system because of the enormous distances involved.

7. *Satellite antenna gain.*

Finally the signal reaches the satellite and the spacecraft antenna intercepts a small part of the radiated power. The antenna can produce a budgetary gain on the uplink by collecting distributed power over its surface and concentrating it into a feedhorn placed at its focus. The larger the antenna, the higher its gain.

To determine the dimensions of the antenna, it is necessary to refer to the parameters which govern the downlink beam, because the satellite antenna is sized and shaped in accordance with its footprint. The WARC plan gives the 3 dB beamwidths for the UK as 1.84° on the major axis and 0.72° on the minor—an elliptical beam. Using the rearranged beamwidth equation (antenna diameter $D = 72\lambda$ /half-power beamwidth) with the downlink wavelength, the antenna is deduced to be a 1×2.5 m ellipse. By substituting the uplink wavelength, it can be seen that the beamwidths of the same antenna as it receives the uplink frequency are 1.2° and 0.5°.

Figure B4.4 compares the beam areas of the spacecraft antenna at uplink and downlink frequencies. With an uplink earth station situated away from the satellite antenna's beam centre or boresight, the budget would not benefit from the peak gain of the spacecraft antenna. The narrower beamwidths at the uplink frequency mean that placing an earth station just 0.5° from boresight could cause a 3 dB gain reduction. Interestingly, the UK's DBS uplink station has a very low probability of being built at the boresight of the spacecraft antenna, since this would put it somewhere in the Irish Sea!

However pessimistic it seems, it is always advisable to make worst-case estimates in the preliminary design of a link to be sure that it is feasible. Adjustments to more favourable situations can easily be made later. The worst possible siting of the earth station (on the edge of the receive coverage area) will therefore be assumed and the gain of the satellite antenna will be calculated at the half-power point. A convenient expression for the half-power gain of an antenna with dissimilar axes is

$$G = 10\log_{10} \eta(27\ 800/\Theta\phi) \tag{B4.3}$$

where η is the antenna efficiency expressed as a decimal fraction, and Θ and ϕ are the orthogonal half-power beamwidths in degrees. (Alternative values for the constant (27 800) may be found in other texts, possibly where a distinction is made between antennas of different types.) Substituting beamwidth figures for the uplink and the downlink into this expression produces gains of 44.4 dB and 41 dB respectively. This illustrates the frequency dependence of antenna gain and shows that, everything else being equal, the 1 × 2.5 m dish produces a gain improvement on the uplink of 3.4 dB over the downlink. An antenna gain of 44.4 dB is thus entered in the uplink budget.

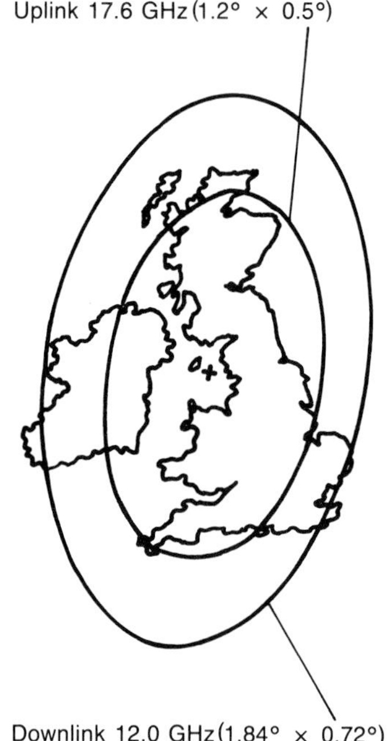

Figure B4.4 Beam area of spacecraft antenna at uplink and downlink frequencies.

8. *Received carrier power at satellite (C).*

A convenient place to take account of the budget is the point where the carrier has reached the satellite antenna (i.e. before it enters the transponder). The received carrier power sum in table B4.1 shows the state of the profit and loss account so far: – 90.4 dBW. (Note: decibel values are always added or subtracted, representing respectively the multiplication or division of the numerical quantities they replace.) From this point in figure B4.1 the power level of the carrier is boosted by the satellite receiver and channel amplifier, which together provide a gain of 60.4 dB. The carrier thus enters the satellite's high-power amplifier (a TWTA) at -30 dBW.

Having obtained a value for the carrier, it is time to consider the noise. The figure at the bottom of the uplink budget (item 14 in table B4.1) will show how far the carrier signal is above the background noise level. This important parameter is called carrier-to-noise ratio, or C/N. It can be likened to the signal-to-noise ratio perceived on an ordinary radio receiver: if the noise is anywhere near the same level as the wanted signal, the quality of the signal suffers an easily perceptible degradation.

9. *Boltzmann's constant*

Just as there is an absolute reference for temperature (absolute zero = -273.15 °C or zero degrees kelvin), there is an absolute reference for noise in a communications system. The very lowest level of background noise is thermal noise, due to the motion of molecules which constitute any physical object. As the absolute temperature on the kelvin scale approaches zero, the molecular motion ceases and brings the thermal noise down to its zero absolute reference. This is represented by Boltzmann's constant (-228.6 dBW $Hz^{-1} K^{-1}$), a constant of proportionality between the molecular kinetic energy and temperature, such that KE = $3/2\ (kT)$. The units of the constant describe, in terms of power (dBW), a level of noise contained in every 1 Hz of bandwidth for every 1 K of temperature. Boltzmann's constant (in SI units, 1.381×10^{-23} J K^{-1}) is the ratio of the universal gas constant (8.315 J K^{-1} mol^{-1}) to Avogadro's number (6.022×10^{23} mol^{-1}) and should not be confused with the Stefan–Boltzmann constant ($\sigma = 5.67 \times 10^{-8}$ W m^{-2} K^{-4}) which is used to define the energy radiated by a black body. As far as the calculation of noise for the uplink budget is concerned, this simply 'sets the counter at zero': the next section evaluates the level of noise.

10. *Satellite noise temperature.*

The first source of noise to consider is that due to the satellite communications equipment. Although specifications are improving continually, there will always be some noise in a given system. An engineering estimate for the noise figure of a DBS receive system is 6.6 dBK (in the same way that power is referenced to the watt in dBW, temperature is referenced to the kelvin in dBK). This includes both the noise introduced by the electronic and RF equipment and that due to the antenna, since it is above absolute zero and contributes an amount of thermal noise. Noise calculations are discussed further in box B4.1.

So far, only the noise contributed by the satellite itself has been considered: the temperature equivalent to the noise of the satellite receive system can be deduced by dividing 6.6 dBK by 10 and taking the antilog, which gives only 4.6 K. A much higher noise temperature is contributed by a noise source external to the satellite—the Earth.

Since the satellite has a part of the Earth within its field of view, it 'sees' the Earth's thermal output as a background source of noise. Its temperature is, from the point of view of a geostationary satellite, an average of 290 K (about 17 °C). This is equivalent to 24.6 dBK, which is the level of noise entering the satellite with the carrier. The effect is

BOX B4.1: NOISE CALCULATIONS

Noise sources

Noise is an undesirable electrical disturbance which, if sufficiently severe, can mask the signal in a communications system. It has a number of origins: the communications equipment itself; other electrical devices such as motors and ignition systems; natural terrestrial or atmospheric noise such as that from electrical storms (significant only below about 30 MHz); and extraterrestrial noise, particularly from the Sun.

A major source of noise from within the communications equipment is that due to the intrinsic random motion of electrons—*thermal noise* (also called Johnson noise or Nyquist noise). The recognised method for reducing this noise is to cool the receiver and other relevant equipment. The greatest benefit is gained by cooling to cryogenic temperatures, close to absolute zero, at which molecular (and therefore electron) motion is reduced and thus so is the noise.

Another source of noise is *shot noise*, which is produced by discrete charges in an electric current. It is therefore impulsive and random, as opposed to the continuum of noise as represented by thermal noise. For example, shot noise may be produced by the electrons in a thermionic diode, a klystron or a travelling wave tube (TWT).

Noise figure and noise factor

A measure of the noise contribution of a device is given by the *noise figure*, which is expressed relative to a theoretical noise-free device at a reference temperature. The noise figure of a device is its *noise factor* expressed in decibels (dBK). Noise factor, F, can be calculated using the expression

$$F = P_N / GkT_aB$$

where P_N is the noise power at the output, G is the gain, and kT_aB is the noise power of the theoretical source resistance at ambient temperature (where k is Boltzmann's constant, T_a is the normal ambient temperature of the source resistor, taken to be 290 K, and B is the effective noise like looking at a candle against the glow of a light bulb: the signal from the candle must be discerned despite the 'noise' emitted by the light bulb. The total satellite noise temperature is the sum of the thermal noise of the satellite receive equipment and that of the Earth: item 10 in table B4.1 shows this to be 31.2 dBK.

11. *System noise power density at satellite (N_0).*
It is now possible to calculate the noise level on the uplink in absolute terms. The noise level is 31.2 dB higher than the absolute reference,

bandwidth of the system in hertz). If no noise is contributed by the device, the noise factor is 1 and the noise figure is 0 dBK.

Noise power

A quantitative measure of noise in a communications system is given by the *noise power* (units dBW), which is given by the expression

$$P_N = kTB$$

where T is the noise temperature of the source, k is Boltzmann's constant and B is the effective noise bandwidth in dBHz.

Noise power is analogous to carrier power; the two quantities are compared in the calculation of the carrier-to-noise (power) ratio (C/N), that is the ratio of the power level of a signal carrier wave to the power level of the noise at the same time and place in a radio transmission system.

Noise power density (NPD: N_0)

A quantitative measure of the level of noise in every 1 Hz of bandwidth is given by the noise power density (units dBW Hz^{-1}), which is given by the expression

$$N_0 = P_N / B = kT$$

where P_N is the noise power (dBW), B is the bandwidth (dBHz), k is as above and T is the absolute temperature (dBK).

The ratio of the power level of a signal carrier wave to the NPD, carrier-to-noise power density ratio (C/N_0), is given by the expressions

$$\frac{C}{N_0} = \frac{P_C}{(P_N/B)} = \frac{P_C}{kT} = \frac{C}{kT}$$

where P_C is the carrier power and P_N, B, k and T are as above.

The difference between C/N and C/N_0 is that C/N_0 considers only the noise in 1 Hz of bandwidth (noise Hz^{-1}), whereas C/N takes account of the total noise in the full bandwidth.

Boltzmann's constant, so the value of -228.6 dBW $Hz^{-1}K^{-1}$ increases to -197.4 dBW Hz^{-1}. This is the level of noise in every 1 Hz of bandwidth—the temperature (K^{-1}) has been accounted for.

12. *Carrier-to-noise power density ratio (C/N_0).*
The level of the carrier in item 8 (at -90.4 dBW) is higher than the noise (at -197.4 dBW Hz^{-1}), so the carrier-to-noise power density ratio, which is simply the difference between the two, comes out positive. Item 12 shows the C/N_0 to be 107.0 dBHz. The units of C/N_0 are dBW divided by dBW Hz^{-1}, that is carrier divided by noise power density. For simplicity, the numbers can be considered as decibel values added together.

To summarise the budget so far, a carrier signal received at the satellite antenna credits the account; the noise received is a debit. The balance is the sum of the carrier and the noise. The C/N_0 of 107 dB means that the carrier is 107 dB higher than the noise.

13. *Bandwidth.*
All communications signals are spread over a band of frequencies, the signal bandwidth, the width of which depends on the complexity of the signal. As discussed in chapter B2, the DBS TV signal requires a bandwidth of 27 MHz.

The units of Boltzmann's constant describe noise for every 1 Hz of bandwidth for every 1 K of temperature. Having ascertained the temperature, what remains is a net level of carrier-to-noise ratio for every 1 Hz of bandwidth (C/N_0). Since the bandwidth of the TV carrier is 27 MHz, the noise associated with the total bandwidth, and the final noise level received at the satellite, will be 27 million times greater. The decibel equivalent of 27 MHz is 74.3 dBHz ($10\log_{10}27 \times 10^6$), so the carrier-to-noise ratio at the satellite is degraded by this amount. Relating the carrier level to the signal bandwidth in this way shows that bandwidth 'costs power': as the bandwidth is increased the noise increases, the carrier-to-noise ratio decreases and the TV signal is degraded.

14. *Carrier-to-noise at satellite (C/N).*
The level of the carrier above the noise is thus reduced to give the margin of C/N at the satellite. The level of 32.7 dB indicates a healthy system, but it is the C/N_0 value which will be important in the downlink leg of the system.

The uphill struggle to the satellite is now complete. The satellite receiver has collected the uplinked signal with a good margin above that needed to transmit a good-quality TV signal back to Earth. The uplink budget shows a profit.

4.3 THE DOWNLINK BUDGET

The downlink budget follows the structure of the uplink quite closely, beginning with the TWTA which amplifies the received low-power carrier to a sufficient level to ensure good reception by the domestic DBS antenna/receiver. The crux of the WARC DBS plan is that it uses spacecraft with high-power amplifiers, with typical RF output powers of about 200 W. The gain of the TWTA in this example is 53 dB, a value which brings the carrier power up to 23 dBW (shown in figure B4.1 and as item 1 in table B4.2). Item 2 in table B4.2 shows estimated output losses of 2 dB, incurred in a similar way to those in the uplink earth station.

Table B4.2 Downlink budget. Frequency f = 12.0 GHz, wavelength λ = 0.025 m (UK channel 16).

Parameter		Value	Notes
1.	TWTA O/P power (200 W TWT)	23.0 dBW	
2.	Output losses	2.0 dB	
3.	Satellite antenna gain (at half-power point)	41.0 dB	†
[4.	Satellite EIRP (3 − 2 + 1)	62.0 dBW]	
5.	Free-space loss	205.2 dB	‡
6.	Atmospheric attenuation (including rain)	2.2 dB	§
7.	Domestic antenna peak gain	38.0 dB	‖
8.	**Carrier (C)**: Received carrier power at ground (C = 1 + 3 + 7 − 2 − 5 − 6)	− 107.4 dBW	
9.	Boltzmann's constant	− 228.6 dBW $Hz^{-1} K^{-1}$	
10.	Receive system noise temperature (500 K)	27.0 dBK	
11.	**Noise (N_0)**: system noise power density at ground (N_0 = 9 + 10)	− 201.6 dBW Hz^{-1}	
12.	Carrier-to-noise power density ratio (C/N_0 = 8 − 11)	94.2 dBHz	
13.	Revised C/N_0 due to contribution of uplink noise	94.0 dBHz	¶
14.	Bandwidth (27 MHz)	74.3 dBHz	
15.	Carrier-to-noise ratio at ground (C/N = 13 − 14)	19.7 dB	††
	Overall system margin:	19.7 − 14 = 5.7 dB	

† Half-power gain: $G = 10\log_{10} \eta\,(27\,800/\Theta\phi)$, η = 60%, Θ = 1.84°, ϕ = 0.72°.

‡ FSL = $20\log_{10} (\lambda/4\pi R)$, R = 36 000 km.

§ Elevation angle 25°: oxygen and water vapour 0.2 dB; rain (6 mm hr^{-1}) 2.0 dB.

‖ Peak gain $G = 10\log_{10} \eta(\pi D/\lambda)^2$, D = 0.9 m, η = 50%.

¶ Power sum of uplink and downlink C/No:

$1/5.01 \times 10^{10}$ + $1/2.63 \times 10^{9}$ = $1/2.49 \times 10^{9}$

↑ ↑ ↓

(107.0 dBHz) (94.2 dBHz) (94.0 dBHz).

†† WARC recommends at least 14 dB for C/N.

The gain of the satellite antenna is calculated, using equation (B4.3) with the downlink beamwidths, as 41 dB (item 3). The half-power gain is used since this is the power level which will be received by earth terminals towards the edge of the service area. Only those near the boresight will benefit from peak power.

Although not a necessary step in the calculation of a downlink budget, a parameter often referred to can be derived at this point, namely the satellite EIRP or equivalent isotropic radiated power (item 4) given by the sum of items 1, 2 and 3. This is a figure of merit for the satellite which shows how powerful a given combination of transponder power and antenna gain make it before its power is dissipated into the downlink beam.

The value of FSL on the downlink (item 5) is slightly less than on the uplink because of the lower frequency (12 GHz), and atmospheric attenuation (item 6) is reduced for the same reason. Taking into account the same factors as for the uplink, the attenuation due to oxygen and water vapour can be easily deduced from figure B4.2(*a*) to be 0.2 dB; that due to rain is 2 dB from figure B4.2(*b*).

If it is assumed that the domestic receive antenna will be pointed reasonably well towards the satellite, a figure for peak gain can be used in the budget (item 7). Although smaller antennas may well be used, the 0.9 m dish suggested by WARC is assumed, and a pessimistic 50% efficiency value has been allocated to it to cover inferior manufacturing. This results in a gain of 38 dB which in turn gives a received carrier power of -107.4 dBW for item 8.

The noise temperature of the receive system (item 10) is estimated at 500 K for a typical domestic unit, which leads to a noise power density on the ground of -201.6 dBW Hz^{-1} (item 11). Receivers with improved noise figures are already available, but it is inadvisable at the initial stage of the calculation to assume the availability of the 'deluxe' model. Apart from this, rainfall causes an increase in earth station antenna noise temperature because the antenna 'sees' warm rain instead of cold sky, and this can further degrade the carrier-to-noise ratio.

Incidentally, given the values calculated for items 7 and 10, the figure of merit, or G/T, of the receiver installation can be deduced. The antenna gain (G = 38 dB) and the receive system noise temperature (T = 27 dBK) give a G/T of 11 dB K^{-1}. This compares favourably with the value of 6 dBK^{-1} recommended by WARC-77 [see table B3.1]. In fact the 5 dB margin shows that a smaller antenna could be substituted for the 90 cm dish assumed here. For example, a 50 cm antenna with a gain of about 33 dB would give a G/T of 6 dB K^{-1} with the same receiver.

Item 12 combines the carrier and the noise to give a C/N_0 of 94.2 dBHz. At this point it is necessary to tie in the value of C/N_0 derived for the uplink. A certain level of noise is of course received by the satellite along with the carrier. The satellite transponder amplifies this uplink noise, as well as the carrier, so the level of the downlinked carrier, with respect to the noise, must be reduced accordingly. The power sum used to combine the uplink and downlink C/N_0 is shown at the foot of table B4.2 (the decibel equivalents are printed below the numbers in the power sum). It can be seen that, at 107 dBHz, the C/N_0 on the uplink is greater than that on the downlink (94.2 dBHz): the noise level on the uplink is consequently lower. This is possible, indeed desirable, on the uplink because large-diameter antennas and high-power HPAs can be used on the ground. The satellite payload, on the other hand, is mass-, power- and size-limited. Any shortfalls in a link budget are therefore best made up on the uplink—indeed there is no point in sending a noisy signal through a satellite transponder only to have that already high level of noise amplified still further.

The same philosophy of the uplink 'bearing the brunt' appears in frequency allocation: since free-space and atmospheric losses are greater at higher frequencies, it is always the uplink section of a satellite's communications service which is assigned the higher frequency of the pair.

The revised C/N_0 for the downlink is 94.0 dBHz (item 13). Item 14 takes into account the transmission of a signal of 27 MHz bandwidth and gives a carrier-to-noise ratio of 19.7 dB (item 15). The WARC plan recommends a carrier-to-noise ratio of 14 dB to produce an acceptable TV picture, so this 'first cut' at the link budget for a hypothetical system has produced an overall system margin of 5.7 dB.

4.4 CONCLUSION

This analysis of the communications link budget for a DBS service has shown how a TV signal can be transmitted via a satellite in geostationary orbit to domestic receivers placed anywhere within the satellite's coverage area. The overall system margin gives an indication that the system will deliver a relatively noise-free TV picture to a subscriber's home. A system with a good overall margin, or good carrier-to-noise ratios on both the uplink and the downlink, is thus able to combat the problems of interference discussed in previous chapters. It has been assumed that there will be no errors in the pointing of the domestic antenna, which will not be so in all cases. This is why a margin is always useful.

The link budget is fundamental to the design of any satellite communications system since it provides the numerical backing for the initial system concept. It is worth noting, however, that a link budget like this is usually just the starting point for discussions on how the actual system will be put together. It shows the overall feasibility of the system and is an irreplaceable design tool, but it can also be manipulated at will. If the margin is too large, it can be reduced by decreasing the power of the HPA or the antenna size (within the constraints of beamwidth). Alternatively, if a wider bandwidth is required these factors can be increased, within reason, to accommodate this.

REFERENCES

Pratt T and Bostian C W 1986 *Satellite Communications* (New York: Wiley) ch. 4
Williamson M 1984 *Satell. Cable TV News* May 53–5, June 46–8, July 56–7

Appendix A Satellites in Geostationary Orbit (as of February 1989)

The following table has been compiled with reference to NASA's *Satellite Situation Report* (vol. 28, no. 4, 31 December 1988), published by Goddard Space Flight Center, and extended to include launches to GEO up to and including the Ariane V28 mission on 26 January 1989.

The table includes all satellites with measured orbital parameters or 'elements' close to nominal for GEO (an apogee and a perigee of about 36 000 km and an inclination within a few degrees of zero). It specifically excludes those entries in the *Satellite Situation Report* listed as 'orbital elements not maintained' or 'elements not available'. Some of these satellites have been moved to a higher 'graveyard orbit', but some may still be in GEO, which means that the total number of objects in GEO may be higher than the table suggests. However, those omitted are either military satellites about which little is known, or defunct satellites which can be regarded as 'space junk'. Amongst those whose orbital elements are 'not maintained' are the early Syncom satellites, Early Bird (Intelsat I) and some of the Intelsat II and III series spacecraft.

Given the above, the total number of objects in GEO (as of 26 January 1989) is 309. However, 79 of these are not satellites but upper stages used to boost satellites into GEO. Loosely termed 'apogee motors', they include a number of Titan Transtage motors, a few inertial upper stage (IUS) second-stage motors and a large number of apogee motors jettisoned from Soviet Cosmos, Ekran, Gorizont and Raduga satellites. The table therefore lists 230 satellites, the majority of which are communications satellites of some description. A handful are meteorological or 'Earth resources' satellites.

No attempt has been made to indicate whether or not the satellites are operational. This depends on a variety of factors including the age and condition of the satellite and whether it has been replaced by a new generation. It is also difficult to obtain information on Soviet systems and military systems. That said, a rough calculation can be made in light of the fact that the majority of western-built satellites launched in the 1980s were designed to operate for between 7 and 10 years.

Geostationary satellite launches, at the time of writing, planned for 1989 are included in tabular form as an addendum.

Launch date	Satellite	Affiliation
1967		
5 November	ATS-3	USA
1968		
26 September	LES 6	USA
1970		
20 March	NATO I	NATO
1971		
3 November	DSCS 1	USA
3 November	DSCS 2	USA
20 December	Intelsat IV-F3	Intelsat
1972		
23 January	Intelsat IV-F4	Intelsat
13 June	Intelsat IV-F5	Intelsat
10 November	Anik A1	Telesat Canada
1973		
20 April	Anik A2	Telesat Canada
23 August	Intelsat IV-F7	Intelsat
13 December	DSCS 3	USA
13 December	DSCS 4	USA
1974		
26 March	Cosmos 637	USSR
13 April	Westar 1	USA
30 May	ATS-6	USA
10 October	Westar 2	USA
21 November	Intelsat IV-F8	Intelsat
23 November	Skynet 2B	UK
19 December	Symphonie-A	France/FRG
1975		
7 May	Anik A3	Telesat Canada
22 May	Intelsat IV-F1	Intelsat

27 August	Symphonie-B	France/FRG
26 September	Intelsat IVA-F1	Intelsat
8 October	Cosmos 775	USSR
16 October	GOES 1	USA
13 December	RCA Satcom I	USA
1976		
17 January	CTS	Canada
29 January	Intelsat IVA-F2	Intelsat
19 February	Marisat 1	USA
15 March	LES 8	USA
15 March	LES 9	USA
26 March	RCA Satcom II	USA
27 April	NATO III-A	NATO
13 May	Comstar 1	USA
10 June	Marisat 2	USA
8 July	Palapa A1	Indonesia
22 July	Comstar 2	USA
14 October	Marisat 3	USA
26 October	Ekran 1	USSR
1977		
28 January	NATO III-B	NATO
23 February	ETS-II (Kiku 2)	Japan
10 March	Palapa A2	Indonesia
12 May	DSCS 7	USA
12 May	DSCS 8	USA
26 May	Intelsat IVA-F4	Intelsat
16 June	GOES 2	USA
14 July	GMS-1 (Himawari-1)	Japan
23 July	Raduga 3	USSR
25 August	Sirio 1	Italy
23 November	Meteosat 1	ESA
15 December	CS-1 (Sakura-1)	Japan
1978		
7 January	Intelsat IVA-F3	Intelsat
9 February	Fltsatcom 1	USA
31 March	Intelsat IVA-F6	Intelsat
7 April	BSE-1 (Yuri-1)	Japan
11 May	OTS 2	ESA
16 June	GOES 3	USA
29 June	Comstar 3	USA
14 July	GEOS 2	ESA
19 November	NATO III-C	NATO

14 December	DSCS 11	USA
14 December	DSCS 12	USA
16 December	Anik B1	Telesat Canada
1979		
21 February	Ekran 3	USSR
4 May	Fltsatcom 2	USA
5 July	Gorizont 2	USSR
10 August	Westar 3	USA
3 October	Ekran 4	USSR
28 December	Gorizont 3	USSR
1980		
18 January	Fltsatcom 3	USA
20 February	Raduga 6	USSR
14 June	Gorizont 4	USSR
9 September	GOES 4	USA
5 October	Raduga 7	USSR
24 October	Fltsatcom 4	USA
15 November	SBS 1	USA
21 November	DSCS 13	USA
21 November	DSCS 14	USA
6 December	Intelsat V-F2	Intelsat
26 December	Ekran 6	USSR
1981		
21 February	Comstar 4	USA
18 March	Raduga 8	USSR
22 May	GOES 5	USA
23 May	Intelsat V-F1	Intelsat
19 June	Apple	India
19 June	Meteosat 2	ESA
25 June	Ekran 7	USSR
30 July	Raduga 9	USSR
6 August	Fltsatcom 5	USA
24 September	SBS 2	USA
9 October	Raduga 10	USSR
20 November	RCA Satcom IIIR	USA
15 December	Intelsat V-F3	Intelsat
20 December	Marecs A	ESA/Inmarsat
1982		
16 January	RCA Satcom IV	USA
5 February	Ekran 8	USSR
26 February	Westar 4	USA
5 March	Intelsat V-F4	Intelsat

15 March	Gorizont 5	USSR
10 April	Insat 1A	India
17 May	Cosmos 1366	USSR
9 June	Westar 5	USA
26 August	Anik D1	Telesat Canada
16 September	Ekran 9	USSR
28 September	Intelsat V-F5	Intelsat
20 October	Gorizont 6	USSR
28 October	RCA Satcom V	USA
30 October	DSCS 15	USA
30 October	DSCS 16	USA
11 November	SBS 3	USA
12 November	Anik C3	Telesat Canada
26 November	Raduga 11	USSR
1983		
4 February	CS-2a (Sakura-2a)	Japan
12 March	Ekran 10	USSR
4 April	TDRS-A	USA
8 April	Raduga 12	USSR
11 April	RCA Satcom VI	USA
28 April	GOES 6	USA
19 May	Intelsat V-F6	Intelsat
16 June	ECS 1	ESA/Eutelsat
18 June	Anik C2	Telesat Canada
18 June	Palapa B1	Indonesia
28 June	Galaxy 1	USA
30 June	Gorizont 7	USSR
28 July	Telstar 3A	USA
5 August	CS-2b (Sakura-2b)	Japan
25 August	Raduga 13	USSR
31 August	Insat 1B	India
8 September	RCA Satcom VII	USA
22 September	Galaxy 2	USA
30 September	Ekran 11	USSR
19 October	Intelsat V-F7	Intelsat
30 November	Gorizont 8	USSR
1984		
23 January	BS-2a (Yuri-2a)	Japan
15 February	Raduga 14	USSR
2 March	Cosmos 1540	USSR
5 March	Intelsat V-F8	Intelsat
16 March	Ekran 12	USSR

29 March	Cosmos 1546	USSR
8 April	PRC 15	China
22 April	Gorizont 9	USSR
23 May	GTE Spacenet 1	USA
22 June	Raduga 15	USSR
1 August	Gorizont 10	USSR
2 August	GMS-3 (Himawari-3)	Japan
4 August	ECS 2	ESA/Eutelsat
4 August	Telecom 1A	France
31 August	SBS 4	USA
31 August	Syncom IV-2 (Leasat 2)	USA
1 September	Telstar 3C	USA
21 September	Galaxy 3	USA
9 November	Anik D2	Telesat Canada
10 November	GTE Spacenet 2	USA
10 November	Marecs B2	ESA/Inmarsat
10 November	Syncom IV-1 (Leasat 1)	USA
14 November	Nato III-D	NATO
1985		
18 January	Gorizont 11	USSR
8 February	Arabsat 1	Arabsat
8 February	SBTS 1	Brazil
21 February	Cosmos 1629	USSR
22 March	Intelsat V-F10	Intelsat
12 April	Syncom IV-3 (Leasat 3)	USA
13 April	Anik C1	Telesat Canada
8 May	GStar 1	USA
8 May	Telecom 1B	France
17 June	Morelos A	Mexico
18 June	Arabsat 1B	Arabsat
19 June	Telstar 3D	USA
30 June	Intelsat VA-F11	Intelsat
8 August	Raduga 16	USSR
27 August	ASC 1	USA
27 August	Aussat 1	Australia
29 August	Syncom IV-4 (Leasat 4)	USA
29 September	Intelsat VA-F12	Intelsat
25 October	Cosmos 1700	USSR
15 November	Raduga 17	USSR
27 November	Aussat 2	Australia
27 November	Morelos B	Mexico
28 November	Satcom Ku-2	USA

1986		
12 January	Satcom Ku-1	USA
17 January	Raduga 18	USSR
1 February	PRC 18	China
12 February	BS-2b (Yuri-2b)	Japan
28 March	GStar 2	USA
28 March	SBTS 2	Brazil
4 April	Cosmos 1738	USSR
24 May	Ekran 15	USSR
10 June	Gorizont 12	USSR
25 October	Raduga 19	USSR
18 November	Gorizont 13	USSR
5 December	Fltsatcom 7	USA
1987		
26 February	GOES 7	USA
19 March	Raduga 20	USSR
20 March	Palapa B2P	Indonesia
11 May	Gorizont 14	USSR
27 August	ETS-V (Kiku-5)	Japan
4 September	Ekran 16	USSR
16 September	Aussat K3	Australia
16 September	ECS 4	ESA/Eutelsat
1 October	Cosmos 1888	USSR
28 October	Cosmos 1894	USSR
21 November	TV Sat 1	FRG
26 November	Cosmos 1897	USSR
10 December	Raduga 21	USSR
27 December	Ekran 17	USSR
1988		
19 February	CS-3a (Sakura-3a)	Japan
7 March	PRC 22	China
11 March	GTE Spacenet IIIR	USA
11 March	Telecom 1C	France
31 March	Gorizont 15	USSR
26 April	Cosmos 1940	USSR
6 May	Ekran 18	USSR
17 May	Intelsat VA-F13	Intelsat
15 June	Meteosat	ESA
15 June	Panamsat 1	USA
21 July	ECS 5	ESA/Eutelsat
21 July	Insat 1C	India
1 August	Cosmos 1961	USSR
18 August	Gorizont 16	USSR

8 September	SBS 5	USA
16 September	CS-3b (Sakura-3b)	Japan
29 September	TDRS-C	USA
20 October	Raduga 22	USSR
27 October	TDF-1	France
10 December	Astra 1A	Luxemburg
10 December	Ekran 19	USSR
10 December	Skynet 4B	UK
22 December	PRC 25	China
1989		
26 January	Intelsat V-F15	Intelsat

Acronyms used in this table (in order of first appearance):

ATS	Applications Technology Satellite
LES	Lincoln Experimental Satellite
NATO	North Atlantic Treaty Organisation
DSCS	Defense Satellite Communication System
GOES	Geostationary Operational Environmental Satellite
CTS	Communication Technology Satellite
ETS	Engineering Test Satellite
GMS	Geostationary Meteorological Satellite
CS	Communications Satellite
BSE	Broadcasting Satellite Experimental
OTS	Orbital Test Satellite
GEOS	Geostationary Earth Observation Satellite
SBS	Satellite Business Systems
TDRS	Tracking and Data Relay Satellite
ECS	European Communications Satellite †
BS	Broadcasting Satellite
PRC	People's Republic of China
SBTS	Sistema Brasileiro de Telecomunicacoes por Satelite
ASC	American Satellite Company
TDF	Telediffusion de France

†European Communications Satellites (ECS) are redesignated Eutelsat 1-F1, F2, etc, once they reach orbit.

Note: Anik spacecraft are alternatively designated 'Telesat' (after their operator Telesat Canada).

Geostationary satellite launches planned for 1989 (as of February 1989).

Ariane	V29 JC-Sat1 & MOP 1 V30 Tele-X V31 SCS-A & DFS-1 V32 Olympus V33 TV-SAT 2 V35 Intelsat VI-F1 V36 SCS-B & Inmarsat II-F1 V37 TDF-2 & DFS-2
Shuttle	STS-29 TDRS-D STS-32 Syncom IV-5 (Leasat 5)
Atlas-Centaur	Fltsatcom 8
Delta	Insat 1D BSB-1
Titan 3	JC-Sat 2
Titan 4	DSP-Block 14
H-I	GMS-4

JC-Sat 1 will be operated by Japan Communications Satellite Co. Inc.
MOP—Meteosat Operational Programme (MOP-1 will be re-designated Meteosat 4 in orbit).
SCS—Space Communications Corporation 'Superbird' satellites (A & B).
DFS—Deutscher Fernmelde Satellit (otherwise known as Kopernikus).
BSB—British Satellite Broadcasting.
DSP—Defense Support Programme.

Appendix B Types of Earth Orbit

Earth orbits can be divided into four main types [see figure B1].

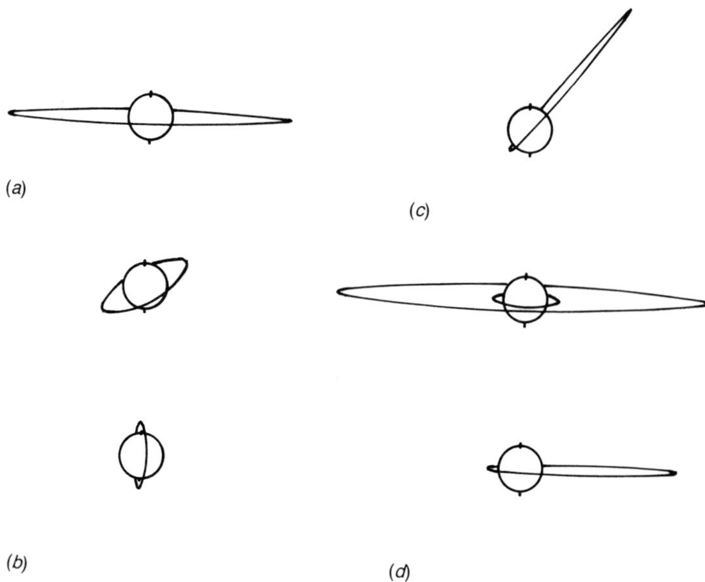

Figure B1 Four main categories of Earth orbit: (a) geostationary; (b) inclined circular geosynchronous; (c) inclined eccentric geosynchronous; (d) non-synchronous.

(a) GEOSTATIONARY ORBIT (GEO)

This unique orbit (which is circular, equatorial and has the same sense and period of rotation as the Earth) is used extensively for communications and constitutes a natural and limited resource. It is ideal for communications with countries where the elevation angle of the satellite is high (i.e. those at low latitudes) but becomes increasingly less attractive for countries at high latitudes (i.e. above about 70°). Apart from this its main disadvantage is that it is fast becoming overcrowded in the available frequency bands. GEO has a radius of 42 164 km (26 141 miles) and a height (above the Earth's surface) of 35 786 km (22 187 miles).

(*b*) INCLINED CIRCULAR GEOSYNCHRONOUS ORBITS

A geosynchronous orbit is one where, after a certain period of time, the sub-satellite point follows an identical ground track to some previous orbit. The repetition occurs on a regular basis: the ground track of a 24 h geosynchronous orbit is a figure of eight, the satellite spending half its orbital period north of the equator and half south; 12 h orbits follow two distinct ground tracks, each of which is repeated on alternate orbits; 8 h orbits have three ground tracks, etc. In the same terminology GEO could be defined as the only non-inclined circular geosynchronous orbit.

These orbits can be useful for coverage of the higher latitudes, but there is an inherent loss in continuity of coverage. An example of a circular inclined orbit with a 12 h period is that used by the Navstar military system, for which a large number of satellites are placed in several different orbits to produce full coverage.

Orbits inclined by 90° to the equator—polar orbits—can also be geosynchronous, with orbital periods which offer total Earth coverage every few days. Remote sensing satellites are usually placed in polar orbits. A type of polar orbit in which the sub-satellite point remains approximately fixed at the same local time on Earth is termed Sun synchronous or heliosynchronous.

(*c*) INCLINED ECCENTRIC GEOSYNCHRONOUS ORBITS

These orbits represent another large family which includes orbits of varying inclination, eccentricity and period: for example periods of 24, 12, 8 and 6 h are possible. These orbits can be inclined to serve high latitudes with an extended period of coverage if the apogee is placed over the latitude of interest. This is because the angular velocity of the satellite about the Earth's axis of rotation is very nearly the same, around apogee, as the angular velocity of the Earth itself. A general problem with orbits which are out of the equatorial plane and have low perigees is that the oblateness of the Earth causes a perigee precession (a rotation of the orbit plane about the Earth) which moves the coverage area progressively westwards. There is, however, a special case which exhibits no significant perigee rotation: the inclination angle of this orbit is about 63.4°.

This type of orbit has been used for a number of years by Soviet spacecraft and is known as a Molniya orbit. Owing to its relative stability and usefulness, it can be considered the elliptical analogue to the circular GEO. As far as the western world is concerned, the Molniya orbit is an unexploited natural resource. The utilisation of orbits of this type could be particularly advantageous for communications with maritime and aeronautical mobiles navigating the polar routes, as well as land-based mobiles and fixed stations in northern Europe, for example.

(*d*) NON-SYNCHRONOUS ORBITS

These orbits, whether circular or elliptical, are not synchronised with the Earth's period of rotation. This category includes the majority of low Earth orbits (LEOs), which are not particularly useful for communications satellites. It also includes the family of circular, equatorial orbits which have periods greater than or less than 24 hours and a group of non-synchronous elliptical orbits which lie in the equatorial plane.

Appendix C Simple Calculations

1 CALCULATION OF THE EARTH'S ROTATIONAL VELOCITY AT THE SURFACE (RELATIVE TO THE 'FIXED STARS')

Earth's equatorial radius, $r = 6377.6$ km (6.3776×10^6 m).
Earth's equatorial circumference, $C = 40\,071.6$ km ($4.007\,16 \times 10^7$ m).
Earth's period of rotation, $t = 23$ h 56 min 4 s (86 164 s) (sidereal day).

Therefore, velocity at equator $v = C/t = 465$ m s^{-1}.

Velocity at higher latitudes is given by the expression $v(\phi) = v \cos \phi$, where ϕ is the latitude in degrees.

The following table includes the world's major launch sites.

Launch Site	Latitude (°N)	v (m s^{-1})
Equatorial	0	465
Guiana Space Centre, French Guiana	5.2	463
Xichang, China	28	410
Kennedy Space Center, USA	28.5	409
Tanegashima, Japan	30.5	401
Vandenberg Air Force Base, USA	35	381
Jiuquan, China	41	351
Baikonur Cosmodrome (Tyuratam), USSR	46	323
Northern Cosmodrome (Plesetsk), USSR	63	211

2 CALCULATION OF THE VELOCITY OF A SPACECRAFT IN GEOSTATIONARY ORBIT (RELATIVE TO THE 'FIXED STARS')

Radius of GEO, $r = 42\,164$ km (4.2164×10^7 m).
Circumference of GEO, $C = 264\,924$ km ($2.649\,24 \times 10^8$ m).

Satellite's orbital period, t = 23 h 56 min 4 s (86 164 s) (sidereal day).

Therefore, velocity of spacecraft $v = C/t = 3074.6\ \text{m s}^{-1}$.

3 DERIVATION OF THE RADIUS OF GEOSTATIONARY ORBIT

A satellite in GEO orbits the Earth in the plane of the equator with the same sense of rotation as the Earth (i.e. anticlockwise as viewed from above the north pole).

The force of gravitational attraction is given by

$$F_G = G(M_E)\,(M_S)/r^2$$

where G is the gravitational constant (6.67×10^{-11} N m^2 kg^{-2}), M_E is the mass of the Earth (5.98×10^{24} kg), M_S is the mass of the satellite, and r is the radius of the orbit.

The centripetal force on a body moving in a circular orbit is given by

$$F_C = M_S\omega^2 r$$

where ω is the angular frequency, a measure of the rotational velocity, and M_S and r are as above.

In a circular orbit the two forces are equivalent, such that

$$GM_EM_S/r^2 = M_S\omega^2 r$$

which can be rearranged in the form

$$r^3 = GM_E/\omega^2.$$

It is required that ω for the satellite orbit should equal the angular frequency, ωE, of the Earth about its axis, so that the satellite will appear stationary. Since $\omega = 2\pi/T$ (where T is the time period in seconds), the value of ωE is given by $2\pi/(1\ \text{day})$:

$$\omega E = 2\pi/8.64 \times 10^4 = 7.3 \times 10^{-5}\ \text{s}^{-1} = \omega.$$

Therefore, $r^3 = (6.67 \times 10^{-11})\,(5.98 \times 10^{24})/(7.3 \times 10^{-5})^2 = 7.48 \times 10^{22}$. Thus, $r = 4.2 \times 10^7$ m.

The radius of GEO is therefore about 42 000 km.

Appendix D A Guide to Using the Decibel (dB)

The decibel is a logarithmic ratio between two signal power levels used, amongst other things, to denote the gain of an antenna, amplifier, etc (a gain of 3 dB is equivalent to a doubling in signal strength). A numerical value (in dB) can be found using the following expression:

$$\text{No. of dB} = 10\log_{10}P_1/P_2$$

where P_1 and P_2 are the output and input powers respectively. A positive decibel value represents a gain; a negative value is a loss (e.g. 0.5 is equivalent to -3 dB, colloquially referred to as '3 dB down'). If two voltages are being compared instead of two powers, V_1 and V_2 replace P_1 and P_2 and '20' replaces '10' in the above expression.

dBi is a measure of gain relative to an isotropic source; dBc denotes a measurement 'relative to the carrier'; dBW is the ratio of power output to a reference signal at 1 W ($P_2 = 1$), expressed in decibels. In the same way that power is referenced to the watt in dBW and the milliwatt in dBm, frequency is referenced to the hertz in dBHz and temperature to the kelvin in dBK.

Conversion from the numerical quantity in SI units to its equivalent in dB . . ., is performed simply by multiplying by $10\log_{10}$ (e.g. as in the bandwidth of a transponder, 36 MHz $\equiv$ 75.56 dBHz, or the noise temperature of a receiver, 4 K $\equiv$ 6.02 dBK).

Appendix E Frequency Bands

The ITU has divided the radio spectrum into the following bands.

Symbol †	Frequency range ‡ (lower limit exclusive, upper limit inclusive)	Corresponding metric subdivision (Note: not widely used)
VLF	3–30 kHz	Myriametric waves
LF	30–300 kHz	Kilometric waves
MF	300–3000 kHz	Hectometric waves
HF	3–30 MHz	Decametric waves
VHF	30–300 MHz	Metric waves
UHF	300–3000 MHz	Decimetric waves
SHF	3–30 GHz	Centimetric waves
EHF	30–300 GHz	Millimetric waves
	300–3000 GHz	Decimillimetric waves

†V = very; U = ultra; S = super; E = extra; L = low; M = medium; H = high; F = frequency.

‡Prefixes are those assigned by the ITU: k = kilo (10^3), >3 kHz, ⩽ 3000 kHz; M = mega (10^6), >3MHz, ⩽ 3000 MHz; G = giga (10^9), >3 GHz, ⩽ 3000 GHz; T = tera (10^{12}), >3000 GHz, that is centimillimetric, micrometric and decimicrometric waves.

Satellite communications frequencies are mainly in the SHF band, but UHF and VHF are also used. For many communications applications the sub-bands are designated by letters. The following example is the IEEE Radar Standard 521:

L-band:	1–2 GHz
S-band:	2–4 GHz
C-band:	4–8 GHz
X-band:	8–12 GHz
Ku-band:	12–18 GHz
K-band:	18–27 GHz
Ka-band:	27–40 GHz

Standard 521 also defines 40–100 GHz as the millimetre waveband (mm-wave), and 0.3 – 1 GHz as UHF (at variance with the ITU definition).

These designations are intended as a guide to common usage in the satellite communications industry and should not be taken as definitive for all applications. Over the years, several frequency band standards have evolved for use in terrestrial radio, radar and spacecraft communications, often differing according to which side of the Atlantic they were devised. Thus readers may also come across, amongst others, P-band (0.23 – 1 GHz), R-band (1.7 – 2.6 GHz), H-band (3.95 – 5.85 GHz), and designations which differ in that Ku-band is shown as J-band and Ka-band as Q-band. It is also common for the 1 – 300 GHz band to be defined as the microwave band.

Although frequency allocation is a complex subject, it can be said that each particular band is dominated by certain services: communications satellites operate at C-band, Ku-band and, increasingly, K-band and Ka-band; X-band is largely reserved for military communications; L-band is used for the mobile – satellite service; and S-band is used, amongst other things, for satellite telemetry and telecommand.

Frequency combinations such as 4/6 GHz, 12/14 GHz, 12/18 GHz, 20/30 GHz are commonly used. These refer to approximate uplink and downlink frequency bands for various satellite services. The uplink is always the higher figure, since the greater losses (due to atmospheric attenuation) at higher frequencies can be more readily overcome by higher power earth station transmitters, as opposed to increasing the spacecraft HPA power [see chapter B4].

Appendix F Physical Constants and Metric/Imperial Conversions

CONSTANTS

Speed of light in a vacuum, $c = 2.998 \times 10^{8}$ m s^{-1}
Gravitational constant, $G = 6.67 \times 10^{-11}$ N m^{2} kg^{-2}
Acceleration due to gravity (standard), $g = 9.807$ m s^{-2}
Absolute zero: -273.15 °C or 0 K
Boltzmann's constant, $k = 1.381 \times 10^{-23}$ J K^{-1}
(-228.6 dBW Hz^{-1} K^{-1})
Stefan–Boltzmann constant, $\sigma = 5.67 \times 10^{-8}$ W m^{-2} K^{-4}
Solar constant (at 1 AU) $= 1.37 \times 10^{3}$ W m^{-2}

OTHER USEFUL VALUES

Earth

Equatorial radius: 6377.6 km (6.3776×10^{6} m)
Equatorial circumference: 40 071.6 km ($4.007\,16 \times 10^{7}$ m)
Sidereal period of rotation: 23 h 56 min 4 s (86 164 s)
Synodic period: 24 h
Mass: 5.98×10^{24} kg
Astronomical unit (average radius of Earth's orbit): 1 AU $= 1.496 \times 10^{11}$ m

Geostationary orbit

Height: 35 786 km (3.5786×10^{7} m)
Radius: 42 164 km (4.2164×10^{7} m)
Circumference: 264 924 km ($2.649\,24 \times 10^{8}$ m)
Satellite's orbital period: 23 h 56 min 4 s (86 164 s)

CONVERSION TABLE

Although the majority of the world's space industry uses the metric system of units, imperial units can still be found, especially in American publications. The following tables should assist the reader in converting between the two.

Length

1 in = 25.4 mm	1 mm = 0.039 in
1 ft = 0.305 m	1 m = 3.281 ft
1 yd = 0.914 m	1 m = 1.094 yd
1 mile = 1.6093 km	1 km = 0.6214 miles
1 nautical mile = 1.852 km	1 km = 0.54 nautical miles

Area

1 square ft = 0.093 m^2	1 m^2 = 10.764 ft^2
1 square yd = 0.836 m^2	1m^2 = 1.196 yd^2

Volume/capacity

1 cubic ft = 0.028 m^3	1 m^3 = 35.315 ft^3
1 cubic yd = 0.765 m^3	1 m^3 = 1.308 yd^3
1 pint = 0.568 l	1 l = 1.76 pints
1 gallon = 4.546 l	1 l = 0.22 gallons
1 US gallon = 3.785 l	1 l = 0.264 US gallons

Mass

1 ounce = 28.349 g	1 g = 0.035 ounces
1 lb = 0.454 kg	1 kg = 2.205 lb
1 ton = 1.016 tonnes	1 t = 0.984 tons

Force

1 lb force (thrust) = 4.448 N	1 N = 0.223 lbf (pounds thrust)

Pressure

1 psi = 6.9 kPa (6.9 kN m^{-2})	1 kPa = 0.145 psi
1 mm Hg = 133.3 Pa	1 Pa = 0.0075 mm Hg
1 atmosphere (atm) = 1.013 bar	1 bar = 0.987 atm

(1 bar = 100 kPa; 1 atm = 760 mm Hg)

Velocity

1 ft/s = 0.305 m s^{-1}

1 mph = 1.6093 km h^{-1}

1 m s^{-1} = 3.281 ft/s

1 km h^{-1} = 0.6214 mph

Temperature conversion formulae

°C = 5(°F − 32)/9

°F = [9(°C)/5] + 32

Acronyms

In common with all technical fields, a great many acronyms are used in the satellite industry. This list may serve as a useful reminder for the reader of this book and may also prove useful for those reading technical specifications and other documents. It includes terms connected with satellite hardware and communications systems as well as other terms in general use in the satellite industry.

ABM	Apogee boost motor (same as AKM)
ACE	Attitude control electronics
ACS	Attitude control system
AEF	Apogee engine firing
AIT	Assembly, integration and test
AKM	Apogee kick motor (same as ABM)
AM	Amplitude modulation
AMF	Apogee motor firing
AND	Active nutation damping
AOCS	Attitude and orbital control system
AOS	Acquisition of signal
APM	Antenna-pointing mechanism
ARM	Automatic reconfiguration mode
ASE	Airborne support equipment (on Space Shuttle)
ASR	Array shunt regulator
ATE	Automatic test equipment
BAPTA	Bearing and power transfer assembly
BCR	Battery charge regulator
BCU	Battery control unit
BDR	Battery discharge regulator
BDR	Baseline design review
BER	Bit error rate
BMU	Battery management unit

BNSC	British National Space Centre
BOL	Beginning of life
BPSK	Binary phase shift keying
BSS	Broadcasting-satellite service
CADAM	Computer-aided design and manufacture
CATV	Community antenna television (cable TV)
CCHP	Constant conductance heat pipe
CCIR	French acronym: International Radio Consultative Committee
CCITT	French acronym: International Telegraph and Telephone Consultative Committee
CDMA	Code division multiple access
CDR	Critical design review
CEPT	French acronym: European Conference of Postal and Telecommunications Administrations
CEU	Control electronics unit
CFRP	Carbon-fibre-reinforced plastic
C(of)G	Centre of gravity
C/I	Carrier-to-interference ratio
C(of)M	Centre of mass
CM	Communications module
CMC	Ceramic matrix composite
CMD	Command
CMOS	Complementary metal-oxide-silicon
C/N	Carrier-to-noise ratio
CNES	Centre Nationale d'Etudes Spatiales (French National Space Centre)
CPS	Combined propulsion system
CPU	Central processing unit
CSG	Centre Spatiale Guyanais (Guiana Space Centre)
DAMA	Demand assignment multiple access
DANDE	De-spin active nutation damping electronics
DBR	Development baseline review
DBS	Direct broadcasting by satellite
DCIU	Data and control interface unit
DM	Development model
DOD	Department of Defense (USA)
DOD	Depth of discharge
EBU	European broadcasting union
EGSE	Electrical ground support equipment
EHF	Extra high frequency

EHT	Electrically heated thruster
EHT	Electrothermal hydrazine thruster
EIRP	Equivalent (or effective) isotropic radiated power
ELA	Ensemble de lancement Ariane (Ariane launch site)
ELV	Expendable launch vehicle
EM	Engineering model
EM	Electromagnetic
EMC	Electromagnetic compatibility
EMI	Electromagnetic interference
ENT	Equivalent noise temperature
EOC	Edge of coverage
EOL	End of life
EPC	Electronic power conditioner
EPROM	Erasable programmable read-only memory
E/S	Earth station
ESA	European Space Agency
ESA	Emergency Sun acquisition
ESD	Electrostatic discharge
ESOC	European Space Operations Centre
ESR	Emergency Sun reacquisition
ESTEC	European Space Research and Technology Centre
ET	External tank (Space Shuttle)
FCC	Federal Communications Commission (USA)
FCHP	Fixed conductance heat pipe
FDM	Frequency division multiplexing
FDMA	Frequency division multiple access
FEC	Forward error correction
FET	Field-effect transistor
FM	Flight model
FM	Frequency modulation
FMA	Failure modes analysis
FMECA	Failure modes, effects and criticality analysis
FOV	Field of view
FSK	Frequency shift keying
FSL	Free-space loss
FSS	Fixed-satellite service
GEO	Geostationary orbit
GMT	Greenwich mean time
GSE	Ground support equipment
GSO	Geostationary satellite orbit (same as GEO)
GTO	Geostationary transfer orbit

HF	High frequency
HLV	Heavy lift vehicle
HLLV	Heavy lift launch vehicle
HPA	High-power amplifier
HPBW	Half-power beamwidth
IC	Integrated circuit
IF	Intermediate frequency
IFRB	International Frequency Registration Board
IOL	Inter-orbit link
I/P	Input
IRES	Infrared earth sensor
ISL	Inter-satellite link
ISS	Inter-satellite service
ITT	Invitation to tender
ITU	International Telecommunications Union
IUS	Inertial upper stage
KRP	Kevlar-reinforced plastic
KSC	Kennedy Space Center
LAE	Liquid apogee engine
LEO	Low Earth orbit
LF	Low frequency
LHCP	Left-hand circular polarisation
LNA	Low-noise amplifier
LNB	Low-noise block downconverter (similar to LNC)
LNC	Low-noise converter
LO	Local oscillator
LOS	Loss of signal
MF	Medium frequency
MGSE	Mechanical ground support equipment
MIC	Microwave integrated circuit
MLI	Multi-layer insulation
MMC	Metal-matrix composite
MMH	Monomethyl hydrazine
MON	Mixed oxides of nitrogen
MOS	Metal – oxide – semiconductor
MSS	Mobile-satellite service
MTBF	Mean time between failures
MTTF	Mean time to failure

NASA	National Aeronautics and Space Administration (USA)
NASDA	National Space Development Agency (Japan)
NC	Numerically controlled (machinery)
NORAD	North American Air Defense
NTO	Nitrogen tetroxide
OBC	Onboard computer
OBDH	Onboard data handling
OMT	Orthogonal mode transducer
O/P	Output
OSR	Optical solar reflector
PA	Product assurance
PAHT	Power-augmented hydrazine thruster
PAM	Payload assist module
PAM-A	Payload assist module—Atlas
PAM-D	Payload assist module—Delta
PCB	Printed circuit board
PCM	Pulse code modulation
PCU	Power control unit
PFD	Power flux density
PGSE	Payload ground support equipment
PKM	Perigee kick motor
P/L	Payload
PM	Phase modulation
PMD	Propellant management device
PROM	Programmable read-only memory
PSK	Phase shift keying
PTT	Posts, Telegraph and Telecommunications (Organisation)
QA	Quality assurance
QM	Qualification model
QPSK	Quadrature (or quaternary) phase shift keying
RAE	Royal Aerospace Establishment (Farnborough)
RAM	Random access memory
RARC	Regional Administrative Radio Conference
RCS	Reaction control system
RF	Radio frequency
RFP	Request for proposals
RFQ	Request for quotations
RFT	Request for tenders
RHCP	Right-hand circular polarisation

RMS	Remote manipulator system (Space Shuttle)
RMS	Root mean square
ROM	Read-only memory
RSS	Root sum square
Rx	Receive/receiver
S/A	Safe and arm device (propulsion system)
SAD	Solar array drive
SADA	Solar array drive assembly
SADE	Solar array drive electronics
SADM	Solar array drive mechanism
S/C	Spacecraft
SCPC	Single channel per carrier
SEE	Single-event effect
SEU	Single-event upset
SHF	Super high frequency
SM	Service module
SM	Structure model
SMATV	Satellite master antenna television
S/N	Signal-to-noise ratio
SOT	Select on test
SOW	Statement of work
SPELDA	Structure porteuse externe pour lancements doubles Ariane (external support structure for Ariane dual launches)
SPF	Single point failure
SRB	Solid rocket booster
SRM	Solid rocket motor
SSM	Second-surface mirror
SSME	Space Shuttle main engine
SSPA	Solid state power amplifier
SSTDMA	Satellite-switched time division multiple access
SSUS	Spinning solid upper stage (same as PAM)
STDN	Space tracking and data network
STS	Space transportation system (Space Shuttle)
SYLDA	Systeme de lancement double Ariane (Ariane dual-launch system)
TAT	Transatlantic telephone (cables)
TC	Telecommand
TCR	Telemetry command and ranging (TC&R)
TDM	Time division multiplexing
TDMA	Time division multiple access
TDRS(S)	Tracking and data relay satellite (system)

TM	Telemetry
TM	Thermal model
TPA	Transistor power amplifier
TTC	Telemetry tracking and command (TT&C)
TTL	Transistor–transistor logic
T/V	Thermal vacuum (spacecraft tests)
TVRO	Television receive only
TWT	Travelling wave tube
TWTA	Travelling wave tube amplifier
Tx	Transmit/transmitter
UDMH	Unsymmetrical dimethylhydrazine
UHF	Ultra high frequency
UT	Universal time (same as GMT)
VCHP	Variable conductance heat pipe
VHF	Very high frequency
VLF	Very low frequency
WARC	World Administrative Radio Conference
WARC-ST	World Administrative Radio Conference on Space Telecommunications
WCA	Worst-case analysis
WDE	Wheel drive electronics (reaction/momentum wheels)
W/G	Waveguide
XPD	Cross-polar discrimination

Glossary

Actuator A device which produces a mechanical action or motion; a servomechanism that supplies the energy for the operation of other mechanisms (e.g. reaction control thrusters, reaction wheels, momentum wheels, nutation dampers).

Altitude-azimuth mount A category of earth station antenna support which includes elevation-over-azimuth and $X-Y$ mounts.

Amplifier chain A number of amplifiers and associated equipment linked together in series.

Amplitude modulation (AM) A transmission method whereby the amplitude of a carrier wave is varied in accordance with the amplitude of the input signal; the frequency of the carrier remains unchanged.

Anechoic chamber A test facility for the evaluation of radio-frequency (RF) equipment on a spacecraft, which simulates the RF propagation characteristics of free space.

Antenna A device for transmitting and receiving radio waves. The four main types are wire, horn, reflector and array antennas.

Antenna farm A collection of antennas on an orbiting platform or smaller satellite, or a collection of earth station antennas.

Antenna radiation pattern A measure of the directional sensitivity of an antenna.

Aperture antenna A horn antenna or a reflector antenna.

Apogee The point at which a body orbiting the Earth is at its greatest distance from the Earth; the opposite of perigee.

Apogee kick motor (AKM) A rocket motor used to transfer a satellite from geostationary transfer orbit (GTO) to geostationary orbit (GEO) fired when the spacecraft reaches apogee; also called an apogee boost motor (ABM).

Array shunt regulator A device which switches sections of a spacecraft solar array on and off in accordance with the demand for power.

Attitude and orbital control system (AOCS). A spacecraft subsystem which controls attitude (pointing direction) and position in orbit.

Axial ratio A measure of polarisation circularity in a cross-polarised antenna.
Balance mass The mass added to a completed spacecraft before launch to ensure static and dynamic balance during all mission phases.
Band-pass filter A filter with both high- and low-frequency cut-offs (e.g. input, output and channel filters).
Bandwidth The width of a frequency band, typically measured in MHz.
BAPTA (bearing and power transfer assembly) A rotating interface between a spacecraft body and a solar array.
Baseband The frequency band that a signal occupies when initially generated (e.g. the output of a telephone, telex, TV camera, computer, etc).
Battery reconditioning The controlled discharging and subsequent recharging of a battery conducted on a regular basis to improve battery performance.
Baud The transmission or signalling rate on a data channel; a signalling element per second.
Beacon A radiated signal used for spacecraft tracking or propagation tests.
Beam area An area on the surface of the Earth, defined for the purposes of satellite communications, which is within the half-power beamwidth of the spacecraft antenna.
Beam waveguide An arrangement of reflectors enclosed in a 'weatherproof tube' used in some earth stations to conduct the signal between the HPA and the antenna, thereby reducing RF losses.
Beamwidth The width of the beam of radiation shaped by a communications antenna (measured in degrees).
Beginning of life (BOL) The beginning of a spacecraft's operational lifetime.
Bipropellant A propellant with two chemical components (fuel and oxidiser), stored separately and combined for combustion to take place.
Bit error rate A concept used to express the accuracy of digital demodulation or decoding; a measure of how truthfully the received signal represents the signal originally transmitted.
Blowdown system A liquid propellant delivery system which uses a fixed mass of pressurant throughout its period of operation.
Boresight The main axis of an antenna: a distinction is made between the radio-frequency (RF) boresight and the geometrical boresight (physical axis) of the antenna, the angle between the two being known as the squint angle.
Boresighting The pointing of an antenna.
Box Jargon: an area in the sky, as viewed from an earth station, which bounds the excursions of a satellite in geostationary orbit from its mean position.

BPSK (binary phase shift keying) A method for modulating the RF carrier in digital satellite communications links: a type of phase modulation.
Broadcasting-satellite service (BSS) A satellite communications system in which signals transmitted by the satellite are intended for direct reception by the general public.
Bus A spacecraft service module or a spacecraft power supply and distribution system (power bus).
Bus voltage The voltage rating of a spacecraft's main electrical power distribution circuit, or power bus.
Carrier An electromagnetic wave of fixed amplitude, phase and frequency which, when modulated by a signal, is used to convey information through a radio transmission system.
Cassegrain reflector A dual-reflector antenna with a convex hyperboloidal subreflector and a paraboloidal main reflector. The feedhorn, subreflector and main reflector are coaxial in the standard design, but the feed can also be offset (offset Cassegrain).
Clarke orbit An alternative name for geostationary orbit.
CODEC An electronic device for coding and decoding digital transmissions (a contraction of coder–decoder).
Cold soak Exposure to an extended period of low temperatures.
Cold welding A process whereby two metal surfaces in a vacuum diffuse into each other as a result of a lack of lubricant or air gap between them.
Combined (bipropellant) propulsion system A satellite propulsion system, using liquid bipropellants, which combines the functions of apogee kick motor and reaction control system. Also known as a unified propulsion system.
Common carrier A national organisation or private company granted the authority to provide public telecommunications services in its parent country; in Europe generally the PTT.
Communications payload An assembly of electronic and microwave equipment in a communications satellite, including both transponders and antennas; also called a communications subsystem.
Constellation A number of satellites with approximately equal spacing around an orbit designed to provide maximum coverage of the Earth (e.g. three satellites in GEO with orbital spacings of about 120°).
Control electronics unit (CEU) A device concerned with the overall control of a spacecraft—its central computer.
Co-polar(ised) Having the same polarisation.
Coverage area An area on the surface of the Earth, defined for the purposes of satellite communications, within which the PFD of the radiated satellite signal is sufficient to provide the desired quality of reception for a telecommunications service in the absence of interference.

In contrast to the service area, protection against interference is not guaranteed.

Cross-polar(ised) Having the opposite polarisation.

Cross-strapping The practice of interconnecting two or more redundant chains of equipment to allow switching between the chains in the event of a failure.

Cryogenic propellant A liquid propellant (fuel or oxidiser) formed by the liquefaction of a gas when cooled to very low ('cryogenic') temperatures.

Declination The angle north or south of the celestial equator, in the equatorial or geographical system of celestial coordinates (north is positive, south is negative).

Degree of freedom An attribute of a body capable of linear or angular motion; an ability to move along an axis or rotate about it. A spacecraft has six degrees of freedom: three linear along each of the three orthogonal axes (roll, pitch and yaw), and three rotational about those axes.

Delta modulation A transmission method using a modulated RF carrier, whereby the amplitude of the original analogue input signal is sampled at discrete time intervals to create a representative digital translation of the signal.

Delta V (ΔV) A change in velocity.

Demand assignment multiple access (DAMA) A coding method for information transmission between a number of users, whereby satellite capacity is assigned according to demand and not on a rigid predefined basis.

Depth of discharge (DOD) The amount by which a battery can be discharged without detrimentally affecting its future performance (measured as a percentage).

De-spun platform An antenna platform on a spin-stabilised spacecraft which remains stationary with respect to the Earth to maintain the pointing of the antenna beams.

Diplexer A two-channel multiplexer.

Directivity The ratio of the radiation intensity from an antenna in a given direction to that available from an isotropic antenna.

Discrimination A measure of the ability of a communications system or a component within that system to separate wanted from unwanted signals.

Double hop A signal route in a communications satellite system which includes a pass through two satellites.

Doubler A structural element which increases the thickness of a panel; local reinforcing for highly stressed areas.

Downconverter A device which reduces the frequency of a signal by mixing it with a local oscillator frequency (a process known as heterodyning).

Downlink The communications path between a satellite and an earth station, 'from space to Earth'.
Drift orbit An orbital path into which a satellite is injected by its AKM en route to its final position in GEO.
Dry mass The mass of a spacecraft or launch vehicle without propellant or pressurant.
Earth segment The terrestrial part of a satellite-based communications system (synonymous with ground segment).
Earth sensor A device used to establish a spacecraft's attitude relative to the Earth by detecting the limits of the Earth's disk at infrared wavelengths.
Earth station An installation on the Earth containing the equipment necessary for communications with a spacecraft.
Earth terminal A small earth station.
Earthshine Thermal energy absorbed from the Sun and re-radiated by the Earth.
Eclipse The partial or total obscuration of one celestial body by another, or of a spacecraft by another body (particularly the eclipse of the Sun by the Earth).
Eclipse-protected A communications service which can remain powered during an eclipse, thereby providing a continuous service.
Ecliptic In astronomy, the apparent annual path of the Sun relative to the stars as seen from the Earth; the plane containing the Sun and the Earth.
Electronic power conditioner (EPC) A device which supplies the operating voltages to the components of a TWT.
Electrostatic discharge (ESD) An instantaneous loss of static electrical charge which can cause a satellite communications system to switch off.
Elevation-over-azimuth A type of earth station antenna mount.
End of life (EOL) The end of a spacecraft's operational lifetime.
Equilibrium point One of four points on the geostationary orbit produced by the variation in the gravitational force around the Earth's equator.
Equivalent isotropic radiated power (EIRP) The level of transmitter power available at an antenna multiplied by the antenna gain.
***F/D* ratio** The ratio between the focal length, F, and the diameter, D, of an antenna, or a lens or mirror of an optical system.
Feed A device designed to illuminate the surface of an antenna reflector when transmitting an RF signal, and to collect radiation reflected from the antenna when receiving.
Figure of merit (*G/T*) A parameter which defines the quality of a receiving installation (antenna and receiver), where G is the gain of the antenna and T is the receive system noise temperature (units: $dB\ K^{-1}$).

Fixed-satellite service (FSS) A satellite communications system which uses earth stations at specified fixed points. This service is restricted to a relatively small set of user organisations and is not directly accessible to the general public.
Footprint The projection of a satellite's antenna beam on the Earth's surface, usually depicted as a contour map of radiated power received at the surface.
Free-drift strategy A satellite station-keeping method for reducing propellant consumption.
Free-space loss A loss in power density of a radiated telecommunications signal because of the distance of free space between transmitter and receiver.
Frequency division multiple access (FDMA) A coding method for information transmission between a potentially large number of users, whereby each user is allocated a relatively narrow section of the total transponder bandwidth to which they have continual access.
Frequency division multiplexing (FDM) The process of placing more than one signal on a carrier, whereby each signal is given its own subcarrier frequency. Frequency multiplexing allows continuous data transmission from all channels.
Frequency hopping The practice of switching a radio transmission between different radio frquencies in order to deter unauthorised reception.
Frequency modulation (FM) A transmission method using a modulated carrier wave, whereby the frequency of the carrier is varied, around its nominal or *centre frequency*, in accordance with the amplitude of the lower frequency input signal. The amplitude of the carrier remains unchanged.
Frequency reuse The practice of using the same radio frequency more than once in the same satellite beam or coverage area by the use of opposite polarisations, or by using a number of spot beams.
Front end A satellite receiver.
Gain The ratio of output power to input power of an amplifier, usually measured in decibels (dB); the power multiplication factor of an antenna.
Gas jet A reaction control thruster.
Gateway station A major earth station which handles international telecommunications traffic.
Geostationary arc The portion of GEO 'visible' from a given place on Earth.
Geostationary orbit (GEO) A circular orbit in the same plane as the Earth's equator and with the same period of rotation as the Earth (radius 42 164 km, height 35 786 km).
Geostationary transfer orbit (GTO) An elliptical Earth orbit used to transfer a spacecraft from a low-altitude orbit or flight trajectory to GEO.

Geosynchronous orbit An orbit whose period of rotation is some multiple or submultiple of the period of rotation of the Earth.
Graveyard orbit An orbit beyond geostationary orbit to which satellites are boosted at the end of their lives.
Gregorian reflector A dual-reflector antenna with a concave ellipsoidal subreflector and paraboloidal main reflector.
Ground segment Equivalent to earth segment.
Ground spare A satellite stored on Earth which is available to replace a defective satellite in orbit or enhance an existing satellite system.
Ground station Equivalent to earth station.
Guardband A band of frequencies separating two channels to obviate interference.
Half-power beamwidth The width of the beam of radiation formed by a communications antenna, measured in degrees, between the points where the radiated power is half of its peak value.
Heat pipe A device for transferring thermal energy from one point to another; part of a spacecraft thermal control system.
Heat spreader plate An item of spacecraft hardware with the properties of a heat sink.
Heterodyning Frequency mixing.
High-pass filter A filter with a low-frequency cut-off.
High-power amplifier (HPA) A generic term for a device that amplifies electronic signals to a high power level, usually for transmission between an earth station and a spacecraft.
Hiphet thruster A propulsive device utilising electrical heating to increase the exhaust velocity of the standard hydrazine thruster.
Hohmann transfer orbit A trajectory linking two orbits which can be traversed by a spacecraft using a minimum amount of propellant (e.g. GTO).
Horizon sensor or **scanner** Types of earth sensor.
Housekeeping The function of a number of subsystems which maintain a spacecraft in an operational condition (e.g. thermal control, telemetry, etc).
Hydrazine thruster A propulsive device utilising the chemical compound hydrazine (N_2H_4) as a propellant.
Hypergolic (propellants) Propellants which ignite spontaneously when mixed.
Imux An abbreviation for input multiplexer.
Inertia wheel A momentum wheel or reaction wheel.
Inertial platform A device which uses a set of gyroscopes to define a frame of reference in a moving vehicle, with respect to which the vehicle's attitude can be controlled.
Inertial upper stage An upper stage used on the Amercian Space Shuttle and Titan ELV, which combines the functions of perigee kick

motor and apogee kick motor by having two solid propellant stages.

In-orbit spare A satellite in orbit which is available to replace a defective satellite or enhance an existing satellite system.

Intermediate frequency (IF) A frequency used in a communications transponder, or other receive/transmit device, between the input and the output.

Inter-satellite link (ISL) A communications link between orbiting spacecraft. When the link is between two spacecraft in different orbits, it may be termed an inter-orbit link (IOL).

I^2R loss The loss of power in a transmission line due to heating.

Isoflux contours Contours on a satellite antenna footprint which represent the same level of power flux density.

Isolation A measure of separation, in terms of RF power, between the carrier waves of two potentially interfering communications systems.

Isotropic Having uniform physical properties in all directions.

Isotropic antenna A hypothetical omnidirectional point source antenna used as a reference for antenna gain measurements.

Kevlar composite A material composed of a polymer matrix reinforced by threads or fibres of Kevlar (a form of the organic polymer polyparabenzamide).

Klystron A type of 'electron tube' used as an HPA in earth stations.

Launch window A limited period of time during which the launch of a spacecraft may be undertaken.

Lifetime The length of time for which a device, vehicle or system is intended to perform its function. The design lifetime is the period over which it is designed to be *capable* of operation, whereas the operational lifetime is the actual period of operation.

Link budget A calculation of signal power in a communications link design which indicates whether or not a signal will be successfully received.

Link margin The amount by which the RF power in a link budget is above that required to achieve the performance specified for the link under nominal conditions.

Liquid apogee engine (LAE) A rocket engine used to transfer a satellite from GTO to GEO, fired when the spacecraft reaches apogee.

Local oscillator (LO) A device which provides a stable fixed-frequency source of RF energy used in telecommunications equipment; alternatively known as a *carrier generator*.

Low Earth orbit (LEO) A nominally circular orbit of low altitude (typically less than 1000 km) and short period (approximately 90 min).

Low-noise amplifier (LNA) An amplifier designed to deliver a good signal-to-noise ratio.

Low-noise converter (LNC) An LNA combined with a frequency converter, usually mounted at the focus of an earth terminal antenna.
Low-pass filter A filter with a high-frequency cut-off.
Luni-solar gravity The collective gravitational attraction of the Moon and Sun; a cause of perturbations of a spacecraft's orbit.
Magneto-torquer A device which makes use of a planet's magnetic field to stabilise a satellite using the principle that a freely suspended current-carrying coil aligns itself with the local magnetic field.
Mass budget A method of accounting for the mass of a spacecraft or subsystem.
Mass ratio The ratio of the mass of a launch vehicle's payload to the mass of the total vehicle at lift-off, or the ratio of a launch vehicle's initial mass at lift-off to its mass without propellant.
Mixer An electrical device in which two or more input signals are combined and filtered to produce a single output signal.
Mobile – satellite service (MSS) A satellite communications system providing links between mobile earth stations and one or more orbiting spacecraft.
Modulation The process whereby a signal is superimposed upon a higher frequency carrier wave which 'carries' the signal until it is received and demodulated.
Momentum dumping The practice of reducing the angular momentum of a momentum wheel or reaction wheel which has reached its maximum allowable spin speed. Also called 'momentum offloading'.
Momentum wheel A wheel mounted in a three-axis-stabilised spacecraft used to provide gyroscopic stability and attitude control.
Monomethyl hydrazine (MMH) A storable liquid bipropellant (fuel) used in rocket engines.
Monopropellant A chemical propellant comprising a single component.
Multi-layer insulation (MLI) Spacecraft thermal insulation comprising a number of reflective layers.
Multipaction An electron avalanche or 'breakdown' effect which can occur in high-power RF equipment operating in a vacuum. Also called 'Multipactor' (chiefly USA).
Multipath An effect of a radiated signal taking more than one path from transmitter to receiver (a line-of-sight path and a reflected path) causing interference at the receiver.
Multiplexed analogue component (MAC) A TV transmission standard developed by the UK Independent Broadcasting Authority (IBA) for use with European DBS.
Multiplexer A device for combining or separating different signal frequencies.

Nitrogen tetroxide (N_2O_4) A storable liquid bipropellant (oxidiser) used in rocket engines.
Noise An undesirable electrical disturbance which can mask the signal in a communications system.
Noise figure (NF) A measurement of the noise contribution of a device expressed relative to a theoretical noise-free amplifier at a reference temperature.
Noise power A quantitative measure of noise in a communications system (units dBW) analogous to carrier power.
Noise power density (NPD; N_0) A quantitative measure of the level of noise in every 1 Hz of bandwidth (units dBW Hz^{-1}).
Noise temperature The ratio of the noise power at the source (or other reference point) of a system to the temperature of a reference resistance from which the same thermal power is available (units dBK).
Nutation A periodic variation in the precession of a spinning body; a 'nodding' motion superimposed on the precession of a satellite's spin axis.
Nutation damper A device designed to reduce the nutation of a spinning spacecraft.
Offset feed An antenna system in which the feedhorn is placed away from the principal axis of the antenna reflector.
Omnidirectional antenna or **Omni** An antenna system which provides coverage in (almost) all directions (e.g. a spacecraft TT&C antenna).
Omux An abbreviation for output multiplexer.
Optical solar reflector (OSR) A spacecraft surface material that reflects the Sun's visible spectrum (particularly second-surface mirrors).
Orbital parameters The parameters which define a spacecraft orbit (i.e. semi-major axis, eccentricity, inclination, argument of perigee, right ascension of ascending node, true anomaly). Also called orbital elements.
Orthogonal mode transducer (OMT) A component of a waveguide circuit which separates or combines two orthogonally polarised signals.
Outgassing The release of a gas from a material when it is exposed to an ambient pressure lower than the vapour pressure of the gas.
Overspill The unavoidable tendency of a satellite beam to cover more than the intended service area.
Parametric amplifier A type of low-noise amplifier (abbreviated to paramp).
Payload assist module (PAM) A PKM designed for use with the American Space Shuttle and Delta launch vehicles. Also known as a spinning solid upper stage (SSUS), since it uses solid propellant and is capable of spin-stabilising its payload during the transfer orbit.
Payload envelope The dimensional constraint on the volume available for a payload within a launch vehicle shroud.

Perigee The point at which a body orbiting the Earth is at its closest to the Earth; the opposite of apogee.
Perigee kick motor (PKM) A rocket motor used to place a satellite in GTO, fired when the spacecraft reaches perigee.
Perturbations Disturbances which cause a deviation in a spacecraft's nominal orbital path or trajectory. The three important perturbation mechanisms for a satellite in GEO are triaxiality, luni-solar gravity and the solar wind.
Photovoltaic cell A solar cell.
Pitch axis The axis of a satellite in an equatorial orbit which is aligned with the Earth's spin axis (also the spin axis for the spin-stabilised satellite).
Plume impingement An undesired contact between a thruster plume and a spacecraft component or surface, which can disturb the spacecraft's attitude.
Pointing loss A loss in signal power in a communications link because of the mispointing of an antenna.
Polar mount A type of earth station antenna support.
Polarisation The restriction of electromagnetic waves to certain directions of vibration (linear: vertical or horizontal; circular: right hand or left hand).
Power budget A method of accounting for the power requirements of a spacecraft or subsystem.
Power bus The main electrical power distribution circuit in a spacecraft.
Power flux density (PFD) The level of radiated power received per square metre (units dBW m^{-2}).
Propellant budget A method of accounting for the amount of propellant in a spacecraft's propulsion subsystem available during its operational lifetime.
Propellant management device (PMD) A device or structure inside a propellant tank which uses surface tension (capillary) forces to position and hold the propellant over the tank outlet, thus allowing propulsion in weightless conditions.
Propulsion module A self-contained section of a modular spacecraft containing the propulsion subsystem.
PTT An acronym for the Posts, Telegraph and Telecommunications organisation; a country's provider of telecommunications and other associated services.
Pyrotechnics A class of devices which use explosive charges as actuators or igniters.
QPSK (quadrature phase shift keying) A method for modulating the RF carrier in digital satellite communications links; a type of phase modulation. Also called quaternary phase shift keying.

Radio frequency (RF) The frequency of a radio carrier wave; any frequency which lies in the range 10 kHz to 300 GHz.
Ranging The determination of the distance between a launch vehicle or a spacecraft in orbit and the ground station tracking it.
Reaction control thruster A small rocket engine used to make fine adjustments to the orbit or trajectory and the attitude of a spacecraft by the expulsion of gas molecules at high velocity (hence the more colloquial term 'gas jet').
Reaction wheel A wheel used to control the attitude of a three-axis-stabilised spacecraft.
Receiver A device used to detect and amplify electrical signals or modulated radio waves; a satellite receiver is also known as a *front end*.
Redundancy An attribute of a system in which a spare or back-up device is available in case of failure of the prime system.
Regulated power supply A spacecraft power supply which provides a constant voltage, both in sunlight and eclipse.
Reliability A branch of engineering which quantifies the likelihood of a failure.
Repeater A device which receives, amplifies and retransmits electrical signals.
Residuals The propellants or pressurants left in a tank at the end of a mission.
RF sensor A spacecraft sensor used to maintain accurate antenna pointing. The most common type is the monopulse RF sensor.
Rocket engine A propulsive device that uses liquid propellants.
Rocket motor A propulsive device that uses solid propellants.
Roll axis The axis of a satellite aligned with the direction of orbital motion.
Saturation The condition of an HPA when driven to its maximum output.
Second-surface mirror (SSM) A thin sheet of glass or quartz, silvered or aluminised on one side, bonded to the exterior surface of a spacecraft as part of the thermal control system.
Semi-major axis One of the halves of the major axis of an orbit.
Service area An area on the surface of the Earth, defined for the purposes of satellite communications, within which the administration originating a service can demand protection against interference from other transmissions.
Service module A self-contained section of a modular spacecraft containing the subsystems which support the payload (e.g. power, thermal and attitude and orbital control systems).
Shear web A structural element which transfers or shares structural (shear) loads between panels or other components.
Shroud A protective fairing which houses and protects the payload of a launch vehicle.

Shunt dump regulator A device which takes excess spacecraft power from the solar arrays, converts it to heat and radiates it into space.
Sidelobe Sensitivity in a direction outside the main beam or main lobe of an antenna radiation pattern.
Simulation heater A heater designed to simulate a deactivated device which produces heat when it is operating; also called a substitution heater.
Single-event effect (SEE) Damage to spacecraft electronic components, particularly semiconductor storage devices, caused by natural radiation; also called *single-event upset* (SEU).
Single point failure (SPF) A concept in reliability engineering whereby a single component can cause the failure of a complete system because there is no redundancy provision.
Soft fail The failure of a component, device or subsystem which does not seriously degrade the performance of a system.
Solar array An area of solar cells designed to provide electrical power to a spacecraft.
Solar cell A device for the conversion of solar energy to electrical energy (also known as a photovoltaic cell, since it operates in accordance with the photovoltaic effect).
Solar generator A solar array.
Solar sailing The use of the solar wind in controlling the position and orientation of spacecraft.
Solar wind A continuous stream of atomic and other particles ejected from the Sun, otherwise known as solar radiation pressure; a cause of perturbations in a spacecraft's orbit.
Solid state power amplifier (SSPA) A device, using semiconductor materials or components, for the amplification of RF signals; an alternative to the TWTA.
Solstice Either of two annual occasions when the Sun is overhead at the Tropic of Cancer (summer solstice, 21 June) or the Tropic of Capricorn (winter solstice, 21 December); the longest and the shortest day of the year, respectively.
Space-qualified An attribute of any component or device considered able to perform its function in the space environment.
Space segment The space-based part of a satellite communications system (as opposed to the earth segment).
Specific energy A figure of merit for a power source; measured in watt hours per kilogram ($W\ h\ kg^{-1}$).
Specific impulse A measure of the efficiency and performance of a rocket engine or motor; the impulse per unit mass of propellant consumed (units $m\ s^{-1}$).
Specific power A quantitative measure of the mass efficiency of a power subsystem or device (units $W\ kg^{-1}$).

Spillover The portion of an antenna feed's radiation pattern not intercepted by the antenna reflector.
Spin-stabilised spacecraft A spacecraft which maintains its stability by rotating about its longitudinal (pitch) axis.
Spreading loss A loss in power density experienced by a telecommunications signal as it radiates from its point of origin (incorporated with a frequency term to give an expression for free-space loss).
Step-track A type of satellite tracking system which moves an antenna in predetermined 'steps' in a direction dependent on the strength of the signal received.
Structure model The version of a spacecraft built for the purposes of structural and component-fit tests.
Subreflector A subsidiary reflector in a 'reflector antenna' system, interposed between the feedhorn and the main reflector in a folded-optics configuration (e.g. Cassegrain and Gregorian reflectors).
Sub-satellite point The point of intersection, on the surface of a planet, of a line drawn between a satellite in orbit and the planet's centre.
Substitution heater Equivalent to simulation heater.
Telecommand An instruction transmitted to a spacecraft from a controlling earth station or control centre.
Telemetry Information or measurements transmitted from a remote source to an indicating or recording device (e.g. from a spacecraft to an earth station).
Thermal blanket Multi-layer insulation.
Thermal model The version of a spacecraft built to test the thermal control system and verify that all components and subsystems will operate in the thermal environment of space.
Three-axis-stabilised spacecraft A spacecraft which maintains its stability using an attitude control system comprising sensors and actuators (as opposed to spin stabilisation).
Time division multiple access (TDMA) A coding method for information transmission between a potentially large number of users, whereby all users utilise the same frequency band but transmit and receive at different times.
Time division multiplexing (TDM) The process in which several signals share the same carrier but at different times.
Tracking The determination of a spacecraft's orbital parameters or the trajectory of a launch vehicle from a ground installation; a function of a spacecraft's TT&C subsystem.
Transponder A chain of electronic communications equipment on a satellite which receives, filters, amplifies and transmits a signal.

Travelling wave tube (TWT) A device for the high-power amplification of radio-frequency (RF) signals.
Travelling wave tube amplifier (TWTA) A combined TWT and EPC.
Triaxiality An attribute of an ellipsoid with three dissimilar orthogonal axes (e.g. the Earth); one of the mechanisms that results in the perturbation of a spacecraft's orbit.
Unified propulsion system A combined (bipropellant) propulsion system.
Unregulated power supply A spacecraft power supply which provides a variable voltage, dependent upon the power source and electrical load.
Upconverter A device for increasing the frequency of a signal, containing a mixer which mixes the incoming signal with a local oscillator frequency.
Uplink The communications path between an earth station and a satellite, 'from Earth to space'.
Upper stage The uppermost stage in a multi-stage launch vehicle or a rocket motor used to launch a spacecraft from the American Space Shuttle (e.g. payload assist module (PAM), inertial upper stage (IUS)).
Van Allen belts Two regions of charged particles in the Earth's magnetosphere which present a potential hazard to satellites because of their degrading effect on materials.
Waveguide A type of transmission line for the conveyance of electromagnetic waves.
X*–*Y A type of earth station antenna mount.
Yaw axis The axis of a satellite which passes through the sub-satellite point.

List of Boxes Included in Chapters

Index

Page numbers in bold refer to illustrations